DESIGNER DRUGS

EMERGING ISSUES IN ANALYTICAL CHEMISTRY

DESIGNER DRUGS

Chemistry, Analysis, Regulation, Toxicology, Epidemiology & Legislation of New Psychoactive Substances

ROY GERONA
University of California San Francisco
San Francisco, CA, United States

Elsevier
Radarweg 29, PO Box 211, 1000 AE Amsterdam, Netherlands
125 London Wall, London EC2Y 5AS, United Kingdom
50 Hampshire Street, 5th Floor, Cambridge, MA 02139, United States

Notices
Knowledge and best practice in this field are constantly changing. As new research and experience broaden our understanding, changes in research methods, professional practices, or medical treatment may become necessary.

Practitioners and researchers must always rely on their own experience and knowledge in evaluating and using any information, methods, compounds, or experiments described herein. In using such information or methods they should be mindful of their own safety and the safety of others, including parties for whom they have a professional responsibility.

To the fullest extent of the law, neither the Publisher nor the authors, contributors, or editors, assume any liability for any injury and/or damage to persons or property as a matter of products liability, negligence or otherwise, or from any use or operation of any methods, products, instructions, or ideas contained in the material herein.

ISBN: 978-0-12-811764-4

For information on all Elsevier publications visit our website at https://www.elsevier.com/books-and-journals

Publisher: Candice Janco
Acquisitions Editor: Maddie Wilson
Editorial Project Manager: Shivangi Mishra
Production Project Manager: Sruthi Satheesh
Cover Designer: Matthew Limbert

Typeset by TNQ Technologies

Dedication

This book is dedicated to my mom, Remedios Laylo Gerona, from whose examples I learned the meaning of hard work, patience, compassion, and kindness. My work on NPS analysis and surveillance that is instrumental to writing this book has constantly leaned on these values.

Contents

4. NPS analysis

5. NPS screening

6. NPS confirmation using targeted analysis

7. NPS analysis using high-resolution mass spectrometry

8. NPS surveillance and epidemiology

9. NPS regulation and legislation

Foreword

One night in March 1983, at about 2:00 am, the night-shift emergency department (ED) physician walked into the San Francisco Poison Control Center[1] at San Francisco General Hospital (SFGH) with a small plastic packet containing a white powder. He told the on-duty poisoning expert that the police had brought to the emergency department a man who had been found running down the street, naked and in a state of extreme agitation. His pupils were dilated and his heart was racing. He was carrying a suitcase that, when opened, was found to contain stacks of 100-dollar bills and hundreds of clear packets of the white powder. The two clinicians examined the packet and concluded that it probably contained cocaine or methamphetamine. About an hour later, both the ED physician and the poisoning expert began having intense visual and auditory hallucinations and had to be relieved by backup staff. They were reportedly incapacitated for the next 24 hours.

Routine screening tests for cocaine and amphetamines were negative. The toxicology lab director and poison center staff discussed the case and decided to pursue additional testing. Thin-layer chromatography was strongly suggestive that the packets contained lysergic acid diethylamide (LSD).

This was one of the earliest collaborations in a years-long working relationship between the San Francisco Poison Control Center and the clinical toxicology laboratory at SFGH. It had started with John Osterloh, MD, and picked up considerable momentum when Alan H.B. Wu, Ph.D., arrived in 2004 as Chief of Clinical Chemistry and Toxicology. He fostered a strong collaborative ethic and expanded ties with our statewide poison center system, which serves a population of over 30 million people and over 300 hospitals in California. Poison control centers are often the first entities to hear about a new drug and can facilitate the transfer of biological specimens or recovered drug samples from the treating hospital to the laboratory at SFGH. He and his postdoctoral clinical chemistry fellows regularly participate in weekly Toxicology Grand Rounds and are always available for consultation on questions about laboratory analysis for puzzling poison center cases. He was an early adopter of novel

[1]The San Francisco Poison Control Center is officially known as the San Francisco Division of the California Poison Control System. Before it moved to its current location, it was housed at the San Francisco General Hospital.

technology, including quadrupole time-of-flight mass spectrometry (QTOF/MS), to help identify unknown substances.

Roy Gerona joined Alan Wu's team as a clinical chemistry postdoctoral fellow in 2009, after earning his Ph.D. at the University of Wisconsin. Roy took collaboration with poison centers to a new level, developing a special interest in new psychoactive substances (NPS) based on chemical alteration of cathinones and cannabinoids. After a successful collaboration with an expanding group of poison control centers across the United States to identify cases and procure samples for analysis, new stakeholders were added to his list of participants, including the US Drug Enforcement Administration (DEA), state public health departments, and medical examiners. In 2016, Roy established the Psychoactive Substances Consortium and Analysis Network (PSCAN), a consortium of ten medical centers collecting cases that may involve a NPS etiology. As readers of this book will learn, he has developed and refined protocols for laboratory analysis as well as procedures for sharing information about emerging trends with clinicians, health departments, and law enforcement.

Roy originally intended the book mainly for chemists interested in NPS, but has expanded the focus to appeal to a broader audience from multiple disciplines. Five of the nine chapters are background information and are written in such a way that anyone with a college-level course in chemistry and biology can understand the main points. Even the more in-depth chapters on laboratory analysis techniques include introductory material suitable for a general medical audience. Roy clearly recognizes that it is difficult to satisfy the needs of all stakeholders, given their diverse backgrounds, but hopefully readers will find that he has achieved a "goldilocks" compromise that serves the largest possible number of users.

In my role as Medical Director of the San Francisco Poison Control Center from the 1980s well into the new century, I have had the privilege to witness the long and successful evolution of collaboration between the laboratory and clinicians in the poison center, starting with that LSD case in 1983. I am awestruck by the expanse and effectiveness of the NPS surveillance system Roy Gerona has nurtured over the past decade, and I look forward to its continued achievements.

Kent R. Olson, MD
Professor Emeritus, University of California San Francisco,
San Francisco, CA, United States

Preface

Towards the end of graduate school, I asked my Ph.D. adviser how one can be certain of the research field one is meant to spend the rest of their career on. He replied without hesitation, "If in the shower most mornings, all you can think about is the next big experiment you will be doing that day, then you know...." I finished graduate school not understanding what he meant. I moved on and decided that basic science is far too demanding to allow me to find balance between my scientific career and personal life.

I pursued postdoctoral studies in clinical chemistry and decided to focus on analytical toxicology. My postdoctoral adviser convinced me that high-resolution mass spectrometry is the future of drug testing. When I started my fellowship at UCSF in 2009, I developed a workflow using liquid chromatography–time-of-flight mass spectrometry (LC-TOF/MS) that allowed identification of known unknowns and unknown unknowns in samples from intoxication or poisoning. A year later, we piloted the workflow in responding to emergency intoxication cases referred to the San Francisco Poison Control Center, where routine drug testing did not provide an answer.

I realized right away the power of the analytical tool I was working on. Its ability to facilitate nontargeted data acquisition allowed searches even for drugs, metabolites, and compounds I was not necessarily looking for when I started my analysis, and provided unending opportunities to query the data it generated to gain almost limitless insights on what substances may be in the sample. At the same time, I was acutely aware of the thrill and exhilaration I feel every time I sleuth for drugs and poisons in cases referred to me. Not long after, I caught myself more and more in the shower most mornings thinking about and looking forward to the cases I will solve each day. Ironically, it was like an addiction: the time and effort required did not matter just so long as I got that rush whenever I produced the result and felt that I had contributed to solving an emergency physician's conundrum and sometimes saved a life. It was then that I understood what my Ph.D. adviser was trying to tell me. I was home. I had found the field where I truly belong, where I was excited to make my contribution.

The training and realization early on in my postdoctoral studies is key to writing this book. Not long after we implemented our workflow, mysterious cases of obvious drug intoxications that tested negative in urine drug screens started flooding emergency departments in the United States. The great microbiologist and chemist Louis Pasteur once said, "In the field of observation chance favors only the prepared mind." The same can be said of serendipitous discoveries in figuring out a vocation one is meant to have. The agents that turned out to be causing the mysterious intoxications belong to a new generation of designer drugs (later called new psychoactive substances, NPS), evolving much faster in the marketplace than the time it takes for hospitals to develop targeted tests for them. They were not only coming out and changing in molecular composition furiously fast, but also their breadth in variety and chemical class was unprecedented. The routine targeted drug tests available in hospitals and reference laboratories were far more comprehensive than those of a few decades ago but could not respond to the onslaught of new designer drugs changing faster than fashion trends in Paris and Milan. This problem could, however, be solved by an analytical platform with the ability to carry out nontargeted analysis. The high-resolution mass spectrometry workflow I developed and became addicted to suddenly provided an answer to one of the most difficult challenges in contemporary analytical toxicology. And my significant training and newfound passion allowed me to seize a niche right away where I could further flourish scientifically.

This book is a culmination of more than a decade's experience in NPS analysis. Everything I learned not only in the laboratory but also in the collaborations I pursued in solving NPS-related cases and mass intoxications informed the format and tone of this book. Anyone who has spent substantial time working on NPS will tell you that solving NPS cases requires a concerted effort with inputs from experts as diverse in their training as the variety of chemical classes that comprise NPS. NPS analysis is certainly one of the cornerstones in that solution. However, what is critical in solving the analytical problem NPS poses and efficiently utilizing the data obtained from its analysis is the ability of the laboratory expert to effectively communicate with various stakeholders from different academic and professional backgrounds.

When NPS intoxications started to appear in 2010, we often did not know the drug we were looking for in a case. I realized right away that getting as much information about the case from first-line responders—emergency medical technicians, law enforcers, and emergency physicians and nurses—significantly informed my analysis. To effectively search for new drugs in these cases requires going beyond chemistry and analytical toxicology. It entails learning about the pharmacology and toxicology of recreational drugs, their market sources, forms, and modes of application, their epidemiology, and the control and regulatory policies that their manufacturers are trying to evade.

An even more important and critical realization that came later is that getting my collaborators to give me the more relevant data that will be helpful in my analysis benefits a lot from explaining to them how I analyze NPS in the laboratory in a manner and language that they can understand. Their understanding of the analysis, no matter how simplified it may be, allows them to triage all information about the case and distill it to those factors most relevant to my analysis. This is the guiding principle in writing this book. It is written for a wide range of NPS stakeholders, from anyone with a background in college chemistry and biology, medical professionals, public health and policy experts, scientists, and students who are either simply curious or looking to learn more about NPS and how they are analyzed in the laboratory, to chemists and toxicologists who want more information on current trends in NPS analysis and surveillance. The book is meant to be neither a scientific review of the latest methodologies nor a comprehensive treatise on recent publications in NPS analysis. If that is what the reader is looking for, they will be better off searching for scientific reviews on NPS analysis in PubMed. With a wider range of readers in mind, I tried to strike a balance between making information about NPS and its analysis accessible to the average reader and providing a discourse on NPS analysis that can impart useful technical information to an expert in analytical, clinical, or forensic toxicology.

The book is divided into three parts: the first three chapters provide background on NPS, the middle four are focused on NPS analysis, and the last two illustrate how NPS analysis informs NPS surveillance, control, and regulation. The first three and last two chapters are accessible to most readers even without extensive background in chemistry. The middle four are more technical. Parts of the middle four are still accessible to the average reader: the first of them describes the overall workflow for NPS analysis, and the early part of each of the other three describes the respective chapter topic in a more general manner.

Chapters 1 to 3 give a background on NPS. The origins of NPS are described comprehensively in Chapter 1, from the time they were being synthesized as medication alternatives to the boom of designer drugs in the 1980 and 1990s to the present-day resurgence. It also gives an introduction to all topics relevant to NPS use—production and marketing, patterns of use, effects and toxicity, detection methods to identify use, and current controls and regulations in place. Chapter 2 provides a survey of classification schemes. With more than 1000 NPS reported worldwide, keeping track of their effects and toxicity is impossible without organization. The more popular schemes are discussed, and a scheme based primarily on their chemistry is proposed. The metabolism of the major classes is also discussed. Metabolism of drugs affects their duration and strength of action, toxicity, and methods of analysis. Common metabolic transformations in the most common NPS classes are presented. Chapter 3 discusses the pharmacology and toxicology of NPS. Because NPS functionally mimic the pharmacology of traditional recreational drugs (common drugs of abuse), they are similarly grouped according to pharmacological action: cannabinoids, stimulants and empathogens, hallucinogens (psychedelics and dissociatives), and depressants (opioids and benzodiazepines). Pharmacology also dictates acute toxic effects, which are discussed for the different pharmacological classes.

Chapters 4 to 7 comprise the meat of NPS analysis. Chapter 4 describes the general workflow of how NPS are analyzed in a clinical or forensic laboratory, from sample documentation and types of samples to sample preparation and analysis to data analysis and reporting of results. This chapter describes the whole process of NPS analysis that is accessible to the general reader. Those interested in the technical details of the major steps of NPS analysis—screening (Chapter 5), confirmation by targeted analysis (Chapter 6), and nontargeted analysis by high-resolution mass spectrometry (Chapter 7)—can carry on reading the entire chapter. Each of these chapters begins with a general description of the topic before details on the analysis of different types of NPS using specific analytical platforms (immunoassay, liquid chromatography–tandem mass spectrometry, and liquid chromatography–quadrupole time-of-flight mass spectrometry) are discussed.

The last two chapters, 8 and 9, illustrate how data from NPS analysis inform surveillance, control, and regulation. Chapter 8 presents current data from different NPS surveillance groups around the world as well as patterns of use and epidemiology. Chapter 9 ends the book with a survey of control policies enacted by different governing agencies.

With the structure of the book, I circle back to the most important lesson I learned in working on NPS analysis: what is critical in solving the analytical problem NPS poses and efficiently utilizing the data obtained is the ability of the laboratory expert to effectively communicate with various stakeholders from different academic and professional backgrounds. The basic information provided in Chapters 1 to 3 empowered me in solving NPS cases, especially those where I found previously unreported drugs. The analytical workflow and tools described in Chapters 4 to 7 provided the means to tackle NPS cases. And, making sure that the data I gathered are communicated through surveillance programs and utilized by control and regulatory agencies as described in Chapters 8 and 9 gave me a platform to effectively communicate and utilize NPS data. I learned most of the information in the first three chapters from communicating with various collaborators along the way, while the privilege to make my data available to agencies described in Chapters 8 and 9 required me to learn how to communicate it effectively.

At the end of each chapter, a closeup article features an interesting story or issue related to the chapter topic. The book is in a way a reflection of my personal journey, so the contributors I invited to write these articles are former mentors, students, and collaborators whom I worked with or am still working with on NPS cases and analysis. Although placed at the end, reading some of these articles ahead of the chapter may provide a taste of what is to be discussed.

I hope you will enjoy reading this book and will learn from it as much as I enjoyed and learned in writing. The first letters in the subtitle—" Chemistry, Analysis, Regulation, Toxicology, Epidemiology, Legislation"—spell out CARTEL. Drug cartels have given the word a bad rap. We often forget that historically a cartel is a cooperative arrangement between political parties intended to promote a mutual interest. By the same token, a concerted effort from experts in each of the fields in the subtitle is required to fully understand a common interest, NPS and its analysis.

Acknowledgments

This book would have not been written without the help, encouragement, and inspiration of several individuals with whom I had the privilege of working.

I thank Dr Gerald Pollard for meticulously reading and reviewing the whole book. I learned a lot though the countless comments, suggestions, and back and forth revisions of each chapter. Big thanks to Dr Anna Wetterberg of RTI Press for the patience, encouragement, and suggestions in our bimonthly check-ins as I wrote, edited, and revised each chapter. Thanks also to Ms Maria Ashbaugh of RTI for accommodating my ideas and requests in redesigning the book cover.

My sincere gratitude to my mentors, colleagues, and students who agreed to write the foreword and closeup articles and gave permission for use: Dr Kent Olson, Dr Axel Adams, Dr Greg Endres, Dr Hallam Gugelmann, Mr Andrew Reckers, Dr Aaron Schneir, Dr Michael Schwartz, Dr Stephen Thornton, and Dr Alan H.B. Wu. Your contributions have added depth to demonstrating the actual work and challenges behind NPS analysis and surveillance. Special thanks to Dr Alan H.B. Wu for introducing me to high-resolution mass spectrometry, the analytical tool that propelled my work on NPS analysis.

I had numerous collaborators as my laboratory pioneered NPS analysis at the start of its resurgence and later on as we built up our capacity to address surveillance across the United States. Much of the result found its way into the book. I am especially indebted to Dr Jordan Trecki, Dr Michael Schwartz, and Dr Samuel Banister in making our work known. Without your contributions, the book would not have happened.

Equally deserving of recognition are the research assistants and students in my laboratory who have continuously improved and adapted our methods to the challenges that NPS has thrown at us. The chapter on high-resolution mass spectrometry was informed by their toils. Special thanks to the hands-on researchers in analytical and surveillance methods: Mr Ross Ellison, Mr Andrew Reckers, Dr Axel Adams, Dr Anita Wen, Dr Killian Casagrande, Mr Igor Zakharevich, Mr Spencer Martin, and Mr Jonathan Melamed.

Finally, thanks to my mentors at the University of the Philippines Los Baños, University of Wisconsin Madison, and University of California San

Francisco for the inspiration that has pushed me to constantly strive to be a better scientific version of myself. Your examples provided the reactants that helped synthesize my scientific career, and this book is one of its major by-products.

CHAPTER

1

NPS origins and use

Definition

Change is the only constant in the universe. This is true for recreational drugs as for our evolving lifestyles. "Designer drugs" in the 1980s has morphed into multiple terms: new psychoactive substances, novel psychoactive substances, legal highs, and research chemicals are currently in use, but even experts cannot agree on a formal collective name. In this book, we will call them new psychoactive substances (NPS).

These come in many forms. Depending on their chemical or pharmacological class, they are sold as pills, powders, blotters, injectables, joints, herbs, inhalants, vaping fluids, and edibles (Fig. 1.1). The more common classes are synthetic cannabinoids, synthetic cathinones, new synthetic opioids (NSO, opioid analogs), designer benzodiazepines, amphetamines, phenethylamines, piperazines, and tryptamines. There is usually a popular form associated with an NPS class. Synthetic cannabinoids are available as herbs and joints and are sometimes incorporated in vaping fluids and edibles, while NSO are typically powders, pills, or injectables.

What are NPS? The United Nations Office on Drugs and Crime (UNODC) defines them as "substances of abuse, either in a pure form or a preparation, that are not controlled by the 1961 Single Convention on Narcotic Drugs or the 1971 Convention on Psychotropic Substances, but which may pose as a public health threat" (see the last section of this chapter for more information on these conventions) (UNODC, 2021a). "New" does not necessarily refer to newly created drugs, but rather to substances that have recently become available in the recreational drug market. Hence, some NPS were synthesized and studied by academic and pharmaceutical laboratories decades ago. For example, methcathinone, an analog of methamphetamine, was first synthesized in the 1920s but became a popular NPS in the United States in the 1990s (Gonçalves et al., 2019).

Designer Drugs
https://doi.org/10.1016/B978-0-12-811764-4.00005-7

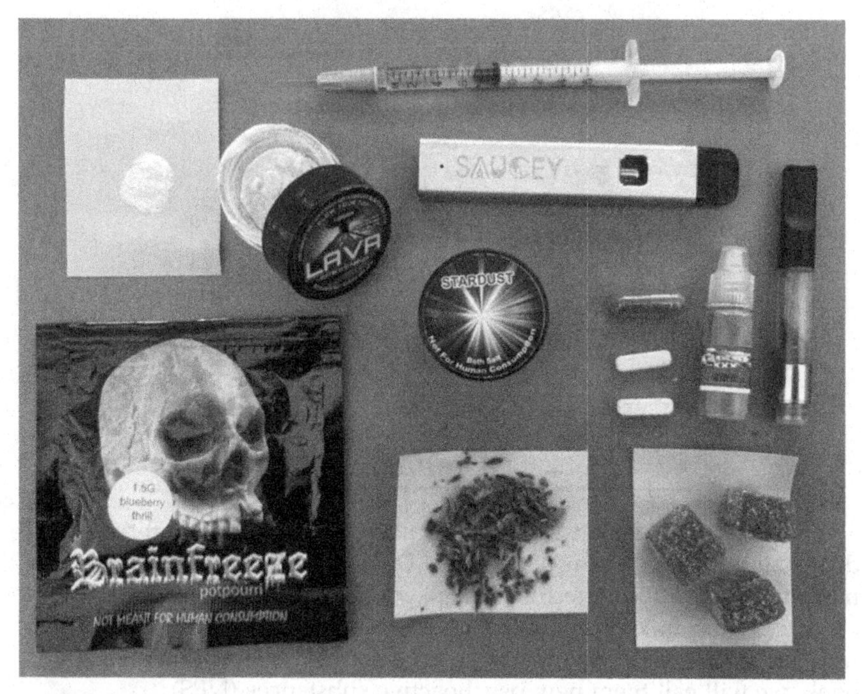

FIGURE 1.1 Common forms of new psychoactive substances (designer drugs).

An NPS can pose a health threat because it exerts pharmacologic effects such as those of traditional recreational drugs (TRD, common drugs of abuse) such as cocaine, methamphetamine, heroin, and cannabis. It does this because it is either a structural or a functional analog of the drug it mimics. Structural analogs have chemical structures very similar to traditional recreational drugs (Fig. 1.2). Acetyl fentanyl, for example, differs from fentanyl only by having one less methyl group. The similarity in structure allows the NPS to bind to the same receptors with which the TRD interacts to facilitate its psychoactive effect. Some receptors can tolerate structural diversity in the molecules they interact with. Hence, an NPS can be a functional analog; i.e., despite its structure not being very similar to that of the TRD, it still binds to the same receptors and facilitates the same functional effect. A lot of synthetic cannabinoids are structurally different from delta-9-tetrahydrocannabinol (THC), the psychoactive component of cannabis, and yet they bind both cannabinoid receptors 1 and 2, sometimes even with stronger affinity than THC. Because of their structural difference, NPS that are functional analogs evade detection by drug screening and legislative control (Banister and Connor, 2018a).

The initial impetus for producing and manufacturing NPS is to provide alternative recreational drugs while evading legislative measures and drug

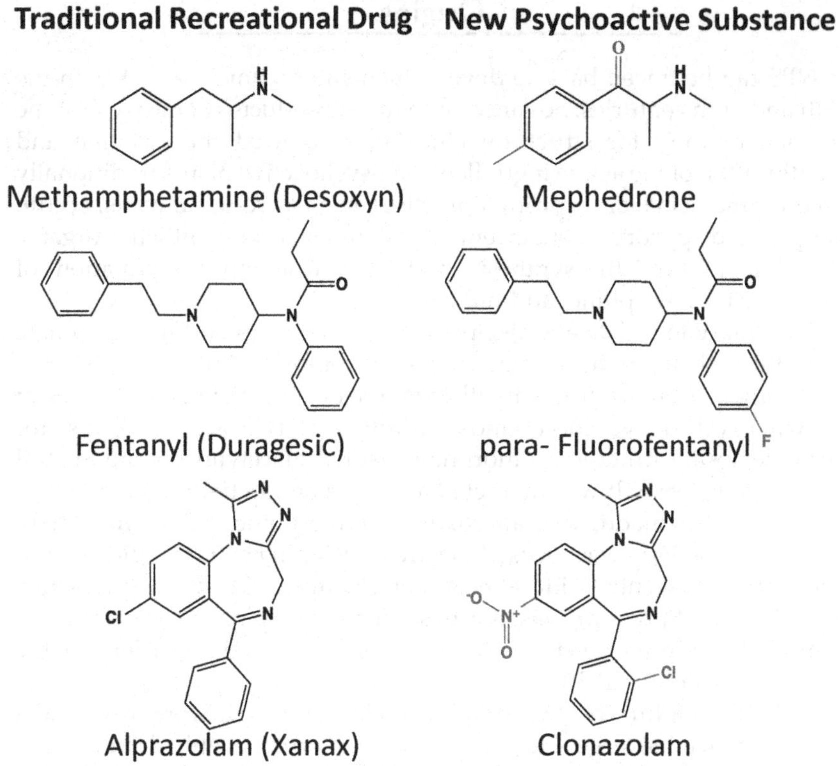

Traditional Recreational Drug | New Psychoactive Substance

Methamphetamine (Desoxyn) | Mephedrone

Fentanyl (Duragesic) | para- Fluorofentanyl

Alprazolam (Xanax) | Clonazolam

FIGURE 1.2 Some new psychoactive substances that are structural analogs of traditional recreational drugs.

screening. Ultimately, most NPS become controlled (Banister and Connor, 2018b). Placing an NPS under schedule triggers clandestine laboratories to hop on other new targets to produce and distribute, thereby creating a vicious cycle of cat and mouse chase between clandestine laboratories and drug regulatory agencies. Not all scheduled NPS disappear from circulation. The synthetic cannabinoid XLR 11 was temporarily scheduled in the United States in July 2012 following its association with several cases of acute kidney injury in the Pacific Northwest (Murphy et al., 2013; Buser et al., 2014). Surveillance studies continued to detect it as late as 2015. In extreme cases, an NPS might transition to becoming a mainstream recreational drug despite being scheduled, as happened with 3,4-methylenedioxymethamphetamine (MDMA), the active ingredient in the party drug Ecstasy. It was originally developed by a pharmaceutical company in 1912 and became popular as a designer street drug in the 1980s (Bernschneider-Reif et al., 2006; DEA, 1988). Even after it was placed under Schedule I by the United States Drug Enforcement Administration (DEA) in 1985, it continued to be one of the most sought-after party drugs. Now it is considered a traditional recreational drug.

Origins

NPS can be traced back to developments in organic chemistry. In the 19th and 20th centuries, advances in natural products chemistry and the creation of tools for structural elucidation allowed the isolation and identification of the active ingredients of psychoactive plants traditionally used for medicinal and spiritual practices, such as coca, cannabis, opium poppy, and peyote. Concurrent developments in synthetic organic chemistry allowed the synthesis, and later structural manipulation, of cocaine, THC, morphine, and mescaline.

The ability to isolate and decipher those structures and to manipulate them through synthetic organic chemistry comprises the first big step in designing new molecules that either mimic or potentiate the activities of known psychoactive compounds. Addition of two acetyl groups, for example, in the structure of morphine created diacetylmorphine in 1898 (Brownstein, 1993). It was marketed as heroin and initially touted to be a more potent, nonaddictive alternative for morphine. Before the 1960s, most of this activity was focused toward developing better medicines and eliminating unwanted side effects. For example, ephedrine (the active ingredient of Ephedra) was the basis for developing amphetamine, a stimulant overprescribed for weight loss and mood disorders in the United States in the 1950s.

Only a handful of the new medicines created were eventually commercialized at large scale (e.g., cocaine, morphine, ephedrine). Many found utility in research as pharmacological tools to probe the body's normal physiology and disease states. Others became chemical building blocks or templates for synthesizing new types of molecules, further expanding the range of NPS.

Many of the synthesized compounds also found their way outside medicine and research and into illicit markets that sprang up for diverted medicines and the products of illicit laboratories. Moreover, when controls were put in place to limit distribution because of the health and social harm they caused, pharmaceutical companies tweaked the structures of controlled psychoactive medicines and boosted the production of their noncontrolled analogs or derivatives. Such was the case for oxycodone in 1918. Hence, creation and distribution of analogs in place of controlled substances is nothing new. What is new is the rapidity by which these analogs evolve in structure, the variety of platforms and markets for their distribution, and the wide range of psychoactive drug classes that can be explored. All these converged with the wide reach of the Internet and catapulted the current public health threat of NPS.

The term designer drugs entered the public consciousness in the late 1970s and early 1980s with the introduction of analogs of the powerful opioid analgesic fentanyl, such as alpha-methylfentanyl, para-fluorofentanyl, and beta-hydroxyfentanyl (King and Kicman, 2011). Produced by illicit laboratories, these compounds were sold as synthetic or fake heroin in the illicit market. Unlike the diverted medicines in the early 20th century, these compounds had little or no history of use. Thus, coupled with their potent effects, fentanyl analogs sometimes proved fatal. The series of deaths in California in 1979 in people who injected drugs were eventually attributed to alpha-methylfentanyl. This was the first recorded novel clandestinely synthesized opioid for which the term designer drug was ascribed.

Following and overlapping with the fentanyl analogs was the creation and production of amphetamine and phenethylamine derivatives in the 1980s. Ring substitution of both drug classes imparted effects other than the typical stimulant effects of amphetamine and its side chain derivatives such as methamphetamine and ethylamphetamine. Instead, the resulting derivatives are either hallucinogens (psychedelics) or entactogens/empathogens. The 2C series of phenethylamines (e.g., 2C–B, 2C–C, 2C–I) are hallucinogens. Most of them were first synthesized by Alexander Shulgin, and their effects were documented in his book *PIHKAL* (Phenethylamines I Have Known And Loved). Another group of hallucinogens are the DOx (4-substituted-2,5-dimethoxyamphetamine) or D series, a class of substituted amphetamines. Both the 2C and Dox/D compounds are based on the structure of mescaline, the psychoactive ingredient of the psychedelic mushroom peyote. MDMA (aka Ecstasy, Molly, E, Beans, Adams) and its closely related analogs MDA (aka Sally, Sass), MDEA (aka Eve), and MBDB (aka Eden, Methyl-J) are the more prominent entactogens. Because they produce experiences of emotional communion (empathy and sympathy) and emotional openness, they became popular in the rave scene, MDMA being the most popular. All these classes were manufactured in clandestine laboratories in the United States and Europe, typically in tablet form bearing characteristic logos (Fig. 1.3). They were usually marketed as Ecstasy, a party drug name that was initially equated to MDMA.

Concurrent with the amphetamine and phenethylamine derivatives, tryptamine derivatives became popular in the 1990s. Like the 2C and DOx compounds, these are hallucinogens, and their structures are based on the alkaloid psilocybin found in magic mushrooms. Shulgin published *TIHKAL* (Tryptamines I Have Known And Loved) in 1997. It systematically documented the effects of various tryptamine derivatives that he synthesized and tried.

Close to the turn of the century, another class of compounds less structurally related to phenethylamine or amphetamine appeared in the

FIGURE 1.3 Different tablets of Ecstasy bearing various types of logos.

illicit market. Piperazines, exemplified by benzylpiperazine, are failed pharmaceuticals; that is, their therapeutic potentials were evaluated by pharmaceutical companies, but they were later abandoned and did not get market authorization. Promoted as safer alternatives to methamphetamine, they are mostly central nervous system stimulants. Soon, various substituted benzylpiperazines and phenylpiperazines appeared in the illicit market; foremost among these are meta-chlorophenylpiperazine (mCPP), meta-trifluoromethylpiperazine (TFMPP), and 3.4-methylenedioxy-1-benzylpiperazine (MDBZP). Like designer drugs in the 1980s, they came first as tablets with markings characteristic of Ecstasy. However, instead of clandestine laboratories, legitimate chemical supply companies, including those in Asia, became the source.

The 21st-century resurgence

At the turn of the century, the primary designer drugs circulating in the United States and Europe were piperazines, phenethylamine and amphetamine derivatives, and a few tryptamine derivatives. Although NPS were continuously being introduced, the pace was relatively slow, and tracking them was manageable. Piperazines introduced a shift in source from small clandestine laboratories to larger chemical supply companies. And because they were not scheduled when they first came out, these drugs were overtly produced and marketed. The growing reach and popularity of the Internet allowed online sale and the creation of chat rooms where their properties and experiences from their use were openly discussed—an online extension of what Shulgin started in *PIHKAL* and *TIHKAL*. Online vendors started to use the term "research chemicals."

Synthetic cannabinoids

Around 2004, a new type started to appear in head shops and on the Internet in Europe. It came in colorful packages containing what appeared to be shredded herbal material laced with a psychoactive drug, usually bearing the warning "not for human consumption." The warning, of course, was meant to skirt regulation and be an added manufacturer's precautionary measure, though the drugs found in the material turned out to be outside the scope of legislative controls. The product came to be popularly known as herbal incense or synthetic marijuana, as it mimicked the effects of cannabis. The first of them was sold under the name Spice in Europe, after the fictional "awareness spectrum drug" in the movie *Dune*. By 2008, the first NPS in Spice was confirmed in Germany to be JWH-018, a synthetic cannabinoid that had been used in cannabimimetic research by John W. Huffman at Clemson University in the 1990s.

Spice did not reach the United States until 2008, where it was popularly known as K2, after the second highest mountain on earth (King and Kicman, 2011). The first detection of synthetic cannabinoid in drug products in the United States was reported by the DEA in November 2008 (Crews, 2013). Because cannabis was very popular but scheduled, while K2 was not legally controlled and mimicked the effects of cannabis, synthetic cannabinoids gained a big following. Furthermore, the structural diversity of synthetic cannabinoids (Fig. 1.4) made them undetectable in routine screening, which added to the appeal by people who wanted to maintain their recreational habit while undergoing regular workplace testing. Spice and K2 were the most popular brands, but hundreds of others were being sold by convenience stores, smoke shops, and Internet retailers from about 2009. Each bag costed between $8 and $20 and typically contained 1 gram of dried herbs. The products evolved from herbal incense to edibles and vaping liquid.

Unlike previous drug classes, where individual drugs were more popular, synthetic cannabinoids are better known as a class, perhaps because there are thousands of possible active structures. More importantly, product composition changes within months, so each drug's presence in the market can be fleeting. Two hundred nine different synthetic cannabinoids were detected in Europe between 2008 and 2020, and roughly 15 subclasses were identified (EMCDDA, 2021a).

Synthetic cathinones

As synthetic cannabinoids were gaining popularity in Europe, cathinone derivatives appeared as alternative stimulants and empathogens among club goers and other enthusiasts of new drugs of abuse. In 2007,

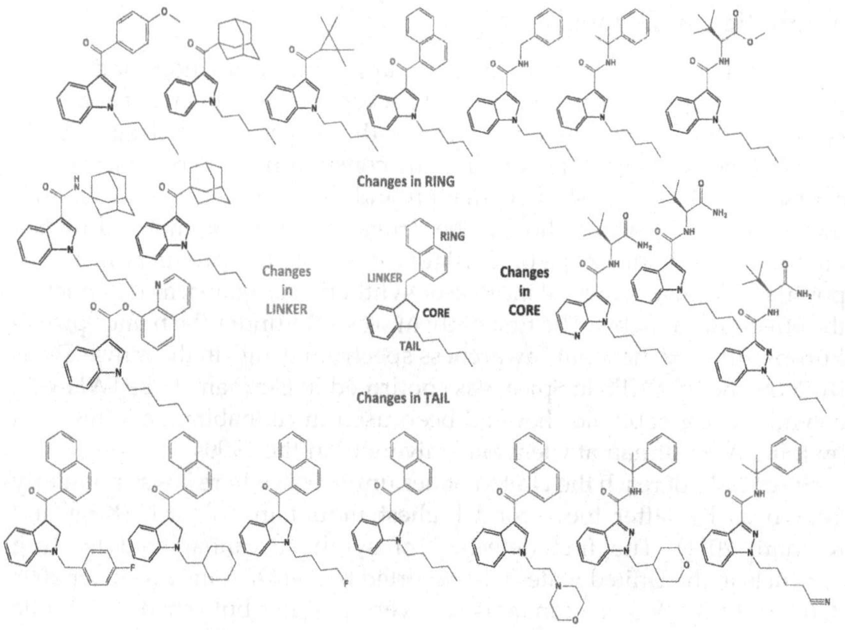

FIGURE 1.4 Structural diversity of synthetic cannabinoids.

cathinones were mentioned in Internet drug forums. They are beta-keto analogs of amphetamine based on the structure of cathinone (Fig. 1.5), a psychoactive ingredient of khat (*Catha edulis*) plant leaves, which are still used in Middle Eastern countries, especially Yemen, for their stimulant properties (Al-Motarreb et al., 2010). Synthesis of the first derivatives, methcathinone and mephedrone, dates to 1928 and 1929, respectively (Hyde et al., 1928; Sanchez, 1929). Methcathinone was used as an anti-depressant in the former Soviet Union in the 1930s and 1940s. As a street drug, it is known as Cat and Jeff. Methcathinone became popular in the United States in the 1990s, but the resurgence of synthetic cathinones did not reach the United States until 2010 where they are called bath salts (Baumann et al., 2014).

Like synthetic cannabinoids, synthetic cathinones are known as a class but a few became popular on their own. In Europe, mephedrone (4-methylmethcathinone) was the first. In the United States, it was mephedrone, methylone, and 3,4-methylendioxypyrovalerone (MDPV). Alpha-pyrrolidinovalerophenone (alpha-PVP, Flakka in the argot) was popular by 2014. Bath salts come in powders or tablets and were sold initially in small tubs containing 0.5–1 g of powder, with colorful designs, priced at $20 to $30. Some products contain a single drug but most have several and can be laced with lidocaine, caffeine, and other stimulants (Schneir et al., 2014). The club drug Molly (slang for

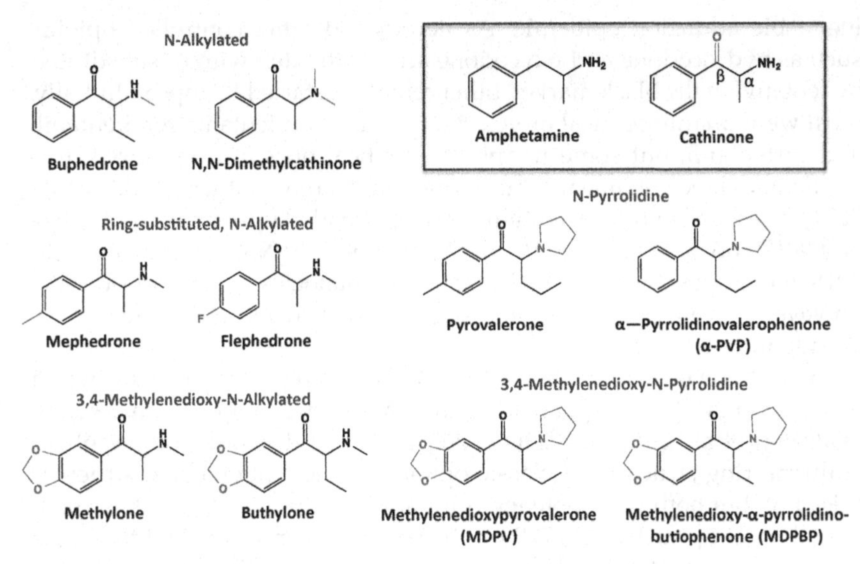

FIGURE 1.5 Different subclasses of synthetic cathinones.

molecular), which is supposed to be the pure crystal powder form of MDMA, has also been found to contain synthetic cathinones. There are hundreds of possible derivatives; between 2008 and 2020, 156 were reported to the Early Warning System (EMCDDA, 2021a).

It is around the time when synthetic cannabinoids and cathinones became popular that the term "legal highs" was introduced. Unlike "research chemicals," this is a more encompassing term that includes both synthetic and plant-derived psychoactive substances and products that are unregulated (EMCDDA and Europol, 2011). Hence, aside from piperazines, amphetamines, synthetic cannabinoids, and synthetic cathinones, the term also applies to plant-based substances such as kratom (leaf of the Southeast Asian plant *Mitragyna speciosa*, with a psychoactive substance that binds the opiate receptors at lower doses but acts as a sedative at higher doses) and *Salvia divinorum* (a hallucinogen grown in Mexico) that have gained popularity since the turn of the century. Like research chemicals, most legal highs are produced mainly in laboratories in Asia and are legally imported as chemicals or packaged products in the United States, Europe, and Oceania (UNODC, 2013a).

New synthetic opioids

By 2012, the resurgence of this class, primarily fentanyl analogs (fentalogs), had begun in Europe. It spread rapidly to the United States, where opiates had long been overprescribed, which set the country up for an inevitable addiction epidemic. As heroin and other controlled opioids such as hydrocodone and oxycodone are adulterated with or substituted by fentanyl in the black market, other fentalogs started to appear. Initially, most were pharmaceutical or veterinary, such as sufentanil, remifentanil, and carfentanil, but some nonpharmaceutical ones also appeared: furanylfentanyl, ocfentanil, and butyrylfentanyl, among others (Lukic et al., 2021). The number of new fentanyl analogs trended upward between 2014 and 2017 and accounted for the highest number of NSO introduced until 2018. Increasing control and regulation of fentalogs as a class worldwide, however, forced the recreational drug market to resort to other types (Vandeputte et al., 2021).

In 2012, the benzamide analog AH-7921 was reported in Norway and Sweden; in 2013, the piperazine analog MT-45 was reported in Sweden (Katselou et al., 2015; EMCDDA, 2021b). These do not share the phenanthrene ring structure of classic opioids or the fentanyl ring structure (Fig. 1.6), but both have potency comparable with morphine. Other benzamide analogs such as U-47700, a structural isomer of AH-7921 that is 7.5 times more potent than morphine, were introduced later. Recently, the number of benzimidazole opioids introduced has overtaken the fentalogs. Benzimidazole opioids were synthesized in the late 1950s in an attempt to find alternative opioid analgesics (Hunger et al., 1957). Although not a

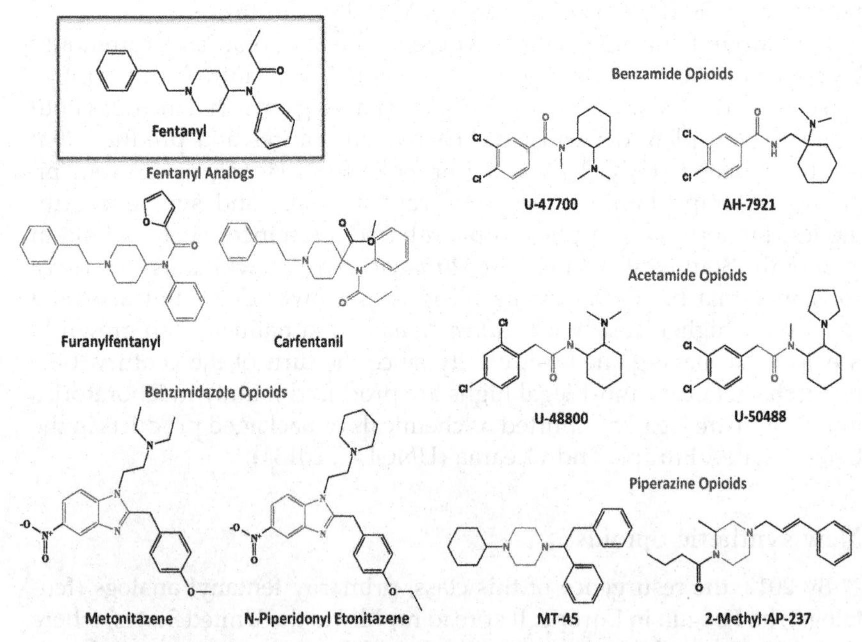

FIGURE 1.6 Diversity in molecular structures of new synthetic opioids.

single drug in the class gained market authorization, extensive structure–activity relationship and pharmacological studies revealed some drugs with potencies orders of magnitude greater than morphine (Ujvary et al., 2021). Isotonitazene was observed to have 500 times the potency of morphine in vitro and was the first member of this class detected in Europe, in April 2019 (WHO, 2020). Soon, other nitazenes appeared, including etonitazine, metonitazene, and butonitazene.

Sixty-seven NSO were detected in Europe between 2008 and 2020 (EMCDDA, 2021a). Although this is fewer than the synthetic cannabinoids and cathinones, the number of potential NSO analogs is in the thousands from fentanyl alone. Moreover, because NSO pose a high risk of life-threatening poisoning from respiratory depression, many countries were quicker to impose regulatory controls on NSO than on other NPS classes. NSO are sold as powder, tablets, and injectables. Recently, they have also been incorporated in vaping liquid and inhalants.

Designer benzodiazepines

An increase in the number of designer benzodiazepines accompanied the rise of NSO in 2012. Structures and clinical effects are similar to those of prescription benzodiazepines such as diazepam (Valium), lorazepam (Ativan), and alprazolam (Xanax), but they have no approved medical uses or have never been formally studied. Although the first designer benzodiazepines, phenazepam and etizolam, came out in 2007 and 2011, respectively, it was not until 2012 that they became readily available in the recreational drug market and were initially distributed by online retailers as research chemicals. Flubromazepam, pyrazolam, and diclazepam were the first ones sold online (Zawilska and Wojcieszak, 2019). In Europe, the most popular until 2017 were phenazepam, etizolam, flubromazolam, and diclazepam (Moosmann et al., 2015).

Designer benzodiazepines were sold as tablets, capsules, powders, and blotters. Some were sold under their own names, marketed as legal versions of scheduled prescription benzodiazepines, but more often as fake versions. Some were combined with other NPS, often an NSO, sometimes a synthetic cannabinoid, counterfeit prescription benzodiazepine, or controlled opioid (Orsolini et al., 2020). Between 2008 and 2020, 30 designer benzodiazepines were detected in Europe (EMCDDA, 2021a).

Other classes

Aside from these four classes, a diverse group of new ones have flooded the recreational drug market since 2008. They range from derivatives of arylcyclohexylamines (e.g., phencyclidine), ketamine, and pipradrol to benzofurans, aminoindanes, and drugs related to cocaine and LSD

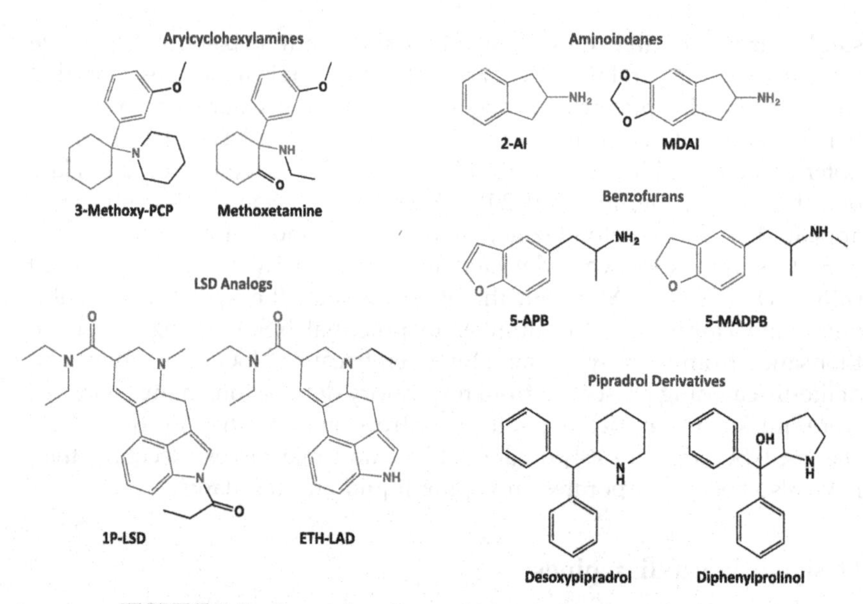

FIGURE 1.7 Some other classes of new psychoactive substances.

(Fig. 1.7) (King and Kicman, 2011). On top of this, drug classes that came out in the last century such as phenethylamines, amphetamines, tryptamines, and piperazines have remained, and chemical motifs from these classes are incorporated into the new classes.

The resurgence of NPS in the 21st century has been furious and alarming. The sheer number of potential structures from the scaffolds of the NPS classes is unprecedented. Along with the availability of information about failed drug candidates in expired patents and the accessibility of synthetic scheme– and/or structure–activity relationship data on tested candidates in the scientific literature, this has allowed rapid evolution since 2010, to a point where drug reference standard manufacturers are not able to keep up. The availability of reference standards is critical in NPS detection, identification, and confirmation in drug products and biological samples. Hence, their scarcity has impeded forensic and clinical laboratories in responding to NPS outbreaks and contributed to proliferation without ample checks. The ease by which information about new drugs and experiences is shared over the Internet, the accessibility of retail markets in the clear web and dark web, the evolution of production in large chemical companies in Asia, and the ease by which the products are transported between continents by couriers and cargo ships comprise synergistic tools that created the perfect storm for NPS to become a worldwide public health threat.

Production, marketing, and supply

One thousand one hundred eighty-two NPS had been reported by national authorities and forensic laboratories in 139 countries by the end of 2022, more than 80% of which were identified in the latter 12 years (UNODC, 2023). This is more than thrice the number (292) that are currently under international control. In each year from 2012 through 2016, more than 60 were reported for the first time. This number has now settled close to 50 (EMCDDA, 2021a). The largest markets for NPS from their resurgence in 2009 until 2017 were the United States, Western Europe, and Oceania (Australia and New Zealand), while the origin of most NPS was (and remains) China. Starting in 2019, the distribution as reflected in NPS seizures has significantly changed. With declining seizures in the United States and Europe, the market is expanding globally, with increasing drug seizures in Asia and Africa (UNODC, 2021a, 2021b).

Unlike designer drugs in the 1980s, which were made by small clandestine laboratories, they are now made by large chemical and pharmaceutical companies employing scientists who are actively consulting the scientific literature for potential targets. Operating on an industrial scale, most of these companies are in Asia, primarily China. Highly pure NPS that range in scale from a few milligrams to hundreds of kilograms are advertised online as research chemicals or dietary supplements. Some companies offer custom synthesis. Some sell ingredients, packaging materials, and formulation equipment (for pressing and compounding tablets and preparing other dosage forms) (Sumnall et al., 2011).

Substances ordered in small amounts are shipped to wholesalers, retailers, and dealers in Europe and the Americas by express mail and courier services. Bulk quantities are shipped by air and sea cargo. To conceal their nature and avoid scrutiny by customs and regulatory agencies, shipments are often misdeclared as foodstuffs, common household items, and other chemical products. This practice has led to a few deadly consequences when misdeclared items were used by unsuspecting buyers for what they were declared to be (EMCDDA, 2016).

After a bulk shipment reaches Europe or the United States, it is processed into branded products as herbal incense, powders, tablets, or other forms that are sold in head shops, smoke shops, and convenience stores openly or under the counter. They can also be advertised and sold online, usually as research chemicals, legal highs, or dietary supplements (Corazza et al., 2014). Sales can also be through existing street-level drug markets. Most products come in colorful packages bearing suggestive names and warnings like "not for human consumption" to sidestep consumer protection laws.

In recent years, NPS have been incorporated in a variety of other products. Synthetic cannabinoids and NSO are added to vaping liquids; fentanyl and its analogs are put into inhalants and patches; synthetic cannabinoids are added to edibles. Dealers are also repackaging smaller quantities as "rocks" or pressed into tablets or other forms that are then sold in the illicit market. Designer benzodiazepines and NSO, for example, are commonly compounded into tablets to produce counterfeit prescribed benzodiazepines or opioids. In the United States, they are often sold as fake Norco, Vicodin, Oxycontin, and Xanax (Compto and Jones, 2019; Orsolini et al., 2020).

Increased monitoring of NPS sales on the clear web has forced online vendors to use the dark web (part of the Internet not accessible to search engines and requiring an anonymizing browser). Silk Road was the first major darknet market for NPS. These markets disappear with law enforcement action, but new ones quickly emerge. The largest after Silk Road are AlphaBay and Dream Market. In the past couple of years, Hydra Market and Cannazon have been leading NPS sites on the dark web (Miliano et al., 2018; UNODC, 2021a).

The increased regulation imposed on fentanyl and fentalogs in China, the United States, and the European Union has led to the direct marketing of their precursors. The profit potential of illegally manufactured fentanyl has attracted the interest of Mexican drug cartels in trafficking fentanyl and fentalogs into the United States. A kilogram of illicit fentanyl from China typically costs $5000 and can be converted to products with a market value upward of $1.5 million (Klobucista, 2021). Precursors are manufactured in China and sent by express mail or courier to these cartels, and then smuggled across the southern border of the United States. Once inside, they are moved to stash houses for NPS production and distribution. The Sinaloa Cartel and the Jalisco New Generation Cartel are known to be most associated with trafficking illegally manufactured fentanyl into the United States (Pergolizzi et al., 2021).

Patterns of use, effects, and toxicity

The wide diversity of NPS classes, the rapid evolution of NPS products' composition, the limited knowledge of users about the substances they actually use, and the uncertainty about which substances are considered NPS all made it initially difficult to acquire data on prevalence and use patterns (Klobucista, 2021). Hence, most of the data were initially limited to specific locations and specialized groups of the population where smaller, more directed surveys were conducted. In the 2011 Eurobarometer survey, 4.8% of the 12,000 randomly selected individuals aged 15 to 24 years (young adults, the age group with the highest drug use) responded that they had experimented with legal substances that imitate the effects of illicit drugs in their lifetime. This trended up in succeeding years. The first formal Eurobarometer survey on NPS, in 2014, showed 8% lifetime and 3% past-year prevalence among that age group. The range for these data is wide among the countries of Europe. As with illicit drugs, NPS use is more widespread among males, those living in metropolitan areas, and the unemployed (EMCDDA and Europo, 2011; UNODC, 2017).

As soon as synthetic cannabinoids and cathinones spread into the United States, the country became the biggest NPS market in the world. In the first 6 months of 2012, for example, 62 NPS were identified, the largest number reported by a single country in this period to UNODC (UNODC, 2013a). In the 2013 Global Drug Survey, close to one in five respondents from the United States indicated use of research chemicals or legal highs in the past 12 months. Second in this survey was the United Kingdom, with a little over 13% (Global Drug Survey, 2014). Synthetic cannabinoids were the most popular with young adults; the prevalence of past-year use among 12th-grade students in 2011 was 11.4% (Johnston et al., 2016).

Some of the motivations for use are similar to those for traditional recreational drugs, including the desire to modify perception, sustain high energy or other functional expectations, enhance sociability, satisfy curiosity, reduce boredom, and increase peer socialization (Greenblatt and Greenblatt, 2019). Motivations unique to NPS use include easier accessibility, lower price, presumed better quality compared with traditional recreational drugs, safety from prosecution with the perceived legal status of NPS, and assumed nondetectability in drug screening assays (UNODC, 2017). Specific social settings such as visiting a bar, pub, or nightclub, attending a rave, concert, or "electro" music scene, and private parties promote the likelihood of NPS use (UNODC, 2013b). Polydrug use is also common, either intentionally or unknowingly because the product consists of multiple drugs.

The methods by which NPS products are used are dependent on how the traditional recreational drugs they mimic are used. Synthetic cannabinoids are usually smoked as herbal incense or e-cigarettes, or orally ingested if incorporated in edibles. Synthetic cathinones, amphetamines, phenethylamines, piperazines, tryptamines, other stimulants, empathogens, and hallucinogens are snorted or smoked as powders or ingested as pills or tablets. NSO are usually smoked, ingested, or injected. Designer benzodiazepines are typically ingested.

Because of the diversity of drug classes that comprise NPS, they facilitate a wide range of effects (Table 1.1). Synthetic cannabinoids functionally mimic THC. Most synthetic cathinones such as amphetamines are stimulants, but some are also entactogens or empathogens. Phenethylamines are primarily hallucinogens, but some are also

TABLE 1.1 Pharmacological actions and effects of different new psychoactive substance classes.

NPS class	Pharmacological action	Effects
Amphetamines, phenethylamines	Serotoninergic receptor agonists Monoamine reuptake inhibitors	Euphoria Increased wakefulness, improved cognitive control Increased muscle strength, fatigue resistance Increased blood pressure, heart rate, body temperature Hallucination, dissociation Sociability, increased libido
Piperazines	Facilitate release of dopamine and noradrenaline Monoamine reuptake inhibitors	Euphoria Increased body temperature Hallucination Sociability
Tryptamines	5HT2A receptor agonists Serotonin reuptake inhibitors	Alteration in sensory perception Visual hallucination Depersonalization
Synthetic cathinones	Act on serotonin, dopamine, noradrenaline pathways	Euphoria and elevated mood Increased energy and wakefulness Increased blood pressure, heart rate, body temperature Aggression, paranoia
Synthetic cannabinoids	CB1 and CB2 receptors agonists	Euphoria, anxiety reduction, and antidepressant-like effects Analgesia (inability to feel pain) Visual/auditory hallucinations
Arylcyclohexylamines	5HT2A receptor agonists NMDA receptor antagonists High affinity for opioid receptors	Sight and sound perception distortion Dissociation, out of body hallucination

TABLE 1.1 Pharmacological actions and effects of different new psychoactive substance classes.—cont'd

NPS class	Pharmacological action	Effects
New synthetic opioids	Mu opioid receptor agonists	Euphoria Analgesia (inability to feel pain) Reduction of consciousness Reduced respiratory rate
Designer benzodiazepines	Facilitate GABA receptor binding	Calming effect Anxiety reduction Increased sedation

stimulants. Piperazines have stimulant effects but in some combinations can also induce empathogenic effects. Tryptamines primarily have hallucinogenic effects. Aminoindanes and pipradrol derivatives have stimulant effects. Like heroin, NSO have pain-relieving (analgesic), sedative, and depressant effects, but most users are after the strong euphoria they temporarily induce. Designer benzodiazepines, like their prescription counterparts, impart a sense of calmness (anxiolysis) and have sedative effects. Arylcyclohexylamines have euphoriant, stimulant, and dissociative effects; some, notably phencyclidine, induce violent behavior (UNODC, 2013a; UNODC, 2013b).

Most NPS have never been thoroughly studied pharmacologically; neither have their toxicological profiles been systematically assessed. Hence, their toxic effects are not known. Most of what we generally know about their toxicity is from reports of intoxication cases and extrapolations from limited in vitro and in vivo pharmacological studies (Table 1.2).

Although most toxicity from NPS use is probably due to misdosing, a significant number of NPS are more potent than the traditional recreational drug they mimic. Some synthetic cannabinoids, for example, including their metabolites, are full agonists of both cannabinoid receptors 1 and 2, while THC is only a partial agonist. Synthetic cannabinoids have been reported to cause seizures, cardiovascular problems including tachycardia and arrhythmia, and psychological disorders. Their use has also been linked to addiction and withdrawal symptoms. A few synthetic cannabinoids are also linked to symptoms that are not often associated with THC toxicity, such as acute kidney injury. As more potent synthetic cannabinoids have come out, fatal intoxications have been reported.

Like amphetamines, synthetic cathinones cause cardiac (tachycardia and hypertension), psychiatric (mild agitation to severe psychosis), and neurological effects (seizures and hyperthermia). In extreme cases of

TABLE 1.2 Toxic effects of different new psychoactive substances classes.

NPS class	Toxic effects
Amphetamines, phenethylamines	Hypertension, hyperthermia; vomiting, convulsion Hallucination, dissociation Liver and kidney failure; rhabdomyolysis; death
Piperazines	Hyperthermia; convulsion Hallucination; kidney failure; death at high doses
Tryptamines	Vomiting, abnormal sweating; headache Tachycardia and irregular heartbeat Hypertension; hyperthermia Hallucination; impaired movement
Synthetic cathinones	Hypertension, hyperthermia Restlessness, agitation, paranoia Abdominal pain, vertigo Convulsion; rhabdomyolysis; death
Synthetic cannabinoids	Psychosis, panic, paranoia Tachycardia Somnolence Convulsion, vomiting, seizures Acute kidney injury
Arylcyclohexylamines	Hypertension, tachycardia Impaired movement Dissociation, catatonia
New synthetic opioids	Bradycardia, hypotension, cyanosis Vomiting, nausea, stoppage of bowel movement Respiratory failure, pulmonary edema; death
Designer benzodiazepines	Drowsiness, coma-like stupor Cognitive impairment Impaired movement and coordination; slow reaction time

hyperthermia, rhabdomyolysis (breakdown of muscles) and renal failure also occur. Deaths related to synthetic cathinone intoxications were reported early on, especially for MDPV. Amphetamines and phenethylamines have similar toxic effects, with some subclasses significantly more potent than others (e.g., DOx/D series, FLY series, NBOMes). Common adverse effects reported for these two classes include hallucinations, seizures, mydriasis (dilatation of the eye pupil), tachycardia, hyperthermia, rhabdomyolysis, severe limb ischemia, and liver and renal failure. Deaths have been associated with the more potent FLY and NBOMe series. Piperazines, being stimulants, also have these effects, specifically toxic seizures, respiratory acidosis, hyperthermia, rhabdomyolysis, and renal failure. Fatal intoxications from BZP and TFMPP have also been reported.

NSO have the toxic effects seen for common opioids such as heroin and morphine. Pinpoint pupils, stupor, cold skin, cyanosis (blue discoloration of the skin), coma, and respiratory depression leading to death are common. Respiratory depression makes NSO significantly more fatal than other NPS classes, especially with the very high potencies of some members such as carfentanil and etonitazene, which are 10,000 and 1000 times more potent than morphine, respectively. Designer benzodiazepines are generally less toxic than NSO. Their toxicity is very similar to that of prescription benzodiazepines and includes slurred speech, lethargy, psychomotor impairment and ataxia (poor balance and coordination from cerebellar damage), slowed reaction time, and memory impairment (Greenblatt and Greenblatt, 2019). Designer benzodiazepines in combination with other drugs such as alcohol and opioids can also be fatal.

Detection and identification of use

Identification of NPS is critical in monitoring developments in the rapidly changing market as well as resolving intoxication cases. However, detection in biological samples is quite challenging given the unique characteristics of NPS. The rapid evolution of product composition poses the biggest challenge. The pace in the past decade has made it impossible for drug reference standard manufacturers to keep up; hence, reference standards for newly introduced NPS are almost always unavailable. Without a standard, a drug cannot be confirmed in samples, nor can a method for it be developed and validated. Rapid evolution also renders traditional targeted testing using liquid chromatography–tandem mass spectrometry (LC-MS/MS) and gas chromatography–mass spectrometry (GC-MS) useless in detecting previously unreported NPS. With 50 to 100 new NPS introduced each year, a significant number will be missed with targeted testing.

The brevity of popularity of most NPS makes it challenging to add them to existing targeted drug assays. Method development and validation for a drug takes about 4–6 months in clinical and forensic laboratories and a couple more months before implementation. By the time a method is available, one or two new generations have come out. A significant number of NPS are rapidly and/or extensively metabolized, adding to the complexity. When the test sample is urine, a metabolite may be a better target; the problem is that, with limited pharmacological studies, the most common metabolite of a newly introduced NPS is usually not known.

Notwithstanding the challenges, several methods for analyzing NPS classes have been developed in multiple biological samples, including urine, serum/plasma, whole blood, oral fluid, hair, and other tissues. As with traditional recreational drug testing, there are two steps: screening and confirmation. Screening qualitatively detects the presence or absence of a class, and confirmation establishes the identity of the NPS and its quantitative level (Table 1.3).

Immunoassays comprise the most common method for drug screening. The standard urine screens for opiates, amphetamines, cocaine, cannabinoids, benzodiazepines, and tricyclic antidepressants are immunoassays. They are developed using antibodies raised against a common motif of drugs in a class. A few NPS—some designer benzodiazepines, synthetic cathinones, and fentanyl analogs—can cross-react with immunoassays for prescription benzodiazepines, amphetamines, and fentanyl, respectively. Generally, however, NPS evade detection by the screen for the traditional recreational drug they mimic. This is one feature that makes NPS more attractive to some users than traditional recreational drugs. It is especially true for synthetic cannabinoids, where there is wide structural diversity. Some immunoassays have been developed specifically for synthetic cannabinoids, synthetic cathinones, and fentanyl

TABLE 1.3 Characteristics of different methods used to analyze new psychoactive substances.

Method	Strength	Limitations	Applications
Colorimetric	Rapid Compatible with on-site testing Low cost	Very low specificity and selectivity Low sensitivity Limited coverage	Screening
Immunoassay	Rapid Moderate sensitivity Lower cost	Low specificity and selectivity Limited coverage	Screening
Chromatography	Moderate sensitivity Quantitative	Low specificity and selectivity Limited coverage	Targeted analysis Quantitative analysis
Mass spectrometry	High sensitivity Quantitative Wide coverage Wide dynamic range of detection	Longer turnaround time Costly Requires more technical expertise	Targeted analysis Nontargeted analysis Quantitative analysis MS/MS spectral collection Structure elucidation

analogs. However, with the rapid evolution and diversity of molecular structures within classes, these assays have limited temporal utility and are fraught with false negatives. Moreover, because a significant number of NPS are extensively metabolized, targeting the metabolites should be given due consideration in developing immunoassays. This is hard to do when the metabolism and pharmacokinetics of most NPS are totally unknown.

LC-MS/MS and GC-MS are the primary platforms for targeted confirmatory testing. LC-MS/MS is the gold standard for quantitative analysis in clinical laboratories. Using selective reaction monitoring, high selectivity and sensitivity for the target analytes are achieved. The detectors are also capable of routinely achieving at least four orders of magnitude in linear dynamic range for most analytes. Various groups have published targeted assays for NPS classes in urine, serum/plasma, and whole blood. A few assays in hair and oral fluid have also been developed. Targeted assays are generally limited to previously reported NPS. Hence, as with screening tests, the fleeting popularity of most NPS and the rapidity with which new ones are introduced limit the useful lifetime of targeted assays.

The need to detect previously unreported NPS forced laboratories to explore alternative platforms and modalities to collect data that can be queried for an unknown that was not targeted a priori. High-resolution mass spectrometry (HRMS) provided this capability; the most common platform is liquid chromatography–quadrupole time-of-flight mass spectrometry (LC-QTOF/MS). Unlike LC-MS/MS, LC-QTOF/MS acquires data on all ions obtained from a sample (nontargeted data acquisition). The total ion chromatogram (TIC) obtained from a sample run can then be queried in either a targeted or nontargeted manner. The targeted way uses databases of target analytes assembled from a set of available reference standards. This is not any different from targeted analysis using LC-MS/MS except that the accurate mass of the parent ion and its fragmentation pattern (if collected) is used for confirmation along with the analyte's retention time. The most useful feature of LC-QTOF/MS is its ability to facilitate suspect screening and nontargeted data analysis. In the absence of a reference standard, a tentative match (suspect) can be assigned to a peak of interest using accurate mass and fragmentation pattern matches along with the retention time plausibility of the suspect. This capability has allowed discovery of previously unreported NPS and has been very useful in surveillance work. Moreover, nontargeted data acquisition allows retrospective data analysis of previously acquired data. This has been very useful in determining whether a newly discovered NPS has in fact been lurking in the past. Various groups have published and assessed LC-QTOF/MS methods for targeted analysis and suspect screening. In contrast with LC-MS/MS assays, most LC-QTOF/MS assays are comprehensive methods that allow the analysis of multiple classes of NPS in one run.

Control and regulation

NPS has challenged the control and regulation of recreational drugs worldwide. International control of drugs is governed by three conventions: (1) the 1961 Single Convention on Narcotic Drugs; (2) the 1971 Convention on Psychotropic Substances; and (3) the 1988 Convention against Illicit Traffic in Narcotic Drugs and Psychotropic Substances. The 1961 convention is focused on cannabis, cocaine, and opioids (UN, 1961), while the 1971 convention is focused on synthetics such as methamphetamine, MDMA, and LSD (UN, 1971). The 1988 convention covers police suppression of illicit markets (money laundering) and control of drug precursor chemicals (UN, 1988; Room and Reuter, 2012).

Addition of new drugs to the 1961 and 1971 conventions is a lengthy process. For a drug to be controlled under either, the risks and abuse potential must be reviewed by the World Health Organization's (WHO, 2020) Expert Committee on Drug Dependence (ECDD). This requires gathering scientific evidence on the potential for harm and the likelihood of causing dependence. These data are usually scant or not available for newly introduced NPS. If the ECDD recommends a drug to be controlled, the recommendation must pass through the WHO Director General to the United Nations Secretary General and ultimately to the Commission on Narcotic Drugs, whose member states vote on scheduling in its annual meeting. Once a drug is scheduled, each member state is required to regulate it. There are currently 186 and 184 member states to the 1961 and 1971 conventions, respectively. As of March 2022, only 71 have been scheduled under the international conventions since the ECDD first examined NPS in 2014 (UNODC, 2023). This is less than 10% of the 1182 NPS that had been reported to UNODC by the end of 2022, a testament to how out of sync international regulation is with evolution of the market.

Without the multilayered requirements intrinsic in placing an NPS under international control, regulation at the regional and national levels happens more quickly and has greater legislative flexibility. Legislation in most countries is based on scheduling of individual substances. In the United States, for example, the Controlled Substances Act signed into law in 1970 established five schedules based on three criteria: (1) potential for abuse, (2) accepted medical use, and (3) safety and potential for addiction. Those with the highest potential for abuse and addiction without medical use are under Schedule I (highest), and those with low potential for abuse and mild potential for addiction that have medical use are under Schedule V (lowest). The United Kingdom's Misuse of Drugs Act of 1971 has four classes: A, B, C, and temporary. Scheduling is done on an individual drug basis.

With the onslaught of NPS starting in 2011, countries used temporary bans to quickly control those thought to pose a significant threat to public health pending decisions on scheduling. This also allowed gathering of substantial evidence. But even this approach could not respond efficiently to the legislative challenge. Placing NPS one by one proved burdensome as more and more were released into the recreational market at an ever-faster pace. Some countries adopted policies that control classes instead of individual agents. Examples are the Synthetic Drug Abuse Prevention Act of 2012 (SDAPA) in the United States and the Psychoactive Substances Act of 2012 (PSA) in the United Kingdom. SDAPA placed 26 types of synthetic cannabinoids and synthetic cathinones into Schedule I, including their specified analogs, while PSA placed "legal highs" as controlled substances. With the use of temporary bans and broader class scheduling, the United States has now put more than 200 NPS under Schedule I since it started temporary scheduling in 2011, evidence that control proceeds faster at the national than at the international level.

At present, NPS control for most parts of the world continues to be reactive. Proactive approaches such as imposing blanket bans on all psychoactive substances remain controversial. On the one hand, this approach would place all future NPS under legislative control, which may deter the rapid pace by which new molecular structures are introduced. However, such all-inclusive bans are not without problems. For one, they can stifle research into the potential therapeutic benefits and legitimate medicinal uses of some NPS that have a low risk of harm. For example, bupropion (Wellbutrin), an FDA-approved antidepressant in the United States, is a synthetic cathinone; other synthetic cathinones may have use as therapeutic agents for trauma and depression. This approach is also difficult to enforce, especially if all member states of the international conventions are required to regulate them. Comprehensive NPS testing is available in only a handful of countries, and experts have yet to agree on an exact definition of "psychoactive" that can be universally adopted.

Close-up: Bath salts

In 2010, clinical toxicologists like myself in the United States became aware of the emergence of a class of NPS being sold as "bath salts." Some burgeoning literature at the time had identified these products to contain primarily various synthetic cathinones. Cathinone, an amphetamine-like sympathomimetic amine, is the main psychoactive compound found in the leaves of khat (*Catha edulis*), a plant native to the Arabian Peninsula and East Africa. Most medical toxicologists were

continued

Close-up: Bath salts (*cont'd*)

likely already familiar with certain populations who regularly chewed the plant, and that sporadic reports had previously detailed recreational use of either pure cathinone or synthetic cathinones. Stimulant-related complications, as expected, were described in these situations. Chewing of khat, its stimulant properties, and the subsequent "crash" that may occur were detailed in the book *Black Hawk Down*, including how US forces considered the timing of their operations in Somalia in relation to these effects (Bowden, 1999).

As has often been the case with NPS, synthetic cathinones appeared first in Europe, and studies there detailed that they were being sold mainly over the internet as "legal highs," "research chemicals," and "plant food." One explanation for the emergence of cathinone derivatives at the time was a demand for alternatives to the drug Ecstasy (MDMA), following a concerted effort in some countries to restrict availability of the plant-derived compound safrole, from which MDMA can be synthesized.

In the United States, although bath salts were available over the Internet, they were also being sold at various stores, including local gas stations, smoke shops, and "head shops." Toxicologists in the Unitd States began to see serious complications from use, such as stimulant-associated cardiovascular, neurologic, and psychopathological problems and even death. It was in this context that colleagues of mine in California became aware that Dr. Roy Gerona at UCSF was using liquid chromatography—time-of-flight mass spectrometry to analyze NPS. As Roy details in this book, the technique is ideal for analyzing NPS.

In collaboration with Roy, multiple colleagues of mine in California embarked on a descriptive study to determine what was in the bath salts being sold over the Internet and at shops in California. A couple of US-based studies had previously analyzed some of these products in different states, but we were intent on doing a more comprehensive study. I was also intent on having quantitative analysis of the products, an aspect that is rarely done in such studies. We purchased samples of these products at various California sites over a specified time period. I also purchased multiple products via the Internet. There was concern among some of the physicians about the potential legal ramifications of buying drugs that were illegal or soon would be. My impression has always been that law enforcement has far greater matters of interest than physicians purchasing NPS for a research study. As it turned out, one physician was in fact stopped by a police officer outside of where he had purchased bath salts. The officer turned out to be concerned about why the respectable looking physician was in the neighborhood, rather than what he had purchased.

Close-up: Bath salts *(cont'd)*

The 35 bath salts we purchased had a variety of colorful names (Fig. 1.8). Most had labeling warning against use, often "not for human consumption." This is not uncommon for NPS products and is undoubtedly an attempt to skirt the US Federal Analog Act (21 USC 813) of 1986 that addressed the synthetic manipulation of previously scheduled drugs. My favorite warning was "This product is not intended to be taken internally or externally by any mammal on planet earth." The majority of products we purchased had synthetic cathinones present, either alone or in different combinations, sometimes in high quantity. We identified multiple synthetic cathinones, along with designer amines that had not been previously identified in US bath salt products. Quantification revealed in some cases dramatic differences in either total synthetic cathinone or stimulant content between products with the same declared weight, and even between identically named and outwardly similar products (Schneir et al., 2014).

By far the most challenging aspect of the entire study for me was summarizing the laboratory technique liquid chromatography–time-of-flight mass spectrometry, which the reviewers insisted I do. Although I was generally aware of how it was performed and its advantages for analyzing NPS, describing it made me acutely aware of how little I understood it. I sent many emails to Roy asking a variety of questions that revealed my ignorance of some of the basic science it derives from. This ended up being a great collaboration between a clinical toxicologist and a laboratory expert. I would have greatly benefitted from having this book Roy has written as a resource (and likely would have simply stated "see book for a description of the technique").

FIGURE 1.8 A sampling of the bath salt products collected for the study.

continued

Close-up: Bath salts (*cont'd*)

The sale and use of bath salts in the United States began to disappear about 2015. Why they were sold under that name was never clear. We received occasional inquiries as to whether these products could be used for an actual bath (bad idea). I always thought that someone would eventually decide to empty many of them in a bathtub and develop toxicity from dermal exposure, although I am not aware of this ever being described. Ironically, what the public most likely may recall about bath salts is a terrifying report that was misattributed to intoxication from their use. This was the highly publicized case of the so-called "Miami Cannibal" who in 2012 was killed by police after chewing off another man's face (Elfrink, 2014).

Aaron Schneir, MD

Professor of Emergency Medicine, University of California San Diego, La Jolla, CA, United States

References

Al-Motarreb A, Al-Habori M, Broadley KJ. Khat chewing, cardiovascular disease and other internal medical problems; the current situation and directions for future research. J Ethnopharmacol 2010;132:540–8.

Banister SD, Connor M. The chemistry and pharmacology of synthetic cannabinoid receptor agonists as new psychoactive substances: origins. Handb Exp Pharmacol 2018a;252: 165–90. https://doi.org/10.1007/164_2018_143. PMID: 29980914.

Banister SD, Connor M. The chemistry and charmacology of synthetic cannabinoid receptor agonist new psychoactive substances: evolution. Handb Exp Pharmacol 2018b;252: 191–226. https://doi.org/10.1007/164_2018_144. PMID: 30105473.

Baumann MH, Solis Jr E, Watterson LR, Marusich JA, Fantegrossi WE, Wiley JL. Baths salts, spice, and related designer drugs: the science behind the headlines. J Neurosci 2014; 34(46):15150–8. https://doi.org/10.1523/JNEUROSCI.3223-14.2014.

Bernschneider-Reif S, Oxler F, Freudenmann RW. The origin of MDMA ("ecstasy")–separating the facts from the myth. Pharm 2006;61(11):966–72.

Bowden M. Black Hawk down a story of modern war. New York: Atlantic Monthly Press; 1999. p. 21.

Brownstein M. A brief history of opiates, opioid peptides and opioid receptors. Proc Natl Acad Sci USA 1993;90:5391–3.

Buser GL, Gerona RR, Horowitz BZ, Vian KP, Troxell ML, Hendrickson RG, et al. Acute kidney injury associated with smoking synthetic cannabinoid. Clin Toxicol 2014 Aug;52(7): 664–73. https://doi.org/10.3109/15563650.2014.932365.

Compton WM, Jones CM. Epidemiology of the U.S. opioid crisis: the importance of the vector. Ann N Y Acad Sci 2019;1451(1):130–43. https://doi.org/10.1111/nyas.14209.

Corazza O, Valeriani G, Bersani FS, Corkery J, Martinotti G, Bersani G, et al. "Spice," "kryptonite," "black mamba": an overview of brand names and marketing strategies of novel psychoactive substances on the web. J Psychoact Drugs 2014 ;46(4):287–94. https://doi.org/10.1080/02791072.2014.944291. PMID: 25188698.

Crews B. Synthetic cannabinoids: the challenges of testing designer drugs. Clinical Laboratory News; February 1, 2013.

DEA. Schedules of controlled substances; scheduling of 3,4-Methylenedioxymethamphetamine (MDMA) into schedule I of the controlled substances act. Remand 1988;53(34):5156.

Elfrink T. Bath salts didn't cause the Miami Cannibal attack, scientists say. Miami New Times December 2, 2014.

EMCDDA. EU drug markets report. In: Depth analysis. Luxembourg: Publications Office of the European Union; 2016. https://doi.org/10.2810/219411. [Accessed 10 October 2021].

EMCDDA. European drug report 2021: Trends and developments. Luxembourg: Publications Office of the European Union; 2021a.

EMCDDA. Risk assessment report of a new psychoactive substance: 1-cyclhexyl-4-(1,2-diphenyl)piperazine (MT-45). Luxembourg: Publications Office of the European Union; 2021b.

EMCDDA and Europol. (2011). EMCDDA-europol 2011 annual report on the implementation of Council decision 2005/387/JHA. Lisbon, Portugal.

Global Drug Survey (GDS). The global drug survey 2014 findings. https://www.globaldrugsurvey.com/past-findings/the-global-drug-survey-2014-findings/. [Accessed 14 October 2021].

Gonçalves JL, Alves VL, Aguiar J, Teixeira HM, Camara JS. Synthetic cathinones: an evolving class of new psychoactive substances. Crit Rev Toxicol 2019:1–18. https://doi.org/10.1080/10408444.2019.1679087.

Greenblatt HK, Greenblatt DJ. Designer benzodiazepines: a review of published data and public health significance. Clin Pharmacol Drug Dev 2019 Apr;8(3):266–9. https://doi.org/10.1002/cpdd.667.

Hunger A, Kebrle J, Rossi A, Hoffmann K. Synthesis of analgesically active benzimidazole derivatives with basic substitutions). Experientia 1957;13(1):400–1.

Hyde JF, Browning E, Adams R. Synthetic homologs of D,L-ephedrine. J Am Chem Soc 1928; 50(8):2287–92.

Johnston L, O'Malley P, Miech R, Bachman J, Schulenberg J. Monitoring the future national survey results on drug use, 1975–2015: 2015 overview – key findings on adolescent drug use. Ann Arbor, Michigan: University of Michigan; 2016.

Katselou M, Papoutsis I, Nikolaou P, Spiliopoulou C, Athanaselis S. AH-7921: the list of new psychoactive opioids is expanded. Forensic Toxicol 2015;33(2):195–201. https://doi.org/10.1007/s11419-015-0271-z.

King LA, Kicman AT. A brief history of 'new psychoactive substances. Drug Test Anal 2011 ;3(7–8):401–3. https://doi.org/10.1002/dta.319. PMID: 21780307.

Klobucista C. The US opioid epidemic. Council on Foreign Relations; September, 2021. http://www.cfr.org/backgrounder/us-opioid-epidemic. [Accessed 7 October 2021].

Lukic V, Micic R, Arsic B, Nedovic B, Radosavljevic Z. Overviews or the major classes of new psychoactive substances, psychoactive effects, analytical determination and conformational analysis of selected illegal drugs. Open Chem 2021;19:60–106.

Miliano C, Margiani G, Fattore L, De Luca MA. Sales and advertising channels of new psychoactive substances (NPS): internet, social networks, and smartphone apps. Brain Sci 2018 Jun 29;8(7):123. https://doi.org/10.3390/brainsci8070123.

Moosmann B, King LA, Auwärter V. Designer benzodiazepines: a new challenge. World Psychiatr 2015;14(2):248. https://doi.org/10.1002/wps.20236.

Murphy T, Van Houten C, Gerona RR, Moran J, Kirschner R, Marraffa J, et al. Acute kidney injury associated with synthetic cannabinoids use, multiple states, 2012. MMWR (Morb Mortal Wkly Rep) 2013;62(6):93–8.

Orsolini L, Corkery JM, Chiappini S, Guirguis A, Vento A, De Berardis D, et al. New/designer benzodiazepines': an analysis of the literature and psychonauts' trip reports. Curr Neuropharmacol 2020;18(9):809–37. https://doi.org/10.2174/1570159X18666200110121333.

Pergolizzi J, Magnusson P, LeQuang JAK. Breve F. Illicitly manufactured fentanyl entering the United States. Cureus 2021;13(8):e17496. https://doi.org/10.7759/cureus.17496.

Room R, Reuter P. How well do international drug conventions protect public health? Lancet 2012;379:84–91.

Sanchez SBJ. Sur un homologue de la Societé. Chimique de France 1929;45:284–6.

Schneir A, Ly BT, Casagrande K, Darracq M, Offerman SR, Thornton S, et al. Comprehensive analysis of "bath salts" purchased from California stores and the internet. Clin Toxicol 2014 Aug;52(7):651–8. https://doi.org/10.3109/15563650.2014.933231. PMID: 25089721.

Sumnall HR, Evans-Brown M, McVeigh J. Social, policy, and public health perspectives on new psychoactive substances. Drug Test Anal 2011 ;3(7–8):515–23. https://doi.org/10.1002/dta.310.

Ujváry I, Christie R, Evans-Brown M, Gallegos A, Jorge R, de Morais J, et al. DARK classics in chemical neuroscience: etonitazene and related benzimidazoles. ACS Chem Neurosci 2021 Apr 7;12(7):1072–92. https://doi.org/10.1021/acschemneuro.1c00037. Epub 2021 Mar 24.

United Nations. Single convention on narcotic drugs. 1961. https://www.unodc.org/pdf/convention_1961_en.pdf. [Accessed 18 October 2021].

United Nations. Convention on psychotropic substances. 1971. https://www.unodc.org/pdf/convention_1971_en.pdf. [Accessed 18 October 2021].

United Nations. Convention against the illicit traffic in narcotic drugs and psychotropic substances. 1988. https://www.unodc.org/pdf/convention_1988_en.pdf. [Accessed 18 October 2021].

UNODC. World drug report 2013. United Nations Publication; 2013a. Sales No. E.13.XI.6.

UNODC. Global SMART Update 2013: the challenge of new psychoactive substances. 2013b.

UNODC. World Drug Report 2017, 2017 (ISBN: 978-92-1-148291-1, eISBN: 978-92-1-060623-3, United Nations publication, Sales No. E.17.XI.6).

UNODC. World drug report 2020, cross-cutting issues : evolving trends and new challenges. United Nations publication; 2021a. Sales No. E.20.XI.6.

UNODC. Early warning advisory on new psychoactive substances, 2021. 2021b. https://www.unodc.org/LSS/Home/NPS. [Accessed 4 July 2021].

UNODC. Early warning advisory on new psychoactive substances, 2023. 2023. https://www.unodc.org/LSS/Page/NPS. [Accessed 28 April 2023].

Vandeputte M, Krotulski A, Papsun D, Logan B, Stove C. The rise and fall of isotonitazene and brorphine: two recent stars in the synthetic opioid firmament. J Anal Toxicol 2021. https://doi.org/10.1093/jat/bkab082. bkab082.

World Health Organization. Critical Review report: Isotonitazene. Geneva, Switzerland: 43rd Expert Committee on Drug Dependence Meeting; 2020.

Zawilska JB, Wojcieszak J. An expanding world of new psychoactive substances-designer benzodiazepines. Neurotoxicology 2019 Jul;73:8–16. https://doi.org/10.1016/j.neuro.2019.02.015. Epub 2019 Feb 23. PMID: 30802466.

CHAPTER

2

NPS chemistry, classification, and metabolism

Current classification systems

Like traditional recreational drugs (TRD), new psychoactive substances (NPS) come in a wide variety, so understanding their chemistry, toxicology, and analysis benefits from organizing them into classes. The task is not straightforward, as stakeholders also come from diverse backgrounds. Pharmacists, pharmacologists, toxicologists, epidemiologists, public health officials, and medical practitioners are most familiar with the pharmacological classification used in medicine. Forensic scientists, law enforcement agents, lawyers, and policy experts view drugs through a legislative lens. Analytical, clinical, forensic, medicinal, and synthetic organic chemists are more at ease with chemical classification. Ethnobotanists and others classify according to origin. Because none of these criteria can comprehensively and systematically cover the variety of NPS, some groups combine schemes. The European Monitoring Centre for Drugs and Drug Addiction (EMCDDA), for example, uses both chemical and pharmacological criteria. Because there is no universal scheme, the use of different schemes creates confusion that further complicates understanding what substances comprise NPS and their impact in public health.

Pharmacological classification

The dominant scheme in the literature is based on the pharmacological effects produced in the human body (Global Commission on Drug Policy,

Designer Drugs
https://doi.org/10.1016/B978-0-12-811764-4.00001-X

2019; DEA, 2020). NPS follow the system of common recreational drugs. There are three general classes:

- **Hallucinogens** alter perception, notably of space, time, colors, and forms, primarily by activating the 5-HT_{2A} serotonin receptor. Traditional examples are lysergic acid diethylamide (LSD), *N,N*-dimethyl tryptamine (DMT), and psilocybin. NPS examples are phenethylamines in the 2C series (e.g., 2C–B) and tryptamines such as diisopropyl tryptamine (DiPT).
- **Stimulants** speed up the physiological activity of the central nervous system, boosting alertness and energy and, at high doses, inducing euphoria. They increase catecholamine levels by stimulation of the D_1 dopamine receptor and/or agonistic activity at the α_{2A} adrenergic receptor. Most common are cocaine, amphetamine, and methamphetamine. Among NPS are synthetic cathinones such as mephedrone and MDPV, and piperazines such as benzylpiperazine (BzP) and 3-trifluoromethylpiperazine (TFMPP).
- **Depressants** slow the physiological activity of the central nervous system and produce feelings of relaxation, sleepiness, and less sensitivity to pain. Opioids such as heroin and morphine and benzodiazepines such as diazepam and alprazolam are common. NPS examples are new synthetic opioids (carfentanil, isotonitazene) and designer benzodiazepines (flualprazolam, clonazolam). Opioids bind opioid receptors, which closes N-type voltage-operated calcium channels and opens calcium-dependent inwardly rectifying potassium channels, resulting in neural cell hyperpolarization and reduction in neuronal excitability. Benzodiazepines bind gamma-aminobutyric acid (GABA)-A receptors, enhancing the activity of the inhibitory neurotransmitter GABA and making neurons less responsive to the excitatory neurotransmitters serotonin, dopamine, norepinephrine, and acetylcholine.

Two other classes are anabolic steroids and antipsychotics (Zapata et al., 2021):

- **Anabolic steroids** mimic the effects of the naturally occurring male hormone testosterone, abused to promote muscle growth and enhance physical ability. Examples are nandrolone and trenbolone. Although not commonly categorized as NPS, they are also easily derivatized or modified to create new drugs that are not under international control.
- **Antipsychotics** treat psychoses and can be recreationally abused. They generally block the dopamine D_2 or the serotonin 5-HT_{2A} receptor. Prescription examples are quetiapine, risperidone, and olanzapine. No NPS that is an analog of common antipsychotics has been identified so far.

Hallucinogens are further distinguished into three classes based on their psychopharmacological effects: psychedelics, dissociatives, and cannabinoids:

- **Psychedelics** alter states of consciousness, often through specific psychological, visual, or auditory changes. There are three common families, which are also examples of psychedelic NPS: phenethylamines, tryptamines, and lysergamides (derivatives of LSD).
- **Dissociatives** are hallucinogenic compounds that produce feelings of detachment from the self and/or environment in addition to causing distortion in perception through sight or sound. Arylcyclohexylamines such as those that are derived from phencyclidine and ketamine are NPS examples.
- **Cannabinoids** are psychoactive compounds derived from cannabis or synthetic compounds that mimic the effects of the psychoactive ingredient of cannabis, delta-9-tetrahydrocannabinol (THC). Pharmacologically, they induce an enhancement of sensory perception, euphoria, and antinociception (blocking the detection of a painful or injurious stimulus). Synthetic cannabinoids comprise one of the largest families of NPS.

Depressants are further divided into two classes: sedative-hypnotics and opioids (also called narcotics):

- **Sedative-hypnotics** are depressants that have a relaxing and calming effect often leading to sleepiness. Barbiturates and benzodiazepines are the most common. Designer benzodiazepines such as clonazolam and etizolam are NPS examples.
- **Opioids** are depressants that induce analgesia, sedation, and a strong sense of euphoria along with respiratory depression. This big family includes psychoactive compounds from the opium poppy and their derivatives, with diverse synthetic classes that bind the opioid receptor, such as fentanyl analogs and benzimidazole opioids. The new synthetic opioids comprise a large class of NPS.

Among the stimulants, some are distinguished into a class called empathogens or entactogens. They increase a feeling of empathy and benevolence toward others. These drugs promote emotional communion, relatedness, and openness. MDMA is the most common example. Some amphetamines (MDEA, MBDB) and synthetic cathinones (methylone, butylone) are NPS examples.

Variations in the pharmacological classification of drugs abound in the literature. Cannabinoids and dissociatives, for example, are often regarded as separate from hallucinogens, and opioids as separate from depressants. This is most likely because they have unique receptor targets. Table 2.1 shows pharmacological classification schemes from three well-known organizations.

TABLE 21 Three common pharmacological classifications of NPS.

Organization	Classes						
Global Commission on Drug Policy	Hallucinogens (*dissociatives, cannabinoids*)			Stimulants (*psychedelics*)	Depressants (*narcotics, hypnotics, sedatives*)		Antipsychotics
US DEA	Hallucinogens			Stimulants	Depressants	Narcotics	Anabolic steroids
UNODC	Hallucinogens	Dissociatives	Cannabinoids	Stimulants	Hypnotics/Sedatives	Opioids	

Legal classification

The de facto scheme for law enforcement agents, lawyers, and most policy makers has three categories (Global Commission on Drug Policy, 2019):

- **Legal** drugs are those whose production, trade, and consumption are allowed by national and/or international law. These include prescription medications and substances without established therapeutic use that are recreationally consumed by custom or tradition. Examples of the latter are alcohol, tobacco, and caffeine.
- **Illicit** drugs are those whose production, trade, and consumption are strictly controlled by law because of their public health threat. International regulation is governed by the three conventions briefly discussed in Chapter 1: the 1961 Single Convention on Narcotic Drugs, the 1971 Convention on Psychotropic Substances, and the 1988 Convention against Illicit Traffic in Narcotic Drugs and Psychotropic Substances. Most countries have adopted these conventions with minor modifications. Examples of illicit drugs are cocaine, methamphetamine, and heroin.
- **Unregulated** drugs have been neither approved for therapeutic use nor declared illicit under international regulations. When they first appear, NPS are unregulated; hence, their consumption is not illegal, one of the major reasons for their popularity.

Classification based on origin

Another commonly used scheme categorizes psychoactive substances according to their origin (Feng et al., 2017). There are three categories:

- **Natural** drugs are extracted directly from plants, fungi, and some animals, with little or no physical processing. Traditional recreational drug examples are psilocin from tropical mushrooms and several opiates such as morphine from the exudate of the opium poppy. Salvia, which contains salvinorin A, and kratom, which contains mitragynine, are natural NPS.
- **Semisynthetic** drugs are obtained from natural sources and transformed with minimal chemical modification, typically by altering one or two functional groups. A classical example is heroin, which is synthesized from morphine in opium poppy by adding two acetyl groups. Cocaine is another.
- **Synthetic** drugs are produced primarily by chemical processes often requiring multiple steps. The structures of many are based on

naturally occurring or legitimate synthetic drugs, which they mimic. Most NPS are synthetic; e.g., JWH-018 (cannabinoid), MDPV (cathinone), carfentanil (opioid), and flualprazolam (benzodiazepine).

Chemical classification

This scheme organizes NPS according to their relatedness in chemical and biochemical properties and is especially useful to forensic, clinical, and analytical chemists. Unfortunately, it is considered too technical and not directly meaningful by other stakeholders. Even among medical practitioners, systematic names are too difficult to remember for practical use. The result is the use of multiple common names to refer to the same NPS and a variety of arbitrary nomenclature rules and classification schemes, creating confusion in the literature and in gathering surveillance data.

A few research groups have proposed the use of chemical classification for NPS. One at the University of Hertfordshire adopted a structure-based clustering approach using the nuclear magnetic resonance (NMR) spectra of 478 representative NPS (Zloh et al., 2017). Computational tools using hierarchical clustering analysis were applied to group representative NPS according to their structural similarity. 78 clusters were obtained, among which are 13 superclusters (Fig. 2.1). However, about 53 clusters contain only one or two NPS. Although this approach has also systematically grouped NPS according to their chemical and biochemical properties and is likely useful in structure–activity relationship predictions, the structural grouping of drugs they came up with is not accessible to NPS stakeholders outside chemistry. The sheer number of clusters is impractical, and even among the superclusters, drugs that are intuitively in one class by having a common functional group are scattered in several superclusters (e.g., benzodiazepines and piperazines).

Another group, at the University of Alcala in Madrid, proposed a chemical classification based on all NPS chemical families that had been published by UNODC and EMCDDA as of December 2020 (Zapata et al., 2021). This scheme used four major classes according to their primary functional group: (1) polycyclic hydrocarbons, (2) amines, (3) alcohols/ethers, and (4) other NPS (Fig. 2.2).

Most NPS are amines, and this class is further divided into arylalkylamines and heterocyclic amines, each of which is subdivided into families and groups. Phenethylamines comprise the largest family of arylalkylamines. It consists of eight groups, including amphetamines, cathinones, aminoindanes, and methylenedioxyphenethylamines. Heterocyclic amines are further divided into two big families, heteromonocyclic and

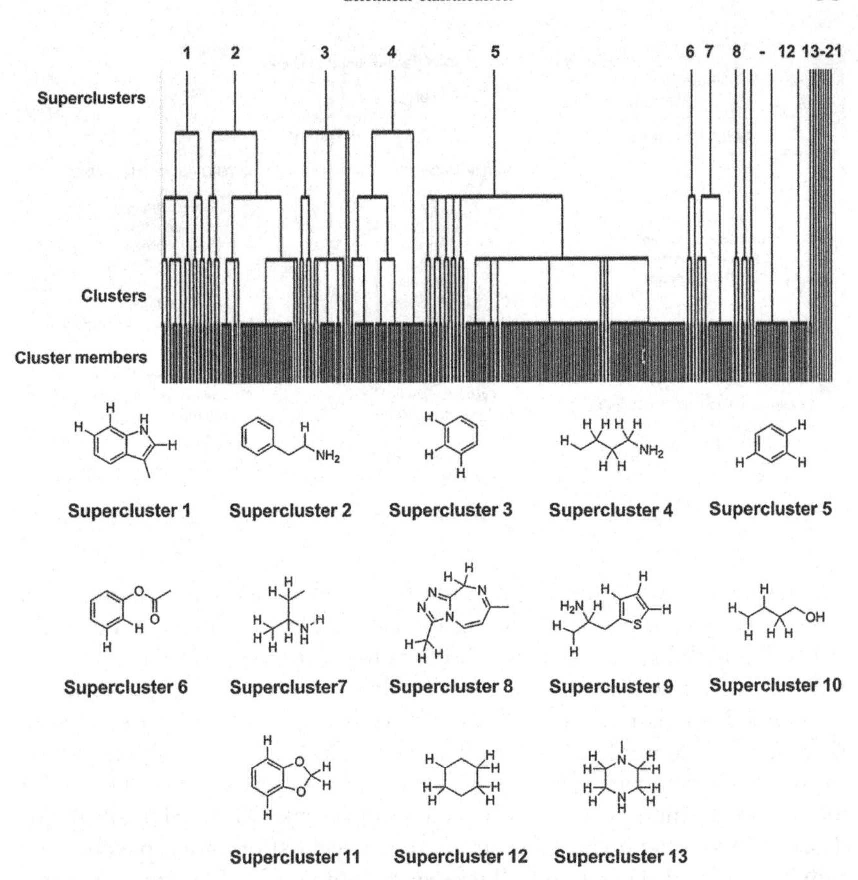

FIGURE 2.1 Dendrogram illustrating the relationship between superclusters, clusters, and cluster membership for the all atom analysis conducted on 478 representative NPS molecules. Fragments of the structure that define each supercluster are given. *Reproduced from Zloh et al. (2017).*

heteropolycyclic. The indoles make up the largest group of heteropolycyclic, consisting of familiar groups such as tryptamines and aminoalkyl indoles. Benzodiazepines also are heteropolycyclic. The heteromonocyclic amines consist of six groups, including piperazines, imidazoles, and piperidines. Fentanyls are piperidines.

Although this system is intuitive and encompasses a significant number of drug groups that are already familiar to NPS stakeholders other than chemists (e.g., piperazines, cathinones, aminoindanes), the large number of groups would still not be easily accessible to NPS experts outside chemistry. All in all, there are 45 groups, not including those classified as other. There are only two groups under alcohol/ethers and one under polycyclic hydrocarbons, but there are 42 groups of amines.

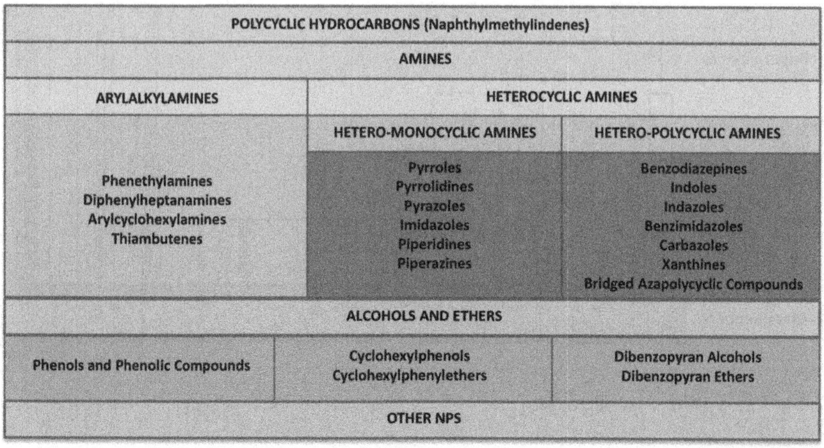

POLYCYCLIC HYDROCARBONS (Naphthylmethylindenes)

AMINES

ARYLALKYLAMINES
Phenethylamines
Diphenylheptanamines
Arylcyclohexylamines
Thiambutenes

HETEROCYCLIC AMINES

HETERO-MONOCYCLIC AMINES
Pyrroles
Pyrrolidines
Pyrazoles
Imidazoles
Piperidines
Piperazines

HETERO-POLYCYCLIC AMINES
Benzodiazepines
Indoles
Indazoles
Benzimidazoles
Carbazoles
Xanthines
Bridged Azapolycyclic Compounds

ALCOHOLS AND ETHERS

Phenols and Phenolic Compounds

Cyclohexylphenols
Cyclohexylphenylethers

Dibenzopyran Alcohols
Dibenzopyran Ethers

OTHER NPS

FIGURE 2.2 Chemical classification of NPS based on functional group similarity.

This number is just impractical to adopt and impose on all stakeholders. What was gained in systematizing the chemical groupings can be easily lost in the inability of most stakeholders to use the system.

Recognizing the problems with a solely chemical scheme, some researchers have combined chemical and other bases of classification. Zawilska's system relies first on binning NPS into six major classes based on their pharmacological classification and refining each class with chemical classification (Zawilska and Wojcieszak, 2018). Her six major classes are synthetic cannabinomimetics, psychostimulants, psychedelic hallucinogens, dissociative hallucinogens, opioids, and benzodiazepines (Table 2.2).

Synthetic cannabinomimetics have the largest number of groups in the Zawilska system, with 11, psychostimulants have 7, and psychedelic hallucinogens, dissociative hallucinogens, and opioids have 2 each. Except for the synthetic cannabinomimetics, the groups in each of these classes are those that most interested parties are already familiar with, such as synthetic cathinones, piperazines, tryptamines, aminoindanes, phenethylamines, arylcyclohexylamines, and fentanyls. This combination scheme is easier to follow for NPS stakeholders outside chemistry. With only 13 groups of drugs plus 10 of the cannabinomimetics, the scheme is easier to remember and handle. One caveat: a group can belong to two different classes, as there is also innate variety in the pharmacological activities of individual drugs in a group. Tryptamines, for example, are classified under psychostimulants and psychedelic hallucinogens.

Other schemes use the combination of pharmacological and chemical classification more freely, that is, without hierarchy. UNODC categorizes NPS into nine classes (UNODC, 2020); some are pharmacological but

TABLE 2.2 Combined pharmacological and chemical classification scheme of NPS.

UNODC	EMCDDA	Zawilska
Plant-based substances Synthetic cannabinoids Synthetic cathinones Phenethylamines Aminoindanes Tryptamines Piperazines Phencyclidine-type substances Others	Plants and extracts Synthetic cannabinoids Cathinones Arylalkylamines Arylcyclohexylamines Phenethylamines Aminoindanes Opioids Tryptamines Piperazines Piperidines and pyrrolidines Benzodiazepines Other substances	Synthetic cannabinomimetics • JWH compounds • Adamantoylindoles • Benzoylindoles • Cyclohexylphenols • Classical cannabinoids • Indazoles • TCMP compounds • Others Psychostimulants • Synthetic cathinones • Piperazines • Piperidines and pyrrolidines • Tryptamines • 2,5-Dimethxyamphetamines • 2-Aminoindanes • Benzofurans Psychedelic hallucinogens • Tryptamines • Phenethylamines Dissociative hallucinogens • Arylcyclohexylamines • Diarylethylamines Opioids • Fentanyl analogs • Others Benzodiazepines

most are chemical. EMCDDA has a similar but extended scheme, with 13 classes (EMCDDA, 2022) (Table 2.2).

A proposed NPS classification

The scheme for this book is based on chemistry except for two big classes, where using pharmacology makes it easier to deal with their toxicology. The number of classes is limited to the fewest that will still allow comprehensiveness. It is similar to published schemes, with a few modifications based on chemistry principles, so it will be familiar to many NPS stakeholders. It is presented in the following 13 bullet points:

- **Phenethylamines** are derived from the naturally occurring trace amine phenethylamine. The core consists of a benzene ring (phenyl) attached to a two-carbon chain (ethyl) and an amine group (Fig. 2.2). Synthetic substitution can happen in all three moieties. The 2C series of compounds that mostly comprise this class were given their name by Shulgin based on the two-carbon atoms that connect the benzene ring and the amine. These are psychedelic hallucinogens, and some are also powerful stimulants. Some 2C derivatives are used in scientific research as selective receptor agonists for neurotransmission.

- **Amphetamines** are closely related to phenethylamines. They differ by having an isopropyl (branched 3C) instead of ethyl (2C) linker between the benzene ring and the amine (Fig. 2.2). Most schemes categorize them as phenethylamines. However, the number and popularity of the compounds in this category merit a separate class. Most amphetamines are stimulants, but the methylenedioxyamphetamines are also empathogens (e.g., MDMA). The large DOx subclass are primarily stimulants.
- **Synthetic cathinones** are derived from amphetamines by oxidation of the carbon beta to the amine (Fig. 2.3). Therefore, they can also be called beta-ketoamphetamines. As their name indicates, they are derivatives of the naturally occurring monoamine alkaloid from *Catha edulis* (khat), cathinone. Next to the synthetic cannabinoids, synthetic cathinones are the second largest NPS class reported to

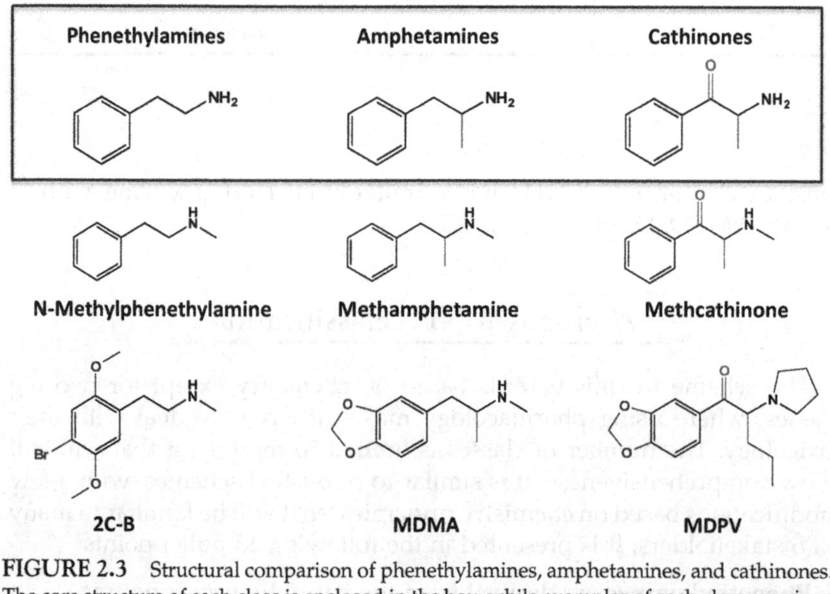

FIGURE 2.3 Structural comparison of phenethylamines, amphetamines, and cathinones. The core structure of each class is enclosed in the box, while examples of each class are given outside.

UNODC and EMCDDA. A variety of subclasses are based on the pattern of substitution in the core structure (ring, alkyl chain, and amine). Like amphetamines, most are either stimulants or empathogens. Some have legitimate medicinal application, such as bupropion (Wellbutrin), an FDA-approved antidepressant.

- **Piperazines** contain the piperazine structural motif. Piperazine itself is a heterocyclic amine that derived its name from piperidine, another heterocyclic amine present in piperine, which is the primary compound responsible for the characteristic taste and smell of pepper. Piperazines are mostly stimulants; some have empathogenic effects. They vary in structure based on the type of aromatic ring attached to the piperazine ring. Two big subclasses are the benzylpiperazines and phenylpiperazines (Fig. 2.4). Piperazines are also used in scientific research and are explored for their therapeutic use as antipsychotics, antidepressants, and anxiolytics (Brito et al., 2019).
- **Piperidines** contain the piperidine structural motif, which is a common motif present in a variety of medications, including cardiovascular drugs, antipsychotics, and opioids (e.g., fentanyl). Derivatives manufactured as NPS are structurally related to the stimulants methylphenidate and pipradrol (Fig. 2.3), so this class of NPS are primarily stimulants.

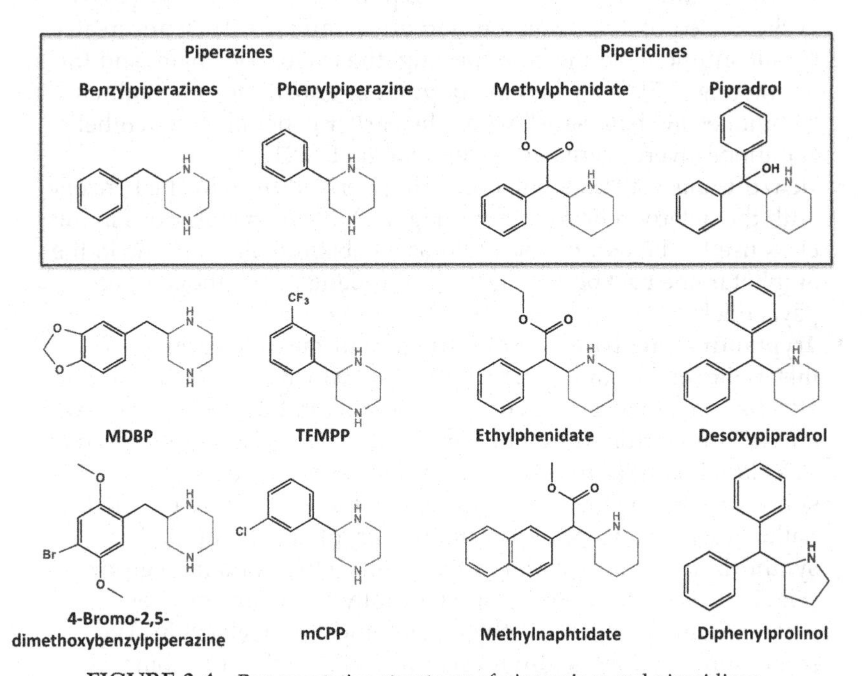

FIGURE 2.4 Representative structures of piperazines and piperidines.

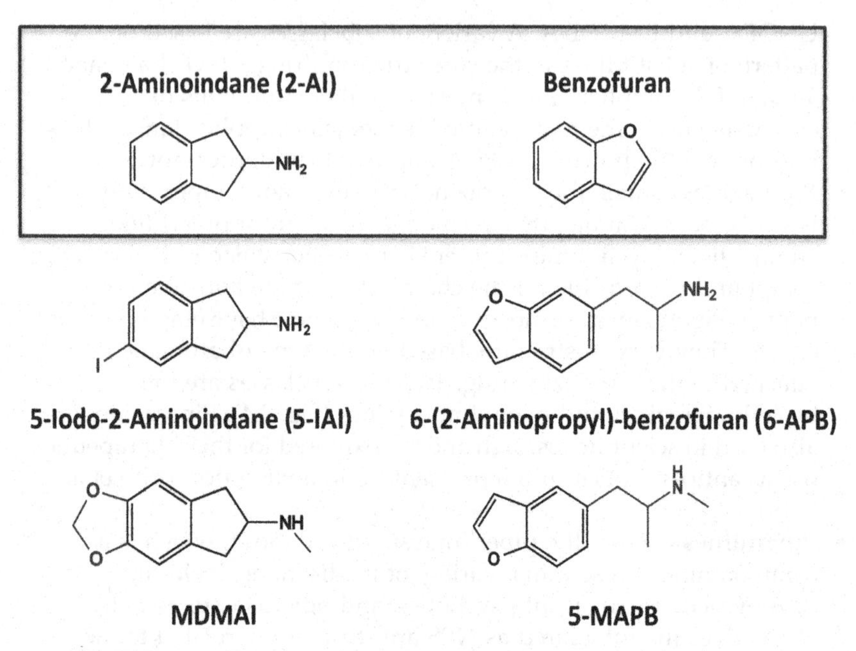

FIGURE 2.5 Representative structures of aminoindanes and benzofurans.

- **Aminoindanes** are based on the compound 2-aminoindane (2-AI), a cyclic analog of amphetamine, and are stimulants. They are created by substitutions on the benzene ring, the cycloalkyl chain, and the amine (Fig. 2.5). They became popular as substitutes for synthetic cathinones in "bath salts" when the earlier generation of synthetic cathinones were regulated (Pinterova et al., 2017).
- **Benzofurans** contain a polycyclic ring derived from fusing benzene with the heterocyclic furan ring (Fig. 2.5). Recreational drugs in this class use the benzofuran scaffold as a substitute for benzene in the amphetamine backbone. They are stimulants, empathogens, or psychedelics.
- **Tryptamines** are based on the structure of the indoleamine metabolite of the amino acid tryptophan, which consists of an indole attached to a two-carbon chain and amine. Indole is a heterocyclic amine formed from fusing benzene and pyrrole. The core scaffold of this class is similar to phenethylamine, where the benzene ring is substituted for by the indole; so most of these drugs are psychedelic hallucinogens. Like phenethylamines, tryptamines can be synthesized by substitution in the indole, ethyl, or amine group (Fig. 2.6). Some tryptamines are naturally occurring, such as serotonin, melatonin, dimethyltryptamine, and psilocybin. The tryptamine scaffold is also present in drugs with more complex

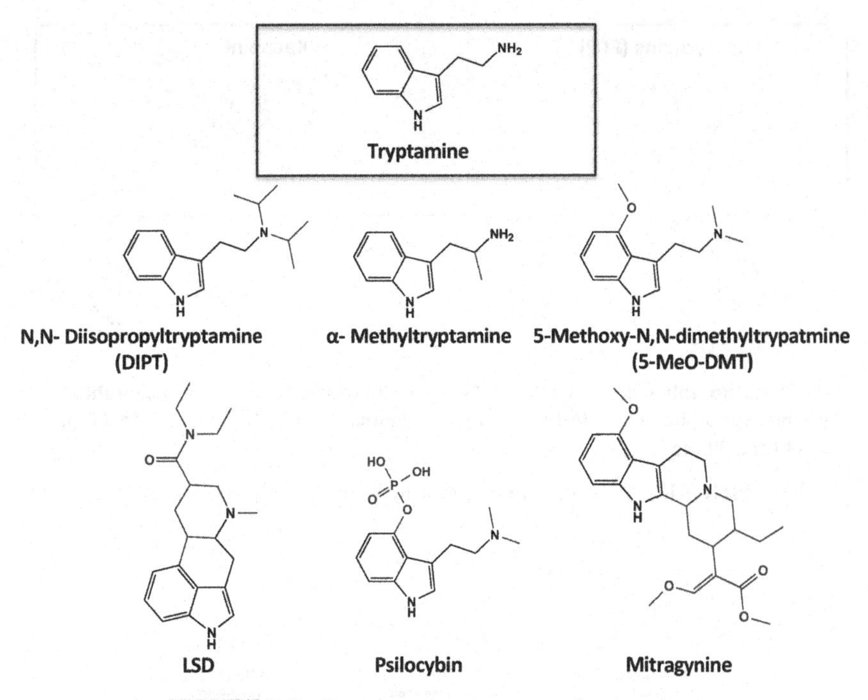

FIGURE 2.6 Representative structures of tryptamines.

structures, such as LSD, mitragynine (from Kratom), ibogaine, and yohimbine.

- **Arylcyclohexylamines** are based on the structure of phencyclidine. The characteristic scaffold of this class consists of a cyclohexylamine with an aromatic ring attached to the same carbon bearing the amine group. The amine group is typically a secondary (e.g., methylamino, ethylamino) or cyclic tertiary (e.g., piperidino or pyrrolidino) amine. The older generation drugs in this class were explored for medicinal applications mostly as anesthetics, such as phencyclidine and its derivatives. The closely related ketamine and its derivatives belong to the class. Most arylcyclohexylamines are dissociative hallucinogens (Fig. 2.7).
- **Synthetic cannabinoids** mimic the effects of cannabis by binding to cannabinoid receptors. There are a wide variety in the chemical structures of this large class. Most of them share a characteristic scaffold consisting of a core structure, linker, pendant chain/ring, and a tail (Fig. 2.8). Based on the types of functional groups that are used for these pharmacophores, synthetic cannabinoids can be classified according to chemical structure into several subclasses: naphthoylindoles, naphthylmethylindoles, naphthylmethylindenes,

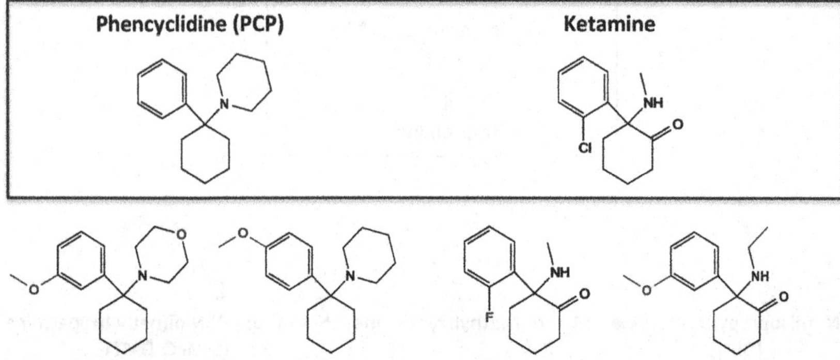

FIGURE 2.7 Representative structures of arylcyclohexylamines.

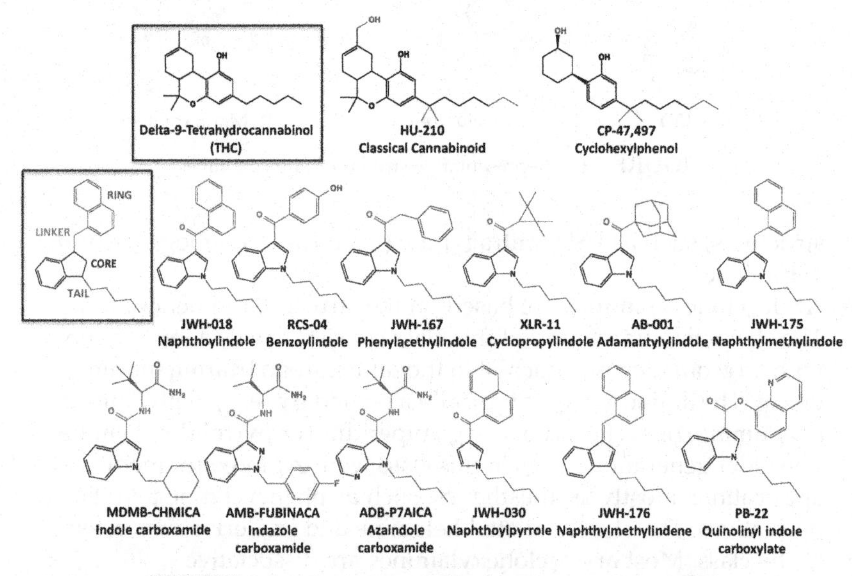

FIGURE 2.8 Common scaffold of synthetic cannabinoids.

benzoylindoles, phenyacetylindoles, cyclopropylindoles, adamantylindoles, aminocarbonylindoles (indole carboxamides), indole carboxylates, indazoles, azaindoles, and naphthoylpyrroles. The number of subclasses is a moving target; new ones are added each year. Synthetic cannabinoids are the most numerous and diverse reported NPS. More than 250 had been reported to both UNODC and EMCDDA as of 2021.

- **New synthetic opioids** have a variety of structures and mimic the effects of heroin by binding the mu opioid receptor. After cannabinoids, this NPS class has the second largest number of potential drugs and is comprised primarily of fentanyl analogs, which follow a common structural scaffold consisting of four pharmacophores, including phenethylamine, piperidine, and aniline rings. Other drugs with structures distinct from heroin and fentanyl have also been reported (Fig. 2.9). A significant number in this class have medicinal application as potent short- or long-acting analgesics.
- **Designer benzodiazepines** are analogs of prescription benzodiazepines but have no approved medical application. They have very similar structures to their ethical counterparts (Fig. 2.10) and produce similar pharmacological effects, but some are significantly more potent.
- **Cocaine analogs** are synthetic and semisynthetic. They are generated by minor modifications of the cocaine structure. Modifications on cocaine's various structural motifs have given rise to at least 17 subclasses. Fig. 2.11 is a sampling of the simple analogs. Other structural and functional analogs exist with significant variation (e.g., piperidines and benztropines).

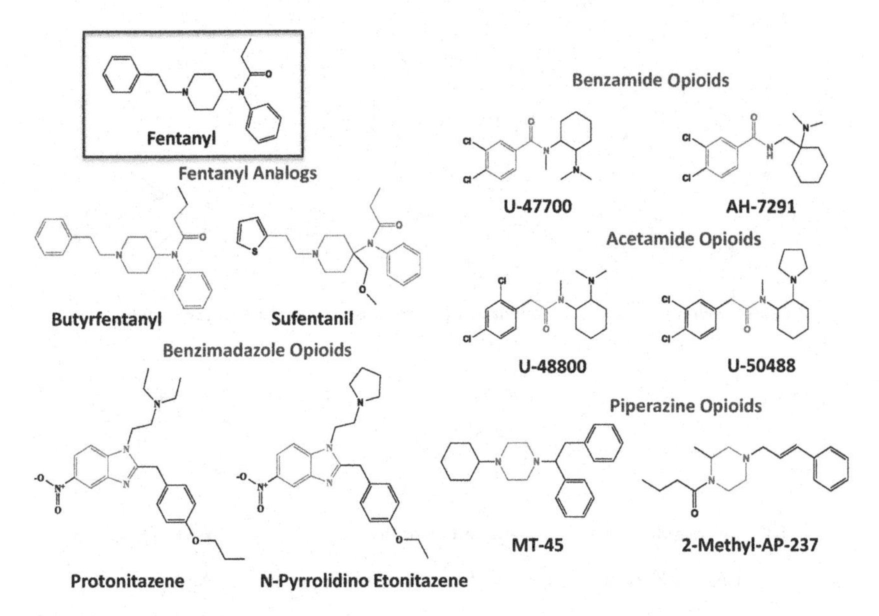

FIGURE 2.9 Representative structures of new synthetic opioids.

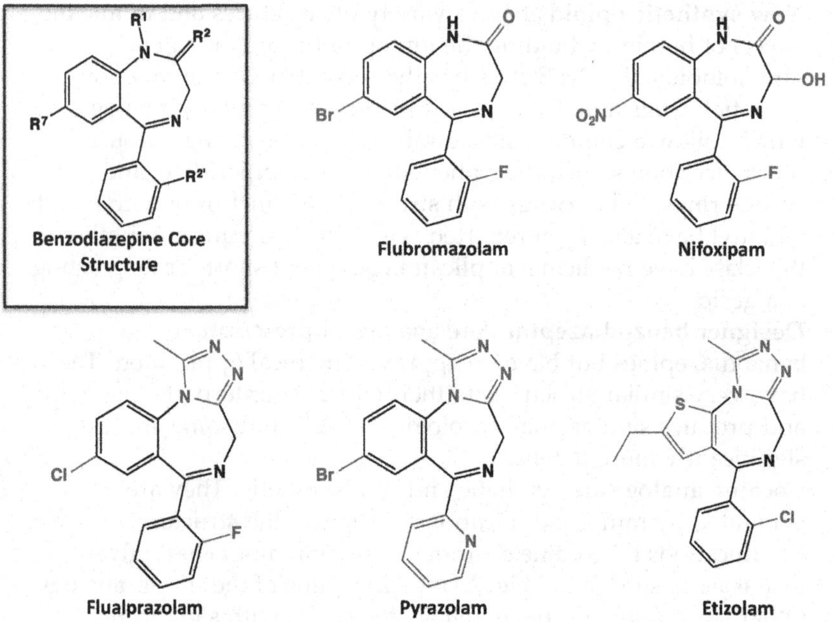

Benzodiazepine Core Structure

Flubromazolam

Nifoxipam

Flualprazolam

Pyrazolam

Etizolam

FIGURE 2.10 Representative structures of designer benzodiazepines.

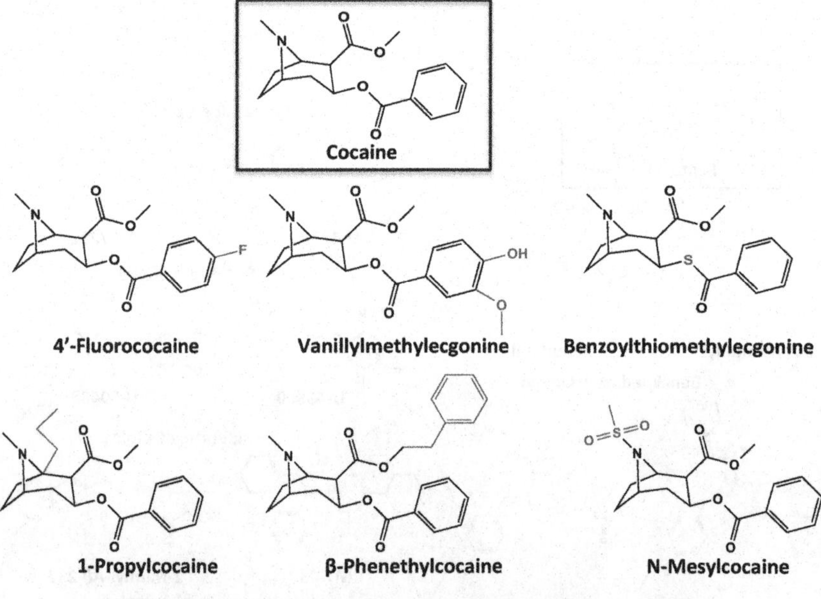

FIGURE 2.11 Representative structures of cocaine analogs.

Metabolic transformations of common NPS classes

Deciphering the metabolic pathway undergone by a therapeutic drug is part of the required pharmacological characterization for its approval as a legal medication. Metabolism determines its half-life, which defines the length of its efficacy. Certain metabolites remain pharmacologically active, while others do not. The half-lives of the intermediate metabolites also affect efficacy and mechanism of action. Moreover, either the parent drug or some of its metabolites can cause toxic or unwanted physiological effects, so metabolism also informs a drug's toxicity and safety.

Metabolism of drugs, like other exogenous compounds, happens primarily in the liver. Two general types of metabolic reactions occur, Phase I and Phase II. Phase I primarily involves reactions that transform lipophilic functional groups to more polar ones, thereby increasing a drug's solubility in biological fluids to promote its excretion. Most of these reactions tend to add or expose $-NH_2$ or $-OH$ functional groups. The resulting metabolites may or may not be biologically active. Three common reactions are employed: oxidation, reduction, and hydrolysis. Oxidative reactions are the most common and are facilitated by hepatic membrane-bound enzymes in the cytochrome P450 system (CYP 450). Phase II metabolic reactions involve conjugation by coupling the drug or its metabolites to another polar molecule, further facilitating their transformation to more polar molecules for excretion. The most common involve conjugation with glucuronic acid, sulfate, acyl group, amino acids (glycine, glutamic acid, and taurine), and glutathione. The resulting metabolites are usually inactive and mostly excreted in the urine. Phase I and II reactions can occur sequentially or simultaneously (Meyer, 1996; Wrighton et al., 1996).

The CYP 450 system consists of more than 50 enzymes that differ in amino acid sequence and target substrate. Each enzyme is assigned a family number (e.g., CYP1, CYP2), a subfamily letter (e.g., CYP2C, CYP2D), and another number to differentiate isoforms in the subfamily (e.g., CYP2C9, CYP2C19) (Nelson, 2009). Despite their variety, only six enzymes are responsible for the metabolism of 90% of drugs: CYP1A2, CYP2C9, CYP2C19, CYP2D6, CYP3A4, and CYP2E1 (Slaughter and Edwards, 1995; Wilkinson, 2005). Each drug can be metabolized by one or multiple CYP 450 enzymes, often in combination with Phase II enzymes. MDMA, for example, is metabolized primarily by CYP 2D6 in combination with catechol-O-methyl transferase (COMT). Aside from being substrates to CYP 450, drugs can also induce or inhibit these enzymes. The antidepressants bupropion, fluoxetine, and paroxetine are strong CYP 2D6 inhibitors, so taking these drugs with MDMA can prolong MDMA's half-life.

The genes that encode CYP P450 enzymes exhibit polymorphisms, that is, differences in genetic sequence that result in slight structural modification and/or lower activity of the encoded enzyme. These alleles (denoted by an asterisk and a number) result in different forms of a CYP enzyme (e.g., CYP 2D6*1 [wild type], CYP 2D6*10 [lower activity variant]) and cause differences in the ability to metabolize drugs. Each person's ability is determined by the pairing of individual alleles he or she inherited (Wilkinson, 2005). Thus, an individual with CYP 2D6*1/*1 genotype will metabolize MDMA faster than an individual with CYP 2D6*1/*10 genotype. Some alleles are expressed at higher frequency in specific ethnic groups compared with others. For example, 75% of East and Southeast Asians carry the CYP 2D6*10 allele, which is rare in Caucasians. Therefore, most of these people metabolize MDMA more slowly, making them more susceptible to MDMA's toxicity at higher concentrations. Genotypic differences that influence drug metabolism and toxicity are the primary focus of pharmacogenomics. It is an important consideration in the metabolism, pharmacokinetics, and toxicity of some drugs, including NPS (Bagheri et al., 2015).

The metabolism of legal drugs is well characterized by the pharmaceutical companies that make them. However, because recreational drugs or NPS may have no therapeutic value, their metabolic characterization is relegated to academic and government institutions. With newly released NPS, especially those with strong potency, metabolic studies are paramount to determining toxicity. Almost always, their metabolic fate is unknown. Even for those revived from expired patents or previous drug candidates, there is often very limited data. This adds to the risk of use. Every drug is a potential poison, depending on the dose, so determining a safe recreational dose can save the lives of those who knowingly or unknowingly use NPS.

From an analytical standpoint, metabolic characterization is useful in developing methods to detect and quantify the drug in different biological matrices. Some drugs are rapidly and extensively metabolized, so efforts to analyze them in urine should be more focused on their metabolites rather than the parent drug. Some are extensively conjugated, so adding a prior deconjugation treatment can increase the sensitivity of the method. Metabolism also dictates the required sensitivity to detect and quantify a drug.

With the rapid turnover of NPS comes the challenge of deciphering the metabolic fate of a large number of previously unreported drugs. Metabolic studies require extensive effort and resources; therefore, it is impossible to study the metabolism of the 50–100 NPS released in the recreational drug market each year in a timeframe that is relevant to their market availability. Alternatively, it is useful to determine the general biochemical transformations that can occur in a class of drugs and thereby make predictions about a new drug based on its class if metabolic studies cannot be done right away. Following are summaries of the general metabolic transformations of major classes of NPS. For other classes, the reader will be referred to reviews.

Amphetamines and phenethylamines

These drugs are targeted by metabolic enzymes at three main sites in their structures: the aromatic ring, alkyl chain, and amine group. Additionally, the methylenedioxy ring is a common site of metabolic reaction in amphetamines that have this moiety, such as MDMA (Staack and Maurer, 2005). Amphetamines and phenethylamines are commonly available as pills or powders to be taken orally or injected. "Ice" (slang for crystallized methamphetamine hydrochloride) is smoked, so inhalation is another route. Phenethylamine itself has a very short half-life of 5–10 min, but the 2C drugs and their derivatives have a wide range, 2–24 h (Stoller et al., 2017). Amphetamines have similar half-lives, 4–24 h.

Amphetamine and phenethylamine derivatives follow metabolic routes very similar to those of methamphetamine and MDMA. Two main reactions occur: N-demethylation (dealkylation) of the amine group and para-hydroxylation of the aromatic ring. A third (usually minor) reaction is the hydroxylation of the beta carbon in the alkyl chain. For MDMA, the analogous reaction to the p-hydroxylation of the aromatic ring is demethylenation of the methylenedioxy ring; this is followed by methylation of the 3-hydroxy group by COMT (Fig. 2.12) (Kreth et al., 2000). Methylenedioxy ring-containing amphetamines follow the same metabolic route. The metabolic products from p-hydroxylation or demethylenation can be

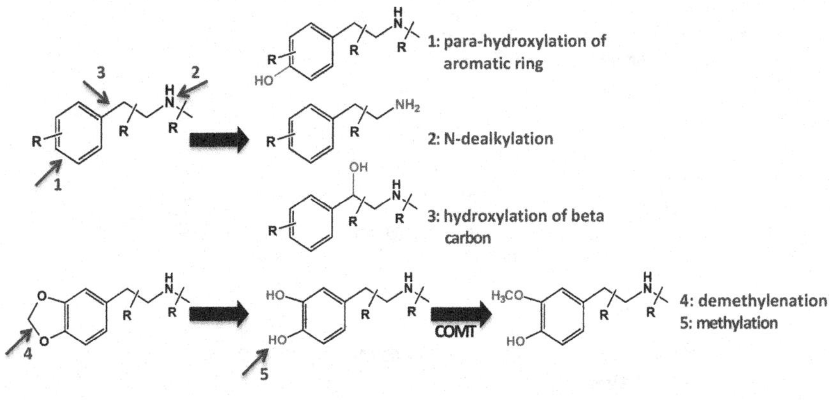

FIGURE 2.12 General metabolic transformations of amphetamines and phenethylamines.

further conjugated with glucuronide or sulfate. N-dealkylation can further proceed to deamination and oxidation to corresponding benzoic acid derivatives; conjugation with glycine can then also follow.

The 2C and DOx series are phenethylamine and amphetamine derivatives with common structural features: both have 2,5-dimethoxy groups and 4-lipophilic groups in their aromatic ring. Their metabolism is not well characterized in humans. Studies in rats and in vitro studies using human liver microsomes suggest that there are slight variations in members of this subclass. Generally, however, the following reactions occur: (1) O-demethylation at the 2C and 5C, (2) oxidative deamination followed by oxidation to carboxylic acid, (3) N-acetylation of the amine group, (4) hydroxylation of the alkyl group and its further oxidation to a keto group, and (5) hydroxylation of the 4-alkyl group and its further oxidation to carboxylic acid (Meyer and Maurer, 2010).

Synthetic cathinones

Like amphetamines and phenethylamines, synthetic cathinones can be substituted in the aromatic ring, alkyl chain, and amine group. Two special subclasses arise from substitution in the aromatic ring (methylenedioxy-substituted) and the amino group (pyrrolidinyl-substituted) (Fig. 2.13). Most synthetic cathinones are sold as "bath salts" in the United States and are commonly snorted or ingested. Sublingual, rectal insertion by enema, and intramuscular and intravascular

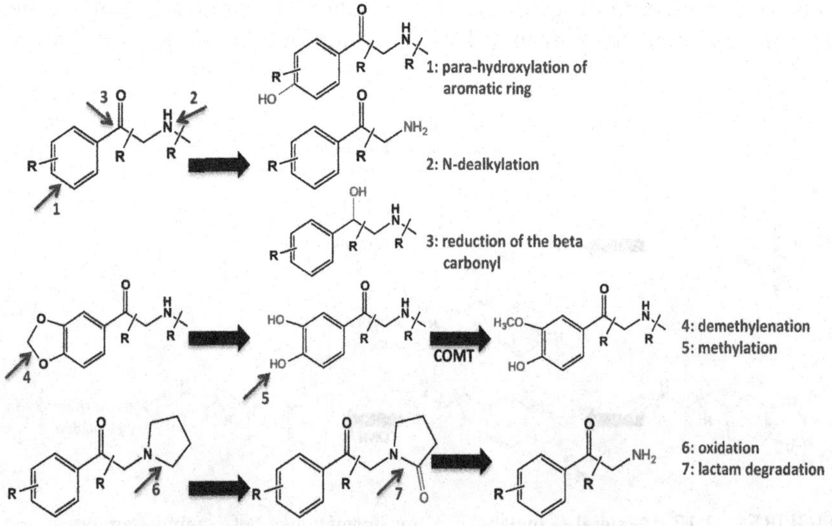

1: para-hydroxylation of aromatic ring
2: N-dealkylation
3: reduction of the beta carbonyl
COMT
4: demethylenation
5: methylation
6: oxidation
7: lactam degradation

FIGURE 2.13 General metabolic transformations of synthetic cathinones.

injection have also been reported. Pharmacokinetic studies on a number of synthetic cathinones indicated that most have short half-lives, typically a couple of hours (mephedrone 126 min, MDPV 78 to 98, alpha-PVP 80). This is consistent with the rapid onset and resolution of effects.

All known subclasses are extensively metabolized in the liver, in both Phase I and II reactions. There are four Phase I reactions common to most synthetic cathinones: (1) N-dealkylation of the amino group, (2) reduction of the carbonyl group to alcohol, (3) hydroxylation of the aromatic ring, and (4) demethylenation of the methylenedioxy group if present (Fig. 2.13). Further reactions can occur in the initial metabolites obtained from these reactions, such as the transfer of a methyl group to a hydroxyl group facilitated by COMT, a reaction analogous to what occurs to MDMA (Zaitsu, 2018). Subsequently, Phase I metabolites undergo Phase II conjugation reactions resulting in addition of either a glucuronide or a sulfate.

There are differences among subclasses in the predominant metabolic reaction. Mephedrone, the first popular synthetic cathinone in bath salts, undergoes all three Phase I reactions in rats and humans. Further oxidation of the hydroxylated tolyl group to a carboxylic acid also occurs in humans. Glucuronidation and sulfation of Phase I metabolites also have been observed in urine samples (Meyer et al., 2010; Pedersen et al., 2013). Ring-halogenated synthetic cathinones (e.g., 3-bromocathinone) undergo the same Phase I reactions, as found in studies using rat urine and human liver microsomes. Additionally, reduction of the N-dealkylated primary amine to an imine occurs significantly in some halogenated synthetic cathinones, such as 3-fluoromethcatinone. The unchanged parent drugs of alkyl- and halogen-substituted synthetic cathinones are also present in urine.

Methylenedioxy-substituted cathinones such as methylone, butylone, and pentylone follow a metabolic pathway observed in MDMA because of their structural similarity (Elmore et al., 2017; Prosser and Nelson, 2012). The major metabolic route in this subclass is the demethylenation of the methylenedioxy ring followed by methylation of either alcohol in the resulting diol by COMT. The resulting methoxy—hydroxy metabolites are then partly conjugated with glucuronide and sulfate. Both N-dealkylation and reduction of the beta-keto group are minor pathways for this subclass, if they occur at all.

In pyrrolidinyl-substituted synthetic cathinones such as MDPV and alpha-PVP, a common pathway is the oxidation of the pyrrolidine ring followed by its dehydrogenation to its corresponding lactam and its eventual degradation to a primary amine (Anizan et al., 2014; Baumann et al., 2017). In some members of the subclass, hydroxylation of the alkyl chain also occurs, as well as oxidation of the 4′-position of the aromatic ring for those that are alkyl- or halogen-substituted. If a methylenedioxy ring is present, such an in MDPV, the characteristic demethylenation followed by methylation by COMT also occurs. Reduction of the carbonyl to alcohol is a minor pathway.

Synthetic cannabinoids

The broad family of cannabinoids is classified into three groups based on origin: phytocannabinoids (produced by plants, such as THC), endocannabinoids (produced within the body and acting as lipid messengers, such as eicosanoids), and synthetic cannabinoids. As noted earlier, synthetic cannabinoids are the most diverse group of NPS based on chemical structure. They differ significantly in structure from classical cannabinoids such as THC. Although they also differ within class, almost all are defined by the same structural scaffold consisting of a heterocyclic ring core (indole, indazole, azaindole, pyrrole), linker (carbonyl, methylene), pendant chain or ring (naphthyl, benzyl, phenylacetyl, cyclopropyl, adamantly, quinolinyl, cumyl, aminoacyl), and tail (alkyl, haloalkyl, cyanoalkyl, fluorobenzyl, cyclohexylmethyl, alkenyl). With this diversity, a wide variety of biochemical transformations occur, and not all can be covered here. We will focus on metabolic transformations in specific structural motifs that are commonly found across subclasses.

Because most synthetic cannabinoids are marketed as ingredients of dried herbs or incorporated in vaping fluids, their most common route of administration is smoking. They are also incorporated in edibles. When smoked, onset of effects occurs within minutes. Taken orally, it is absorbed in the gut and distributed within the body to reach the nervous system, which of course is slower. Smoked JWH-018 reached peak blood concentration within 5 min, followed by a rapid decline over 3 h, and was almost undetectable after 21 h (Toennes et al., 2017). In comparison, AM-2201 orally reached peak blood concentration at 1.3 h (Carlier et al., 2018). The rapid decline after peak blood concentration suggests rapid metabolism. Indeed, synthetic cannabinoids are rapidly and extensively metabolized in the liver following either inhalation or oral ingestion. The estimated half-lives for most JWH compounds are between 30 and 75 min.

Synthetic cannabinoids undergo extensive Phase I and II metabolism. For Phase I, there are three sites of metabolic transformation: the tail, pendant chain or ring, and heterocyclic ring core (Fig. 2.14). Combinations of these metabolic transformations happen within a single synthetic cannabinoid, creating a large number of metabolites (Hutter et al., 2018; Diao and Huestis, 2019).

The tail is a common target of oxidative enzymes in synthetic cannabinoids. Hydroxylation of the terminal and penultimate carbon is the most common reaction observed in alkyl and haloalkyl tails. The terminal hydroxy alkyl metabolite formed in this reaction further undergoes

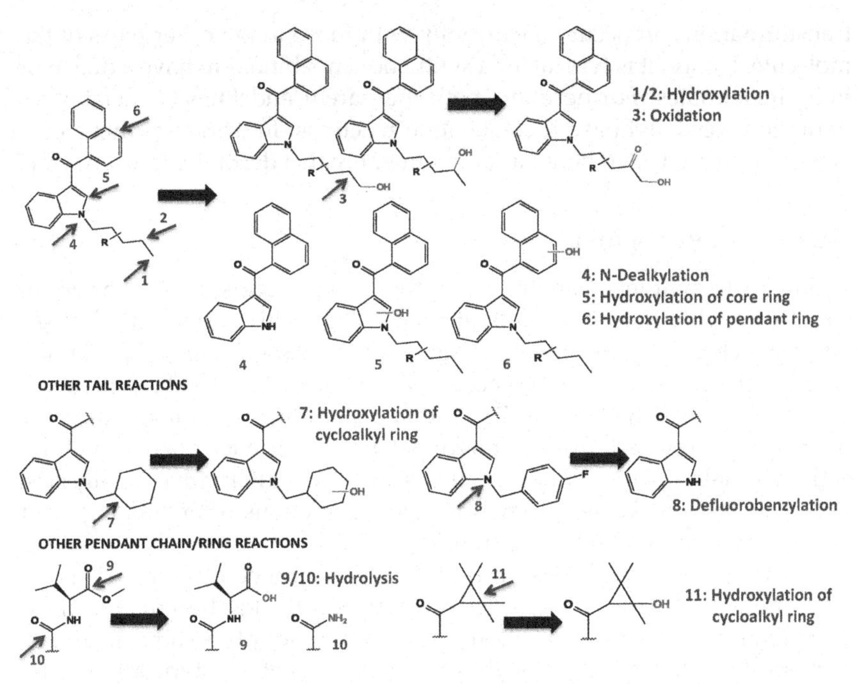

FIGURE 2.14 General metabolic transformations of synthetic cannabinoids.

oxidation into carboxylic acid. For JWH-018, its N-4-hdroxypentyl, N-5-hydroxypentyl, and N-pentanoic acid metabolites are most commonly reported. In haloalkyl tails, oxidative dehalogenation commonly occurs as well. AM-2201, the fluoropentyl analog of JWH-018, is metabolized in this manner to have the same N-5-hydroxypentyl and N-pentanoic acid metabolites as JWH-018. Other reactions that can occur in the tail include N-dealkylation, hydroxylation of the cycloalkyl chain, and defluorobenzylation (Diao and Huestis, 2017; Fabregat-Safont et al., 2022).

Metabolic transformations in the pendant chain or ring depend on the functional group present. The most common metabolic reaction is the hydrolysis of esters or amides in aminoacyl chains. In fact, for esters containing a pendant chain such as in AMB-CHMINACA and AMB-FUBINACA, the acid metabolites are detected in urine samples instead of the parent drugs (Brandon et al., 2021; Franz et al., 2019). For aromatic pendant rings such as in naphthyl, benzyl, phenylacetyl, and cumyl groups, hydroxylation occurs, whereas for the tetramethylcyclopropyl group in XLR-11 and UR-144, extensive hydroxylation and carboxylation are major routes (Diao and Huestis, 2017; Hutter et al., 2018).

The heterocyclic ring core (indole, indazole, azaindole, pyrrole) is oxidized (Diao and Huestis, 2017; Franz et al., 2019). For JWH-018, for example, 5-, 6-, and 7-hydoxyindoles are the observed metabolites. This transformation can occur concurrently with those in the other parts of the molecule. Hence, it is typical for a synthetic cannabinoid to have a dozen or more metabolites. Furthermore, both the parent and Phase I metabolites undergo extensive Phase II conjugation reactions. In urine, synthetic cannabinoids are detected better after treatment with a deconjugation enzyme.

New synthetic opioids

Similar to synthetic cannabinoids, these compounds have a variety of structures. Although most are fentanyl analogs with a common scaffold, other subclasses have appeared since 2012. These include benzamide analogs (U-47700, AH-7921), acetamide analogs (U-48800), piperazine analogs (MT-45, 2-methyl-AP-237), and benzimidazoles (nitazenes). Most of these are revived from expired patents or discontinued structure-activity relationship studies in the search for alternative analgesics. Each subclass has some common metabolic reactions with fentalogs, but also its own unique metabolic transformations.

Because fentanyl analogs and other new synthetic opioids are either knowingly or unknowingly used as a substitute for heroin, the most common routes of administration are smoking and intravenous injection. The onset for both routes is within minutes, with effects typically lasting 1–2 h. Elimination half-life of fentanyl analogs varies widely; remifentanil's is half an hour, sufentanil's is 6–9 h. Fentanyl's is 8–10 h.

The scaffold for fentanyl analogs consists of four structural motifs: the piperidine ring, the anilinophenyl ring, the 2-phenethyl substituent, and the carboxamide moiety linked to the anilino nitrogen. Each motif provides multiple sites of substitution that can generate thousands of analogs. Likewise, they provide multiple sites of metabolic transformation. Similar to fentanyl, the major metabolite of fentalogs arises from N-dealkylation of the piperidine ring. In fentanyl, this forms norfentanyl. However, the presence of the other pharmacophores also facilitates the following metabolic transformations (Fig. 2.15): (1) hydroxylation of the aromatic rings, (2) hydroxylation of the piperidine ring, (3) hydroxylation of the alkyl group of the carboxamide, (4) hydroxylation the ethyl group, (5) N-hydroxylation of the piperidine nitrogen, and (6) hydrolysis of the carboxamide moiety (Wilde et al., 2019; Patel and Parveen, 2022). These transformations can happen sequentially or concurrently. For example, hydroxylation of the piperidine ring and hydrolysis of the carboxamide moiety occur in norfentanyl after its formation from the N-dealkylation of fentanyl. Ring opening is also facilitated when hydroxylation of the piperidine ring happens at the 2-position. Furthermore, a metabolic transformation can split the molecule into two metabolites. N-dealkylation also produces phenylacetaldehyde, which immediately oxidizes to phenylacetic acid. Most

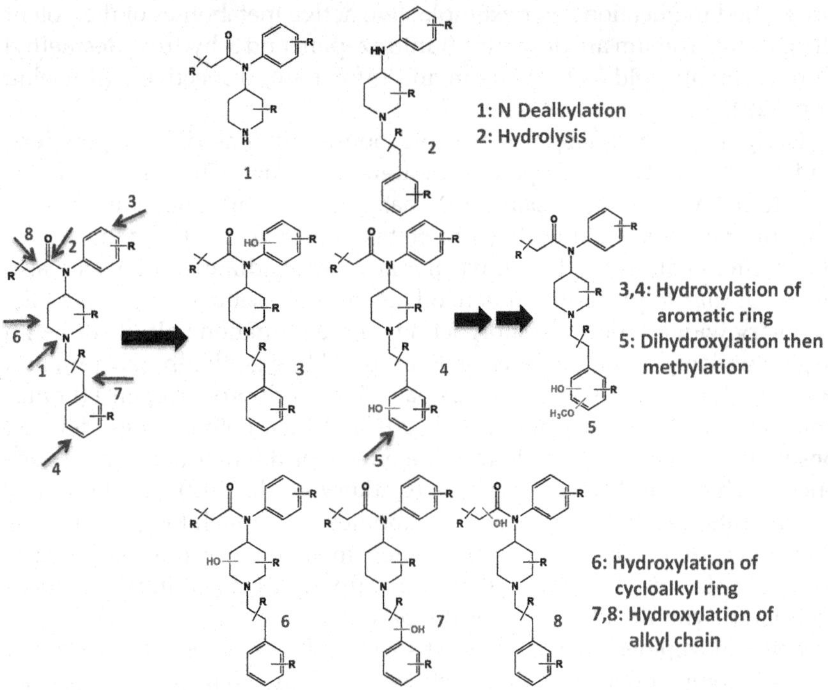

FIGURE 2.15 General metabolic transformations of new synthetic opioids.

of the metabolites formed from these transformations are minor. Of note, some fentanyl analogs share the same metabolite after some of these transformations. Alfentanil and sufentanil, for example, form the same N-dealkylated metabolite.

Not much is published on the metabolism of other subclasses. N-demethylation and hydroxylation seem to be the major metabolic routes for the benzamide analogs U-47700 and AH-7921 (Wohlfarth et al., 2016; Krotulski et al., 2018). For the piperazine analog MT-45, N-dealkylation, hydroxylation, dihydroxylation, and glucuronidation of the hydroxylated metabolites are all reported (Montesano et al., 2017). For benzimidazole opioids, N-dealkylation, O-dealkylation, and reduction of the nitro group have been suggested based on metabolites observed in human urine samples (Ujvary et al., 2021).

Designer benzodiazepines

The majority of these substances are pharmaceutical drug candidates that have never been approved for medical use. Hence, they are structural analogs of prescription benzodiazepines. Some are active metabolites of prescribed medications. For example, two active metabolites of the potent drug flunitrazepam are desmethyflunitrazepam and 3-hydroxydesmethyl flunitrazepam, sold as fonazepam and nifoxipam, respectively (Katselou et al., 2017).

Designer benzodiazepines are sold online often as pills or powders, and are also available as pellets, blotters, or liquid. They may come as counterfeit Valium (diazepam) or Xanax (alprazolam). The main route of administration is oral, but they can be taken intravenously for the "rush" (Carpenter et al., 2018). Intramuscular and rectal administrations are also done. Similar to prescription benzodiazepines, onset and duration of action vary widely. Some are short acting with elimination half-life of 2–4 h (e.g., etizolam), some are intermediate (6–24 h, e.g., flualprazolam), and some are long (e.g., 103 for phenazepam, 106 for flubromazepam) (Lomas and Maskell, 2015; Moosmann et al., 2013). As with other drugs, delayed onset encourages redosing before the effects of the first dose are experienced, which can lead to overdose (Corkery et al., 2012). An important consideration is that some active metabolites have much longer half-lives than the parent. The monohydroxylated metabolite of flubromazepam, for example, can be detected in urine up to 28 days after ingestion, compared with about 7 days for the parent.

Benzodiazepines come in several structural subclasses. Most designer forms belong to one of three subclasses: 1,4-benzodiazepines (e.g., phenazepam and flubromazepam, similar to diazepam), triazolobenzodiazepines (e.g., bromazolam, clonazolam, flualprazolam, flubromazolam, and pyrazolam, similar to alprazolam), and thienotriazolobenzodiazepines (e.g., etizolam and deschloroetizolam, similar to brotizolam). All subclasses undergo Phase I and II metabolism. For 1,4-benzodiazepines, the typical metabolic reactions include N-dealkylation, hydroxylation at position 3 of the diazepine ring, reduction of the nitro group (if present), and its subsequent acetylation. Triazolobenzodiazepines also undergo hydroxylation at the diazepine ring; hydroxylation of the methyl group attached to the triazolo ring (alpha-hydroxylation) as well as ring cleavage of the diazepine ring also occur. Thienotraizolobenzodiazepines undergo hydroxylation at the diazepine ring and the alkyl groups attached to the triazolo and thiophene rings (Fig. 2.16). All Phase I oxidation products, and the parent drugs undergo glucuronidation (Meyer, 2016a; Meyer et al., 2016b; El Balkhi et al., 2017; Zawilska and Wojcieszak, 2019).

Space limitation precludes discussion of the metabolism of all other NPS classes. The interested reader is referred to review articles in Table 2.3.

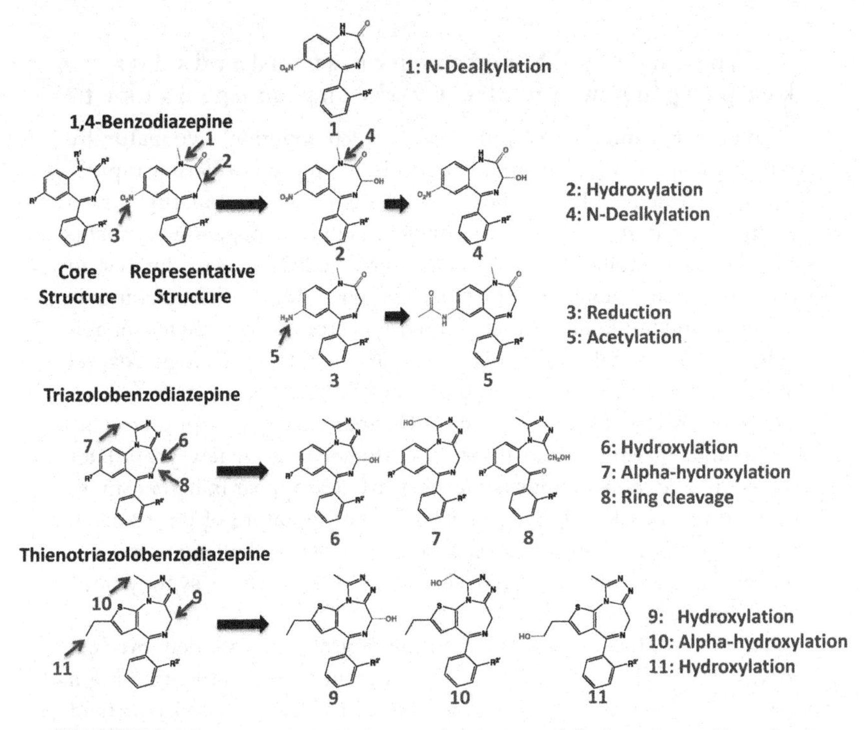

FIGURE 2.16 General metabolic transformations of designer benzodiazepines.

TABLE 2.3 Suggested references on the metabolism of other NPS classes.

NPS class	Suggested reviews
Tryptamines	Katagi et al. (2010), Michely et al. (2015), Caspar et al. (2018), Malaca et al. (2020)
Arylcyclohexylamines	Ho and Dargan (2016), Wallach and Brandt (2018)
Piperazines	Elliott (2011), Schep et al. (2011)
Piperidines	Negreira et al. (2016), White and Archer (2013)
Aminoindanes	Mestria et al. (2021), Zidkova et al. (2017), Manier et al. (2020)
Benzofurans	Welter et al. (2015a, 2015b)

Closeup: NPS reference standards for keeping up with the evolving drug market

Cayman Chemical was founded in 1980, offering five naturally sourced prostaglandin research chemicals. Its product offering rapidly expanded in the next two decades, with an emphasis on bioactive lipids and the arachidonic acid cascade, which include endogenous cannabinoids (endocannabinoids). I joined Cayman in 2004 with an interest in expanding their cannabinoid product line, including endocannabinoids such as anandamide and 2-arachidonyl glycerol (2-AG), phytocannabinoids such as cannabidiol (CBD), and various related analogs. Several published synthetic cannabinoid CB1 and CB2 receptor agonists and antagonists were also introduced into the Cayman catalog to support our academic and pharmaceutical research customer base. A few years later when synthetic cannabinoids (syncans) started to appear in herbal smoking mixtures (aka K2, Spice, Incense), the entire nature of the research chemical business changed, presenting a new set of challenges and opportunities. The term "research chemical" itself became synonymous with "designer drug."

Historically at Cayman Chemical, the typical path to a new product introduction involved reading the scientific literature and identifying a newly reported substance as having potential value in a particular field of research. For example, JWH-018, the dual CB1/CB2 receptor agonist now infamously known for being one of the first widely abused synthetic cannabinoid, was synthesized and introduced based on the work of Dr. John W. Huffman, a prominent cannabinoid research scientist at Clemson University (Wiley et al., 1998). However, after 2009, the primary customer base for this research area changed from academics to forensic scientists, who rely on the independently synthetized reference standards to confirm unknowns in their casework. Likewise, medical examiners and toxicologists rely on the metabolites and internal standards to quantify the presence of these drugs in human specimens.

After JWH-018 and four other syncans were made illegal by a temporary emergency scheduling in 2011, several more analogs with subtle structural differences entered the illicit market to circumvent the new rules. Unscrupulous entrepreneurs were continually formulating smokeable herbs, such as damiana leaf, with new analogs, the intent being to stay one step ahead of law enforcement. Subsequent bans accompanied by waves of new chemical entities with unknown potency and toxicological profile expanded the illicit syncan market at an alarming rate (Seely et al., 2013). At the same time, more new psychoactive substances in other drug classes were showing up in forensic casework. These included amphetamine-based stimulants and synthetic cathinones (aka bath salts),

Closeup: NPS reference standards for keeping up with the evolving drug market (*cont'd*)

novel tryptamine and phenethylamine-based hallucinogens (such as 5-methoxy DALT and 25-NBOMe), as well as new benzodiazepines and opioids. The sudden need for analytical reference standards for the influx of these substances presented challenges: (1) how to keep up with the latest trends to help forensic scientists with drug confirmations, (2) how to determine what metabolites to introduce to help toxicologists and medical examiners confirm and quantify these drugs in human specimens, and (3) what additional quality, safety, and regulatory controls are needed.

The sheer number of new reference standards, and the rate at which they were needed, presented a formidable challenge. Timing was of vital importance for the introduction of these substances because the life span in the illicit market was often as short as a few months between the time of initial detection in forensic casework and being phased out and replaced by another. Therefore, it was critical to synthesize and introduce the new reference standards at a fast enough pace so that criminal laboratories were able to keep up with casework. Without a reference standard, forensic laboratories are unable to positively identify and confirm the unknown chemical entities showing up in new products known as Spice, K2, bath salts, and countless other names. Keeping up with the latest trends required, the insight to know what to make next and the resources to synthesize each new lead. Fortunately, Cayman's chemistry department had the skills and resources to adapt and handle this large volume of work.

As the number of NPS introductions grew, Cayman created spectral libraries which were shared with our forensic science customer base. This was an effective means for forensic chemists to find a presumptive match for the unknowns, then purchase the reference standard (synthesized based on the lead) for the confirmatory test in their casework. Likewise, spectral data from crime samples, with no spectral library match, were often provided to us for assistance in the structural elucidation of these unknowns. Fortunately, the fragmentation data provided by GC-MS often give enough structural insight so that reasonable predictions may be made. This information enabled Cayman scientists to synthesize the predicted substance, and if it proved to be a match based on the GC-MS data, a sample of the newly synthesized reference material was sent to the crime laboratory for confirmation. This data-sharing process became routine and was one of the most effective means of identifying, confirming, and introducing new NPS reference standards. Another

continued

Closeup: NPS reference standards for keeping up with the evolving drug market (*cont'd*)

unconventional method of obtaining new leads that provided opportunities to introduce a reference material much earlier in the NPS life cycle was to monitor websites of suspicious vendors offering the latest "research chemical with discreet packaging" in bulk quantities; however, all introductions were completed through independent synthesis rather than procurement. Online discussion boards of NPS drug users providing "trip reports" proved to be another similarly effective method of monitoring NPS trends for potential leads.

Filling the need for NPS metabolites was the next challenge and required external partnerships. Fortunately, several collaborations were formed, pairing expertise in synthetic organic chemistry with pharmacology to elucidate the primary NPS metabolites. An early partnership with Dr Jeffery Moran at the Arkansas Department of Public Health led to the first published report confirming the JWH-018 metabolites, using biological samples from a fatality case attributed to the substance. While the JWH-018 reference material was available, providing the drug confirmation in the K2 packet found on the deceased, mass spectrometry results of the blood and urine specimens indicated multiple unknown oxidative products. To confirm the metabolites in the human specimen, the potential oxidative products at the alkyl sidechain and aryl indole positions of the molecule were synthesized for comparison (Moran et al., 2011; Chimalakonda et al., 2011). These studies confirmed the predominant oxidative products of this substance, giving forensic toxicologists valuable insight. This work also provided a sound basis for the presumptive metabolites for the second- and third-generation compounds, such as the predicted metabolites of MAB-CHMINACA, which were synthesized in collaboration with Dr Roy Gerona at UCSF in response to a 2015 outbreak in Mississippi (Kasper et al., 2019).

Beyond the synthetic challenges, additional adaptations were needed to create a successful forensic science division. For example, the rigorous quality and traceability demands of this new customer base prompted Cayman to obtain ISO17025 and ISO17034 accreditations. Analytical and manufacturing registration with the DEA was required, along with regulatory approval for items. Additional safety precautions were implemented due to known and unknown hazards in the production of NPS derivatives. Risk mitigation of the high-potency fentanyl analogs, from the start of the synthesis to the final packaging, is just one example. Lastly, it was necessary to implement additional protocols to ensure that products do not fall into the hands of unintended recipients, regardless of

Closeup: NPS reference standards for keeping up with the evolving drug market (*cont'd*)

the relatively small quantities that we provide to our customers. This is of vital importance pertaining to high-potency illicit substances, like carfentanil, which have a potentially lethal dose of 1 milligram.

After working at Cayman Chemical to help create the successful forensic science department that it is today, my current pursuits are in the areas of expert witness testimony and the development of new assays for drug quantitation by mass spectrometry. This work is done through Pinpoint Testing, LLC, where we continue to provide new methods and technologies to assist and enable forensic toxicology laboratories to keep up with their casework in this continually evolving field.

Gregory W. Endres, Ph.D.
Chief Scientific Officer, Pinpoint Testing, LLC, Little Rock, AR, United States.

References

Anizan S, Ellefsen K, Concheiro M, Suzuki M, Rice KC, Baumann MH, et al. 3,4-Methylenedioxypyrovalerone (MDPV) and metabolites quantification in human and rat plasma by liquid chromatography-high resolution mass spectrometry. Anal Chim Acta 2014 May 27;827:54–63. https://doi.org/10.1016/j.aca.2014.04.015. Epub 2014 Apr 12. PMID: 24832995; PMCID: PMC4150738.

Bagheri A, Kamalidehghan B, Haghshenas M, Azadfar P, Akbari L, Sangtarash MH, et al. Prevalence of the CYP2D6*10 (C100T), *4 (G1846A), and *14 (G1758A) alleles among Iranians of different ethnicities. Drug Des Dev Ther 2015 May 13;9:2627–34. https://doi.org/10.2147/DDDT.S79709. PMID: 25999696; PMCID: PMC4435087.

Baumann MH, Bukhari MO, Lehner KR, Anizan S, Rice KC, Concheiro M, et al. Neuropharmacology of 3,4-methylenedioxypyrovalerone (MDPV), its metabolites, and related analogs. Curr Top Behav Neurosci 2017;32:93–117. https://doi.org/10.1007/7854_2016_53. PMID: 27830575; PMCID: PMC5392131.

Brandon AM, Antonides LH, Riley J, Epemolu O, McKeown DA, Read KD, et al. A systematic study of the in vitro pharmacokinetics and estimated human in vivo clearance of indole and indazole-3-carboxamide synthetic cannabinoid receptor agonists detected on the illicit drug market. Molecules 2021;26(5):1396. https://doi.org/10.3390/molecules26051396.

Brito AF, Moreira LKS, Menegatti R, Costa EA. Piperazine derivatives with central pharmacological activity used as therapeutic tools. Fundam Clin Pharmacol 2019 Feb;33(1): 13–24. https://doi.org/10.1111/fcp.12408. Epub 2018 Sep 17. PMID: 30151922.

Carlier J, Wohlfarth A, Salmeron BD, Scheidweiler KB, Huestis MA, Baumann MH. Pharmacodynamic effects, pharmacokinetics, and metabolism of the synthetic cannabinoid AM-2201 in male rats. J Pharmacol Exp Therapeut 2018 Dec;367(3):543–50. https://doi.org/10.1124/jpet.118.250530. Epub 2018 Sep 28. PMID: 30266766; PMCID: PMC6246978.

Carpenter JE, Murray BP, Dunkley C, Kazzi ZN, Gittinger MH. Designer benzodiazepines: a report of exposures recorded in the national poison data system, 2014–2017. Clin Toxicol Phila (Phila) 2018. https://doi.org/10.1080/15563650.2018.1510502.

Caspar AT, Gaab JB, Michely JA, Brandt SD, Meyer MR, Maurer HH. Metabolism of the tryptamine-derived new psychoactive substances 5-MeO-2-Me-DALT, 5-MeO-2-Me-ALCHT, and 5-MeO-2-Me-DIPT and their detectability in urine studied by GC-MS, LC-MS^n, and LC-HR-MS/MS. Drug Test Anal 2018 Jan;10(1):184–95. https://doi.org/10.1002/dta.2197. Epub 2017 May 10. PMID: 28342193.

Corkery JM, Schifano F, Ghodse AH. Phenazepam abuse in the UK: an emerging problem causing serious adverse health problems, including death. Hum Psychopharmacol 2012;27(3):254–61. https://doi.org/10.1002/hup.2222.

Chimalakonda KC, Moran CL, Kennedy PD, Endres GW, Uzieblo A, Dobrowolski PJ, et al. Anal Chem 2011;83(16):6381–8. https://doi.org/10.1021/ac201377m.

Diao X, Huestis MA. Approaches, challenges, and advances in metabolism of new synthetic cannabinoids and identification of optimal urinary marker metabolites. Clin Pharmacol Ther 2017;101:239–53. https://doi.org/10.1002/cpt.534.

Diao X, Huestis M. New synthetic cannabinoids metabolism and strategies to best identify optimal marker metabolites. Front Chem 2019;7:1–15.

Drug Enforcement Administration (DEA) - Drugs of Abuse. A DEA resource guide. 2020. https://www.dea.gov/documents/2017/06/15/drugs-abuse. [Accessed 18 October 2020].

El Balkhi S, Chaslot M, Picard N, Dulaurent S, Delage M, Mathieu O, et al. Characterization and identification of eight designer benzodiazepine metabolites by incubation with human liver microsomes and analysis by a triple quadrupole mass spectrometer. Int J Leg Med 2017 Jul;131(4):979–88. https://doi.org/10.1007/s00414-017-1541-6. Epub 2017 Feb 4. PMID: 28160051.

Elliott S. Current awareness of piperazines: pharmacology and toxicology. Drug Test Anal 2011 ;3(7–8):430–8. https://doi.org/10.1002/dta.307. Epub 2011 Jul 11. PMID: 21744514.

Elmore JS, Dillon-Carter O, Partilla JS, Ellefsen KN, Concheiro M, Suzuki M, et al. Pharmacokinetic profiles and pharmacodynamic effects for methylone and its metabolites in rats. Neuropsychopharmacology 2017 Feb;42(3):649–60. https://doi.org/10.1038/npp.2016.213. Epub 2016 Sep 23. PMID: 27658484; PMCID: PMC5240186.

European Monitoring Centre for Drugs and Drug Addiction. New psychoactive substances: 25 years of early warning and response in Europe. An update from the EU Early Warning System (June 2022). Luxembourg: Publications Office of the European Union; 2022.

Fabregat-Safont D, Mata-Pesquera M, Barneo-Muñoz M, Martinez-Garcia F, Mardal M, Davidsen AB, et al. In-depth comparison of the metabolic and pharmacokinetic behaviour of the structurally related synthetic cannabinoids AMB-FUBINACA and AMB-CHMICA in rats. Commun Biol 2022;5(1):1–13.

Feng L, Battulga A, Han E, Chung H, Li J. New psychoactive substances of natural origin: a brief review. J Food Drug Anal 2017;25:461–71. https://doi.org/10.1016/j.jfda.2017.04.001.

Franz F, Jechle H, Wilde M, Angerer V, Huppertz LM, Longworth M, et al. Structure-metabolism relationships of valine and tert-leucine-derived synthetic cannabinoid receptor agonists: a systematic comparison of the in vitro phase I metabolism using pooled human liver microsomes and high-resolution mass spectrometry. Forensic Toxicol 2019; 37:316–29. https://doi.org/10.1007/s11419-018-00462-x.

Global Commission on Drug Policy. Report. Classification of psychoactive substances – when science was left behind. 2019. 2019, https://www.globalcommissionondrugs.org/reports/classification-psychoactive-substances.

Ho JH, Dargan PI. Arylcyclohexamines (ketamine, phencyclidine, and analogues). In: Brent J, Burkhart K, Dargan P, Hatten B, Megarbane B, Palmer R, editors. Critical care toxicology. Cham: Springer; 2016. https://doi.org/10.1007/978-3-319-20790-2_124-1.

Hutter M, Broecker S, Kneisel S, Franz F, Brandt SD, Auwarter V. Metabolism of nine synthetic cannabinoid receptor agonists encountered in clinical casework: major in vivo phase I metabolites of AM-694, AM-2201, JWH-007, JWH-019, JWH-203, JWH-307, MAM-2201, UR-144 and XLR-11 in human urine using LC-MS/MS. Curr Pharmaceut Biotechnol 2018;19(2):144–62. https://doi.org/10.2174/1389201019666180509163114. PMID: 29745330.

Kasper AM, Ridpath AD, Gerona RR, Cox R, Galli R, Kyle PB, et al. Severe illness associated with reported use of synthetic cannabinoids: a public health investigation (Mississippi, 2015). Clin Toxicol 2019 Jan;57(1):10–8. https://doi.org/10.1080/15563650.2018.1485927. Epub 2018 Jul 10. PMID: 29989463.

Katagi M, Kamata T, Zaitsu K, Shima N, Kamata H, Nakanishi K, et al. Metabolism and toxicologic analysis of tryptamine-derived drugs of abuse. Ther Drug Monit 2010 Jun;32(3): 328–31. https://doi.org/10.1097/FTD.0b013e3181dcb40c. PMID: 20418800.

Katselou M, Papoutsis I, Nikolaou P, Spiliopoulou C, Athanaselis S. Metabolites replace the parent drug in the drug arena. The cases of fonazepam and nifoxipam. Forensic Toxicol 2017;35(1):1–10. https://doi.org/10.1007/s11419-016-0338-5.

Kreth K, Kovar K, Schwab M, Zanger UM. Identification of the human cytochromes P450 involved in the oxidative metabolism of "Ecstasy"-related designer drugs. Biochem Pharmacol 2000 Jun 15;59(12):1563–71. https://doi.org/10.1016/s0006-2952(00)00284-7. PMID: 10799653.

Krotulski AJ, Mohr ALA, Papsun DM, Logan BK. Metabolism of novel opioid agonists U-47700 and U-49900 using human liver microsomes with confirmation in authentic urine specimens from drug users. Drug Test Anal 2018 Jan;10(1):127–36. https://doi.org/10.1002/dta.2228. Epub 2017 Jul 26. PMID: 28608586.

Lomas EC, Maskell PD. Phenazepam: more information coming in from the cold. J Forensic Leg Med 2015;36:61–2. https://doi.org/10.1016/j.jflm.2015.08.017.

Malaca S, Lo Faro AF, Tamborra A, Pichini S, Busardò FP, Huestis MA. Toxicology and analysis of psychoactive tryptamines. Int J Mol Sci 2020 Dec 4;21(23):9279. https://doi.org/10.3390/ijms21239279. PMID: 33291798; PMCID: PMC7730282.

Manier SK, Felske C, Eckstein N, Meyer MR. The metabolic fate of two new psychoactive substances - 2-aminoindane and N-methyl-2-aminoindane - studied in vitro and in vivo to support drug testing. Drug Test Anal 2020 Jan;12(1):145–51. https://doi.org/10.1002/dta.2699. Epub 2019 Nov 25. PMID: 31667988.

Mestria S, Odoardi S, Federici S, Bilel S, Tirri M, Marti M, et al. Metabolism study of N-methyl 2-aminoindane (NM2AI) and determination of metabolites in biological samples by LC-HRMS. J Anal Toxicol 2021 May 14;45(5):475–83. https://doi.org/10.1093/jat/bkaa111. PMID: 32860694.

Meyer MR. New psychoactive substances: an overview on recent publications on their toxicodynamics and toxicokinetics. Arch Toxicol 2016a;90(10):2421–44. https://doi.org/10.1007/s00204-016-1812-x. Epub 2016 Sep 24. PMID: 27665567.

Meyer MR, Bergstrand MP, Helander A, Beck O. Identification of main human urinary metabolites of the designer nitrobenzodiazepines clonazolam, meclonazepam, and nifoxipam by nanoliquid chromatography-high-resolution mass spectrometry for drug testing purposes. Anal Bioanal Chem 2016b;408(13):3571–91.

Meyer MR, Maurer H. Metabolism of designer drugs of abuse: an updated review. Curr Drug Metabol 2010;11:468–82.

Meyer MR, Wilhelm J, Peters FT, Maurer HH. Beta-keto amphetamines: studies on the metabolism of the designer drug mephedrone and toxicological detection of mephedrone, butylone, and methylone in urine using gas chromatography-mass spectrometry. Anal Bioanal Chem 2010;397(3):1225–33. https://doi.org/10.1007/s00216-010-3636-5.

Meyer UA. Overview of enzymes of drug metabolism. J Pharmacokinet Biopharm 1996 Oct; 24(5):449–59. https://doi.org/10.1007/BF02353473. PMID: 9131484.

Michely JA, Helfer AG, Brandt SD, Meyer MR, Maurer HH. Metabolism of the new psychoactive substances N,N-diallyltryptamine (DALT) and 5-methoxy-DALT and their detectability in urine by GC-MS, LC-MSn, and LC-HR-MS-MS. Anal Bioanal Chem 2015 Oct; 407(25):7831–42. https://doi.org/10.1007/s00216-015-8955-0. Epub 2015 Aug 22. PMID: 26297461.

Moosmann B, Huppertz LM, Hutter M, Buchwald A, Ferlaino S, Auwärter V. Detection and identification of the designer benzodiazepine flubromazepam and preliminary data on its metabolism and pharmacokinetics. J Mass Spectrom 2013;48(11):1150–9. https://doi.org/10.1002/jms.3279.

Montesano C, Vannutelli G, Fanti F, Vincenti F, Gregori A, Togna AR, et al. Identification of MT-45 metabolites:in silico prediction, in vitro incubation with rat hepatocytes and in vivo confirmation. J Anal Toxicol October 2017;41(8):688–97. https://doi.org/10.1093/jat/bkx058.

Moran CL, Le V-H, Chimalakonda KC, Smedley AL, Lackey FD, Owen SN, et al. Anal Chem 2011;83(11):4228–36. https://doi.org/10.1021/ac2005636.

Negreira N, Erratico C, van Nuijs AL, Covaci A. Identification of in vitro metabolites of ethylphenidate by liquid chromatography coupled to quadrupole time-of-flight mass spectrometry. J Pharm Biomed Anal 2016 Jan 5;117:474–84. https://doi.org/10.1016/j.jpba.2015.09.029. Epub 2015 Oct 23. PMID: 26454340.

Nelson DR. The cytochrome P450 homepage. Hum Genom 2009;4:59–65. Retrieved from, http://drnelson.uthsc.edu/CytochromeP450.html.

Patel JC, Parveen S. In vitro and in vivo analysis of fentanyl and fentalog metabolites using hyphenated chromatographic techniques: a review. Chem Res Toxicol 2022 Jan 17;35(1): 30–42. https://doi.org/10.1021/acs.chemrestox.1c00225. Epub 2021 Dec 25. PMID: 34957817.

Pedersen AJ, Reitzel LA, Johansen SS, Linnet K. In vitro metabolism studies on mephedrone and analysis of forensic cases. Drug Test Anal 2013;5(6):430–8. https://doi.org/10.1002/dta.1369.

Pinterova N, Horsley RR, Palenicek T. Synthetic aminoindanes: a summary of existing knowledge. Front Psychiatr 2017;8:236. https://doi.org/10.3389/fpsyt.2017.00236. Published 2017 Nov 17.

Prosser JM, Nelson LS. The toxicology of bath salts: a review of synthetic cathinones. J Med Toxicol 2012;8(1):33–42. https://doi.org/10.1007/s13181-011-0193-z.

Schep LJ, Slaughter RJ, Vale JA, Beasley DM, Gee P. The clinical toxicology of the designer "party pills" benzylpiperazine and trifluoromethylphenylpiperazine. Clin Toxicol 2011 Mar;49(3):131–41. https://doi.org/10.3109/15563650.2011.572076. PMID: 21495881.

Staack RF, Maurer HH. Metabolism of designer drugs of abuse. Curr Drug Metabol 2005 Jun; 6(3):259–74. https://doi.org/10.2174/1389200054021825. PMID: 15975043.

Seely KA, Patton AL, Moran CL, Womack ML, Prather PL, Fantegrossi WE, et al. Forensic investigation of K2, Spice, and "bath salt" commercial preparations: a three-year study of new designer drug products containing synthetic cannabinoid, stimulant, and hallucinogenic compounds. Forensic Sci Int 2013 Dec 10;233(1–3):416–22. https://doi.org/10.1016/j.forsciint.2013.10.002. Epub 2013 Oct 14. PMID: 24314548.

Slaughter RL, Edwards DJ. Recent advances: the cytochrome P450 enzymes. Ann Pharmacother 1995;129(6):619–24. https://doi.org/10.1177/106002809502900612. PMID: 7663035.

Stoller A, Dolder PC, Bodmer M, Hammann F, Rentsch KM, Exadaktylos AK, et al. Mistaking 2C-P for 2C-B: what a difference a letter makes. J Anal Toxicol 2017 Jan;41(1):77–9. https://doi.org/10.1093/jat/bkw108. Epub 2016 Oct 6. PMID: 28130544.

Toennes SW, Geraths A, Pogoda W, Paulke A, Wunder C, Theunissen EL, et al. Pharmacokinetic properties of the synthetic cannabinoid JWH-018 and of its metabolites in serum after inhalation. J Pharm Biomed Anal 2017 Jun 5;140:215–22. https://doi.org/10.1016/j.jpba.2017.03.043. Epub 2017 Mar 24. PMID: 28365515.

Ujváry I, Christie R, Evans-Brown M, Gallegos A, Jorge R, de Morais J, et al. DARK classics in chmical neurosciences: etonitazene and related benzimidazoles. ACS Chem Neurosci 2021;12(7):1072–92. https://doi.org/10.1021/acschemneuro.1c00037.

UNODC. Early warning advisory on new psychoactive substances. Groups of NPS; 2020. https://www.unodc.org/LSS/SubstanceGroup/GroupsDashboard?testType=NPS.

Wallach J, Brandt SD. Phencyclidine-based new psychoactive substances. Handb Exp Pharmacol 2018;252:261–303. https://doi.org/10.1007/164_2018_124. PMID: 30105474.

Welter J, Brandt SD, Kavanagh P, Meyer MR, Maurer HH. Metabolic fate, mass spectral fragmentation, detectability, and differentiation in urine of the benzofuran designer drugs 6-APB and 6-MAPB in comparison to their 5-isomers using GC-MS and LC-(HR)-MS(n) techniques. Anal Bioanal Chem 2015b;407(12):3457–70. https://doi.org/10.1007/s00216-015-8552-2. Epub 2015 Feb 25. PMID: 25711990.

Welter J, Kavanagh P, Meyer MR, Maurer HH. Benzofuran analogues of amphetamine and methamphetamine: studies on the metabolism and toxicological analysis of 5-APB and 5-MAPB in urine and plasma using GC-MS and LC-(HR)-MS(n) techniques. Anal Bioanal Chem 2015a;407(5):1371–88. https://doi.org/10.1007/s00216-014-8360-0. Epub 2014 Dec 4. PMID: 25471293.

White M, Archer J. Pipradrol and pipradrol derivatives. New Psychoactive Substances; 2013. p. 233–59. https://doi.org/10.1016/B978-0-12-415816-0.00010-9.

Wilde M, Pichini S, Pacifici R, Tagliabracci A, Busardò FP, Auwärter V, et al. Metabolic pathways and potencies of new fentanyl analogs. Front Pharmacol 2019 Apr 5;10:238. https://doi.org/10.3389/fphar.2019.00238. PMID: 31024296; PMCID: PMC6461066.

Wiley JL, Compton DR, Dai D, Lainton JAH, Phillips M, Huffman JW, et al. Structure-activity relationships of indole- and pyrrole-derived cannabinoids. J Pharmacol Exp Ther 1998; 285(3):995–1004.

Wilkinson GR. Drug metabolism and variability among patients in drug response. N Engl J Med 2005 May 26;352(21):2211–21. https://doi.org/10.1056/NEJMra032424. PMID: 15917386.

Wohlfarth A, Scheidweiler KB, Pang S, Zhu M, Castaneto M, Kronstrand R, et al. Metabolic characterization of AH-7921, a synthetic opioid designer drug: in vitro metabolic stability assessment and metabolite identification, evaluation of in silico prediction, and in vivo confirmation. Drug Test Anal 2016 Aug;8(8):779–91. https://doi.org/10.1002/dta.1856. Epub 2015 Sep 1. PMID: 26331297; PMCID: PMC4562414.

Wrighton SA, VandenBranden M, Ring BJ. The human drug metabolizing cytochromes P450. J Pharmacokinet Biopharm 1996 Oct;24(5):461–73. https://doi.org/10.1007/BF02353474. PMID: 9131485.

Zaitsu K. Metabolism of synthetic cathinones. In: Zawilska J, editor. Synthetic cathinones. Current topics in neurotoxicity, vol 12. Cham: Springer; 2018. https://doi.org/10.1007/978-3-319-78707-7_5.

Zapata F, Matey JM, Montalvo G, Garcia-Ruiz C. Chemical classification of new psychoactive substances. Microchem J 2021;163:1–13.

Zawilska JB, Wojcieszak J. Novel psychoactive substances: classification and general information. In: Zawilska J, editor. Synthetic cathinones. Current topics in neurotoxicity, vol 12. Cham: Springer; 2018. https://doi.org/10.1007/978-3-319-78707-7_2.

Zawilska JB, Wojcieszak J. An expanding world of new psychoactive substances-designer benzodiazepines. Neurotoxicology 2019 Jul;73:8–16. https://doi.org/10.1016/j.neuro.2019.02.015. Epub 2019 Feb 23. PMID: 30802466.

Židková M, Linhart I, Balíková M, Himl M, Váňa L, Vetýška M, et al. Study on the metabolism of 5,6-methylenedioxy-2-aminoindane (MDAI) in rats: identification of urinary metabolites. Xenobiotica 2017 Jun;47(6):505–14. https://doi.org/10.1080/00498254.2016.1199919. Epub 2016 Jul 12. Erratum in: Xenobiotica. 2017 Jun;47(6):x. PMID: 27401914.

Zloh M, Samaras EG, Calvo-Castro J, Guirguis A, Stair JL, Kirton SB. Drowning in diversity? A systematic way of clustering and selecting a representative set of new psychoactive substances. RSC Adv 2017;7:53181–91.

CHAPTER

3

NPS pharmacology and toxicology

Acute toxicity and long-term effects

The Global Drug Survey conducted in 2013 showed that new psychoactive substances (NPS) were more likely to send users to emergency department than traditional recreational drugs (TRD). Synthetic cannabinoid users, for example, were 30 times more likely than cannabis users. Several reasons account for this (Winstock et al., 2015). The pharmacology of most NPS is hardly known; hence, avoiding overdosing can be difficult (Hill and Dargan, 2018). The composition of NPS products is unpredictable. With hundreds if not thousands available, and given the lack of quality control and regulatory oversight in manufacture, a product may contain a drug other than what it purportedly contains. To complicate this, products often are a mixture of drugs either from the same class or, worse, from different classes. Some contents work synergistically, thereby increasing potency. Some of the newer cutting agents are active pharmaceutical or veterinary products that can also add potency (Brandt et al., 2011; Schneir et al., 2014; Brunt et al., 2017). Furthermore, the potency of newer NPS has increased. For the first 2 years when synthetic cannabinoids gained popularity in the United States, 2008 to 2010, there was hardly a report of fatal intoxication. Since then, hundreds have occurred (Trecki et al., 2015; Kraemer et al., 2019).

Like TRD, NPS have acute toxic effects. Because the drugs are psychoactive, these effects are commonly neurological and psychological. And because most of them upregulate excitatory neurotransmitter levels, cardiovascular effects are also common. For depressants such as new synthetic opioids (NSO), respiratory effects are significantly reported. A few drugs within a class can also have renal, hepatic, gastrointestinal, musculoskeletal, and metabolic effects. Table 3.1 shows types of acute toxic effects grouped by organ system.

Designer Drugs
https://doi.org/10.1016/B978-0-12-811764-4.00013-6

TABLE 3.1 NPS adverse effects grouped by organ system.

Organ system	Symptoms
Neurological	Seizure, hyperthermia, mydriasis (dilation of the eye pupil), diaphoresis (excessive sweating), hyperkinesis, insomnia, headache, nausea, drowsiness, sluggishness, stupor, somnolence, reduced consciousness, tremor, shivers, jaw clenching, bruxism (teeth grinding), sedation, coma, short-term memory loss, psychomotor impairment, blurred vision, ataxia (loss of movement control), CNS depression, brain edema, ataxia, nystagmus (rapid involuntary movement of the eye), catatonia, miosis (excessive constriction of the eye pupil), hypothermia
Psychological	Anxiety, agitation, confusion, irritability, aggression, hysteria, mania, disorientation, psychosis, panic, paranoia, delusion, excited delirium, agitated delirium, auditory hallucination, visual hallucination, dysphoria (uneasiness), depression, self-harm, suicidal ideation
Cardiovascular	Tachycardia (abnormally rapid heart rate), hypertension, peripheral vasoconstriction (manifested as cold extremities, skin discoloration, rash), chest pain, palpitation, arrhythmia, QT prolongation (delayed contraction and relaxation of the heart muscle), acute coronary syndrome, myocardial infarction, heart enlargement, disseminated intravascular coagulopathy (DIC, abnormal blood clotting throughout blood vessels), bradycardia (abnormally slow heart rate), hypotension
Respiratory	Apnea (not breathing), bradypnea (abnormally slow breathing), tachypnea (abnormally rapid breathing), respiratory depression, pulmonary congestion, pulmonary edema
Renal	Acute kidney injury
Gastrointestinal	Vomiting, diarrhea, constipation
Hepatic	Hepatic failure
Musculoskeletal	Rhabdomyolysis (muscle breakdown), dystonia (abnormal muscle tone resulting to spasm), hypertonia (abnormally high muscle tone resulting to rigidity), tetany (intermittent muscle spasm), hyporeflexia (decreased reflex response), muscle weakness, incoordination
Metabolic	Hyponatremia, hypokalemia, hyperglycemia, metabolic acidosis
Multiorgan system	Serotonin syndrome, sympathomimetic toxicity, multiorgan failure

Because of the fleeting half-life of some NPS and the rapid changes in composition, long-term use is hard to define. It is more easily applied to a class than to a specific NPS. After more than a decade, a literature is accumulating. Dependency and withdrawal symptoms for synthetic cannabinoids and cathinones were reported early on. Dependence on fentalogs has long been known, and dependence on other NSO subclasses has also been reported or predicted. Other NPS classes vary in their ability to cause dependence and withdrawal symptoms.

Synthetic cannabinoids

These compounds mimic the effects of delta-9-tetrahydrocannabinol (THC) by binding to the cannabinoid receptors, of which there are two subtypes. CB1, found in the central nervous system (CNS), facilitates the psychoactive effects of THC; some are also found in the lungs, liver, and kidneys. CB2, mostly in the peripheral nerve terminals, immune system, and hematopoietic cells, plays a role in antinociception (response to toxic stimuli, mechanical injury, or adverse temperature) and is associated with the analgesic effects of cannabinoids. CB2 receptors are also found in the microglial cells of the brain (Wiley et al., 2011, 2013).

After these subtypes were cloned in 1990, most cannabimimetic research was directed toward finding THC-like compounds that had potent analgesic effects without the psychoactive effects, that is, selective CB2 agonists. However, this effort also generated hundreds of compounds with potent CB1 activity through the first decade of the current century, providing research and patent materials that served as rich references for the synthesis and manufacture of NPS. More than 250 synthetic cannabinoids have been reported to the United Nations Office on Drugs and Crime (UNODC) and the European Monitoring Centre for Drugs and Drug Addiction (EMCDDA). The previous chapter shows the structural evolution of popular ones detected in the United States (Fig. 2.7).

Because they target the same receptors as THC, adverse effects from synthetic cannabinoid intoxication are similar to those from THC—headache, drowsiness, sluggishness, disorientation, memory impairment, increased heart rate, nausea, and vomiting. However, a significant number of synthetic cannabinoids are more potent CB1 agonists and so are more severely toxic, including some effects that have never been reported for cannabis (e.g., acute kidney injury) (Wiley et al., 2015). THC and its active metabolite, 11-hydroxy-THC, are partial agonists of both CB1 and CB2. In contrast, some synthetic cannabinoids and their metabolites (e.g., AB-PINACA, 5F-AMB, MAB-CHMINACA, AMB-FUBINACA) are full

CB1 agonists. Some of these compounds have 10–100 times higher CB1 binding affinity in vitro and show correspondingly greater effects in animal studies (Huffman et al., 2005; Wiley et al., 2015; Tai and Fantegrossi, 2016).

The first generation of synthetic cannabinoids, such as naphthoylindoles (e.g., JWH-018), phenylacetylindoles (e.g., JWH-250), and benzoylindoles (e.g., AM-2233), produced acute toxic effects equal to or somewhat greater than THCs, as reflected in emergency department visits (Hermanns-Clausen et al., 2013; Kronstrand et al., 2013). JWH-018, for example, is three times more potent than THC. The much higher potencies of later-generation synthetic cannabinoids are evident in a Global Drug Survey report published in 2015, where a sample of 2176 users of synthetic cannabinoids were observed to be more likely to report having experienced panic and anxiety (81% vs. 60%), paranoia (62% vs. 35%), agitation (47% vs. 22%), chest pain (33% vs. 20%), visual hallucinations (33% vs. 16%), and seizures (19% vs. 8%) compared with 19,024 users of cannabis (Winstock et al., 2015).

In the thousands of intoxications now available in case reports, conference proceedings, and trip reports (time-stamped personal experience of drug users posted online in drug blogs), the most common adverse effects are agitation and tachycardia. Seizures and delirium, sometimes agitated or excited, usually accompanied by hyperthermia, also started to become common around 2013. CNS effects included psychosis, anxiety, irritability, aggression, panic, paranoia, hallucinations, depression, coma, and self-harm. Other cardiovascular effects were hypertension, chest pain, arrhythmia, acute coronary syndrome, myocardial infarction, and stroke (Fattore, 2016). Acute diffuse alveolar hemorrhage, tachypnea, respiratory depression, and acute kidney injury were reported. Metabolic effects such as hypokalemia, hyperglycemia, and metabolic acidosis have been associated with synthetic cannabinoids. Table 3.2 summarizes acute adverse reactions.

Some mass intoxications are notable. In 2012, the cyclopropyl indole XLR-11 was associated with several cases of acute kidney injury in the Pacific Northwest (Murphy et al., 2013; Thornton et al., 2013; Buser et al., 2014). In vitro studies years later showed that XLR-11 induced nephrotoxicity by impairment of the endocannabinoid-mediated regulation of mitochondrial function in human proximal tubule cells (Silva et al., 2018). The growing concern about the health threat was well captured in a 2015 report on mass intoxication surveillance conducted by the United States Drug Enforcement Administration. It showed agitated delirium associated with ADB-PINACA in Georgia, acute delirium and seizures associated with AB-CHMINACA in Florida, and severe illnesses in 721 patients (11 fatalities) associated with MAB-CHMINACA in Mississippi (Trecki et al., 2015; Tyndall et al., 2015; Schwartz et al., 2015; Kasper et al., 2015).

TABLE 3.2 Acute adverse effects in synthetic cannabinoid intoxication.

Organ system	Symptoms
Neurological	Seizure, CNS depression, coma, hyperthermia, headache, drowsiness, sluggishness, disorientation, nausea, memory impairment, psychomotor impairment
Psychological	Agitation, anxiety, irritability, aggression, psychosis, panic, paranoia, agitated or excited delirium, visual hallucination, depression, self-harm
Cardiovascular	Tachycardia, hypertension, chest pain, arrhythmia, acute coronary syndrome, myocardial infarction, stroke
Respiratory	Tachypnea, respiratory depression, diffuse alveolar hemorrhage
Gastrointestinal	Vomiting
Renal	Acute kidney injury
Metabolic	Hypokalemia, hyperglycemia, metabolic acidosis

The epidemic reached Alaska, with 1351 ambulance transports between July 2015 and March 2016 (Springer et al., 2016). Some cases required endotracheal intubation and admission to the intensive care unit. Most severe cases were associated with 5F-AMB, AB-CHMINACA, or MAB-CHMINACA. "Zombie-like" effects from severe CNS depression were associated with AMB-FUBINACA in New York City in 2016; these cases were unique in that the toxidrome did not involve tachycardia, arrhythmia, seizures, hyperthermia, or acute kidney injury (Adams et al., 2017). The same effects were independently observed in laboratory-confirmed AMB-FUBINACA cases from New Zealand (Ong et al., 2020) and parts of Europe in 2017 (Adamowicz et al., 2019). CNS depression as manifested by reduction in consciousness was reported for MDMB-CHMICA in the United Kingdom; these cases were also characterized by respiratory depression and bradycardia (Hill et al., 2016; Meyyappan et al., 2017).

Fatal intoxication directly associated with cannabis is extremely rare, as was the case for synthetic cannabinoids in the first few years of their popularity. However, the increasing potency of later synthetic cannabinoids, especially the indole and indazole carboxamides, had changed this by 2013. Most of the fatal cases directly linked to synthetic cannabinoids were accompanied by seizures, arrhythmia, agitated delirium, and multiorgan failure; those indirectly resulting from synthetic cannabinoids involved hypothermia, trauma, and self-harm.

Synthetic cathinones

Being derived from amphetamines, synthetic cathinones are psychostimulants that act as monoamine (dopamine, serotonin, norepinephrine) reuptake inhibitors as well as effectors of transporter-mediated monoamine release. Both actions have the effect of augmenting the presynaptic concentrations of these excitatory neurotransmitters (Baumann et al., 2013; Simmler et al., 2013). Like amphetamines, synthetic cathinones can have either predominantly psychostimulant or empathogenic properties. The relative selectivity of a synthetic cathinone toward the dopamine transporter (DAT), serotonin transporter (SERT), and norepinephrine transporter (NET) determines its ability to facilitate stimulant or empathogenic activity. In turn, this determines the type of acute toxic effects.

Stimulants such as cocaine, methylphenidate, methamphetamine, and 3,4-methylenedioxymethamphetamine (MDMA, "Ecstasy") differ in their pharmacological effects based on differences in mechanism of action (reuptake inhibition, release, or both) on the monoamine transporter (MAT) (Rothman and Baumann, 2003). MATs are proteins found in the plasma membrane of neuronal cells that synthesize monoamine neurotransmitters; they facilitate the termination of neurotransmitter action in the synaptic cleft by reuptake into the neuronal cytoplasm. Reuptake inhibitors bind the neurotransmitter-binding site on the transporter (orthosteric site), blocking reuptake from the extracellular medium. Other stimulants also act as substrates after binding to the same orthosteric site in the transporter. As substrates, they are translocated through the transporter channel into the neuronal cytoplasm, inducing inward depolarizing currents that reverse the flux of neurotransmitters in the transporter, which triggers the nonexocytotic release or efflux of intracellular neurotransmitters (transporter-mediated release) (Reith et al., 2015; Sitte and Freissmuth, 2015).

Cocaine blocks monoamine reuptake but does not affect transporter-mediated monoamine release. Although it blocks all three MAT nonselectively, the behavioral and reinforcing effects depend primarily on the inhibition of the DAT. Methylphenidate acts primarily as a potent DAT and NET inhibitor, with a thousand-fold less inhibitory activity on the SERT. Like cocaine, it does not promote the release of monoamines and primarily modulates the levels of dopamine (Luethi et al., 2018). Methamphetamine is most potent in inhibiting the NET while being 5- to 10-fold less potent at the DAT and 200- to 500-fold less potent at the SERT. Unlike cocaine, methamphetamine promotes monoamine release through several pathways on top of monoamine reuptake inhibition. MDMA has moderately higher inhibition potency for the SERT and NET than for the DAT. It also promotes release of serotonin and norepinephrine through a pathway that is different from methamphetamines. MDMA primarily modulates serotonin levels, which is associated with its empathogenic activity (Liechti et al., 2000; Hysek et al., 2012).

A useful parameter to differentiate the pharmacological activities of these stimulants is the ratio of their inhibitory potencies between the DAT and SERT (DAT:SERT ratio). Inhibition of the DAT and NET or activation of the dopamine and norepinephrine systems is associated with psychostimulant property and enhanced abuse liability (Rothman et al., 2001). In contrast, increase in serotonergic activity is associated with reduced potential for addiction (Rothman and Baumann, 2006; Baumann et al., 2011) and enhanced empathogenic effect (Liechti et al., 2000). MDMA induces serotonin release and has a DAT:SERT ratio of typically 0.01–0.1. This property is typical of empathogens. Cocaine, being nonselective, has a DAT:SERT ratio close to unity. Methamphetamine is more selective for DAT and has a DAT:SERT ratio >10, mostly exerting psychostimulant effects. A DAT:SERT ratio >1 is generally associated with high abuse potential.

DAT:SERT ratio is calculated by taking the ratio between the inverse of the half-maximal inhibitory concentration (IC_{50}) of a drug toward the respective transporters. IC_{50} values are experimentally derived from in vitro or in vivo functional assays specific for the transporter activity. Table 3.3 presents DAT:SERT ratios of these representative stimulants along with some of the more popular cathinones.

The DAT:SERT ratios of synthetic cathinones align well with their pharmacological activity. Mephedrone, naphyrone, methylone, ethylone, and butylone are nonselective monoamine uptake inhibitors, similar to cocaine. Methylone, ethylone, and butylone also induce serotonin release,

TABLE 3.3 DAT:SERT ratios of selected stimulants and synthetic cathinones.

Drug	DAT:SERT	Drug	DAT:SERT
PMA	0.01–0.07	Cocaine	3–6
MDMA	0.06–0.2	Flephedrone	5–10
Methedrone	0.06–0.8	Methamphetamine	5–200
MDEA	0.08–0.4	Amphetamine	30–60
MDA	0.3–0.6	Methcathinone	30–70
Mephedrone	0.6–2	MDPBP	40–800
Ethylone	0.8–3	MDPPP	60–600
Butylone	1–6	MDPV	70–900
Naphyrone	2–5	Pyrovalerone	90–1000
Methylone	2–8	Methylphenidate	1000–5000

similar to MDMA and other empathogens. The DAT:SERT ratios of these cathinones are close in value to cocaine and MDMA (Hadlock et al., 2011; López-Arnau et al., 2012). Cathinone, methcathinone, and flephedrone act primarily as preferential dopamine and norepinephrine reuptake inhibitors and induce the release of dopamine, similar to methamphetamine. Their ratios are close to those of methamphetamine and amphetamine (Cozzi et al., 1999; Fleckenstein et al., 1999; Simmler et al., 2013). Pyrovalerone and MDPV are highly potent and selective dopamine and norepinephrine reuptake inhibitors, but, like methylphenidate, they do not promote the release of monoamines (Meltzer et al., 2006). Thus, based on their ratios, synthetic cathinones can be grouped pharmacologically as follows:

- MDMA-like (DAT:SERT <1)
- mixed MDMA- and cocaine-like (DAT:SERT ~1)
- methamphetamine-like (DAT:SERT ~5–100)
- methylphenidate-like (DAT:SERT >100)

Curiously, two special subclasses of synthetic cathinones have their own characteristic pharmacology. Methylenedioxycathinones (methylone, ethylone, butylone), like methylenedioxyamphetamines, facilitate empathogenic effects; they have either MDMA or mixed MDMA- and cocaine-like pharmacology. Pyrrolidinyl-substituted cathinones (pyrovalerones) are potent inhibitors of dopamine and norepinephrine reuptake and have methylphenidate-like pharmacology.

The acute toxic effects of a synthetic cathinone depend on its pharmacology. For mixed MDMA- and cocaine-like and methamphetamine-like members, the sympathomimetic toxidrome is the most common. This is characterized by agitation (sometimes accompanied by aggression), tachycardia, hypertension, palpitations, chest pain, mydriasis (dilated pupils), and seizures. In some cases, it can precipitate psychosis, paranoia, hallucinations, and, much less commonly, excited delirium (Backberg et al., 2015a). Hyperthermia is a common sympathomimetic toxidrome presentation, but it is less common in this subgroup of synthetic cathinones. Other adverse effects observed are headache, nausea, vomiting, hyponatremia, jaw clenching, bruxism, diaphoresis, and peripheral vasoconstriction that leads to cold or numb extremities (Boulanger-Gobeil et al., 2012). Mephedrone, the most popular ingredient of "bath salts" and "plant food" when they first became popular, is an archetype of this group. Other ring- and alkyl-substituted synthetic cathinones such as naphyrone, buphedrone, pentedrone, and 4-ethylmethcathinone exhibit similar acute toxicity. Generally, synthetic cathinones have higher dopaminergic effects than their corresponding amphetamine analog, even among mixed MDMA- and cocaine-like synthetic cathinones. This suggests higher stimulant-type effects and greater risk for dependence (Liechti, 2015).

The sympathomimetic toxidrome is also the major adverse effect of pyrrolidinyl-substituted or pyrovalerone-based cathinones. MDPV and alpha-PVP are the most popular (Spiller et al., 2011; Thornton et al., 2012; Umebachi et al., 2016). They potently inhibit DAT and NET and have significantly lower potency at SERT. The STRIDA project (Sweden) has reported the sympathomimetic toxidrome in laboratory-confirmed MDPV, alpha-PVP, and alpha-PBP cases. Typical effects are tachycardia, agitation, hypertension, and delirium. Hyperthermia is seen less often with these drugs (Beck et al., 2015, 2016; Franzen et al., 2018).

There are a few synthetic cathinones with primarily MDMA-like pharmacology. Methedrone is one. Its DAT:SERT ratio of 0.14 is similar to those of MDMA (0.08) and potent amphetamines such as para-methoxymethamphetamine (PMMA, 0.04) and 4-methylthioamphetamine (4-MTA, 0.02). Overdose of these drugs produces a severe toxidrome involving hyperthermia and seizures and usually leads to rhabdomyolysis, acute kidney injury, and multiorgan failure (Liechti, 2015; Simmler et al., 2013, 2014). These effects are most likely a result of severe serotonin syndrome, a condition caused by excess production of serotonin. Milder symptoms have been reported as adverse effects of MDMA-like synthetic cathinones and include agitation, confusion, disorientation, tachycardia, hypertension, mydriasis, diaphoresis, headache, nausea, vomiting, tremor, and shivers. Table 3.4 is a summary of common adverse effects from synthetic cathinone intoxication.

Amphetamines and phenethylamines

Amphetamine NPS have the same pharmacology as methamphetamine and MDMA. They facilitate monoamine reuptake inhibition and/or MAT-mediated release and act as either stimulants or empathogens. Two common subclasses are ring-substituted and methylenedioxy-substituted. The former are usually stimulants (Dolder et al., 2017) that cause the sympathomimetic toxidrome; the latter are usually empathogens that cause serotonin syndrome (Nichols, 1986), although there are a few exceptions (Simmler and Liechti, 2018). Fig. 3.1 illustrates the structural relationships of amphetamines within the class and to other NPS classes. Table 3.5 summarizes the acute toxic effects of amphetamines and phenethylamines.

A ring-substituted example is para- or 4-fluoroamphetamine (4-FA). Like other stimulants, it can cause tachycardia, mydriasis, and hyperkinesis. Severe cardiotoxicity and hyperthermia along with severe headache are commonly reported (Hondebrink et al., 2017; Wijers et al., 2017). Some ring-substituted structures—para-methoxyamphetamine (PMA), PMMA, and 4-MTA—cause greater morbidity and mortality than others. There are

TABLE 3.4 Acute toxic effects of cathinones.

Organ system	Symptoms
Neurological	Seizure, hyperthermia, mydriasis, headache, nausea, jaw clenching, bruxism, diaphoresis, tremor, shivers, insomnia
Psychological	Agitation, aggression, confusion, disorientation, psychosis, paranoia, hallucinations, excited delirium
Cardiovascular	Tachycardia, hypertension, palpitations, chest pain, peripheral vasoconstriction (manifested as cold or numb extremities), disseminated intravascular coagulopathy
Renal	Acute kidney failure
Musculoskeletal	Rhabdomyolysis
Gastrointestinal	Vomiting
Metabolic	Hyponatremia
Multiorgan system	Sympathomimetic toxidrome, serotonin syndrome, multiorgan failure

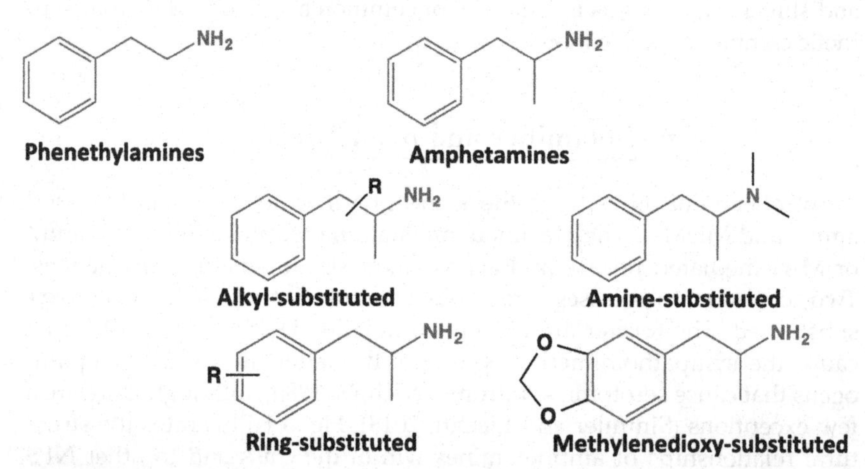

FIGURE 3.1 Structural relationships within amphetamine subclasses and between amphetamines and phenethylamines.

several reports of severe hyperthermia, leading to multiorgan failure and death (De Letter et al., 2001; Vevelstad et al., 2012; Lurie et al., 2012). The hyperthermia is most likely associated with serotonin syndrome. The presence of methoxy and methylthio groups at the 4 (para) position of the aromatic ring imparts the ability to stimulate marked serotonin release and strong monoamine oxidase inhibition.

TABLE 3.5 Acute toxic effects of amphetamines and phenethylamines.

Organ system	Symptoms
Neurological	Seizure, hyperthermia, mydriasis, hyperkinesis, headache, nausea, reduced consciousness, coma
Psychological	Hallucinations, agitation, aggression, excited delirium, confusion, paranoia
Cardiovascular	Tachycardia, hypertension, vasoconstriction, peripheral numbness
Musculoskeletal	Rhabdomyolysis
Renal	Acute kidney failure
Gastrointestinal	Vomiting, diarrhea
Metabolic	Metabolic acidosis
Multiorgan system	Sympathomimetic toxidrome, serotonin syndrome, multiorgan failure

Phenethylamines are psychedelic hallucinogens. Pharmacologically, they are full or partial agonists of 5-HT_2 (5-hydroxytryptamine, serotonin) receptors, usually 5-HT_{2A}. 5-HT_{2A} is thought to underlie hallucinogenic activity, and 5-HT_{2B} and SERT to facilitate empathogenic activity (Nichols, 2004; Fantegrossi et al., 2008). The best known are the 2C series, characterized by the presence of two methoxy groups at positions 2 and 5 and a variable substituent at position 4 of the aromatic ring. Like other classic psychedelics (LSD, dimethyltryptamine, and psilocybin), these drugs induce euphoria, mild stimulation, visually appealing distortions, altered perception of time and space, enhanced appreciation of music, and intensification of sensual or sexual feelings. There is wide variability in the type of sensory and emotional stimulation, and the experience of users is very subjective depending on mood, expectation, personality type, and baseline emotional state. Regardless, the presence of the two methoxy groups in the 2C compounds imparts stronger hallucinogenic activity. Some are also entactogenic or empathogenic (Dean et al., 2013; Nelson et al., 2014). 2C-B first became popular as an alternative to MDMA in the late 1980s and early 1990s, when the latter was scheduled. Psychotherapeutic use was explored but later dropped because of gastrointestinal effects and weaker empathogenic effects than MDMA.

Acute adverse effects from 2C intoxication usually present as a sympathomimetic toxidrome, serotonin toxicity, hallucinations, or a combination. The commonly observed mild effects include hallucinations, agitation, nausea, vomiting, and diarrhea. At high concentrations, unpleasant hallucinations, tachycardia, hypertension, hyperthermia, seizures, and delirium occur (Sanders et al., 2008; Topeff et al., 2011; Bosak et al., 2013; Dean et al., 2013). Excited delirium leading to death from sudden cardiopulmonary arrest has been reported.

A subclass of 2C became popular in the 2010s: the *N*-benzoylmethoxy derivatives (NBOMe). They have significantly stronger 5-HT_{2A} activity than LSD. They were first synthesized in 2003 as a research tool to map the 5-HT receptors in the brain. The three most popular are 25I-NBOMe, 25B-NBOMe, and 25C-NBOMe (Schifano et al., 2015). Intoxication is usually more severe than with other 2C compounds because of their potent serotonergic activity. The most common adverse effects are tachycardia, agitation with aggression, hypertension, and seizure. Other effects are confusion, paranoia, mydriasis, nausea, vomiting, vasoconstriction, and peripheral numbness. Fatal cases typically involve hyperthermia, seizures, rhabdomyolysis, acute kidney injury, metabolic acidosis, multi-organ failure, and coma (Suzuki et al., 2015; Wood et al., 2015).

A subclass of amphetamines has the same 2,5-dimethoxy substitutions in the aromatic ring as the 2C phenethylamines: the DOx or D series. The 2,5-dimethoxy groups impart a hallucinogenic property on top of the stimulant activity (Nichols, 2004), and the result is both dopaminergic and serotonergic stimulation in animal studies. Onset is slow, greater than an hour. This increases the risk of early repeat dosing by inexperienced users, which has led to a lot of intoxications. Duration is also long, 15–30 h. Although they have less stimulant activity than other amphetamines, they have potent 5-HT_{2A} receptor stimulant activity, which is most likely responsible for their ability to induce vasoconstriction. The latter has caused this group to produce greater morbidity and mortality than other amphetamines. Agitation, seizures, reduced consciousness, and metabolic acidosis have been observed.

Other stimulants and psychedelics

Aminoindanes and benzofurans are based on the amphetamine structure, tryptamines on the phenethylamine structure, and they share the respective pharmacological effects. Piperazines and piperidines are stimulants that are structurally related but are not direct derivatives of amphetamines. Fig. 3.2 shows the structural relationships of various classes of NPS stimulants and psychedelics. Table 3.6 summarizes their acute toxic effects.

Aminoindanes

In these cyclic analogs of amphetamine, the isopropyl group is cyclized to form a five-membered ring attached to an amino group. As such, they

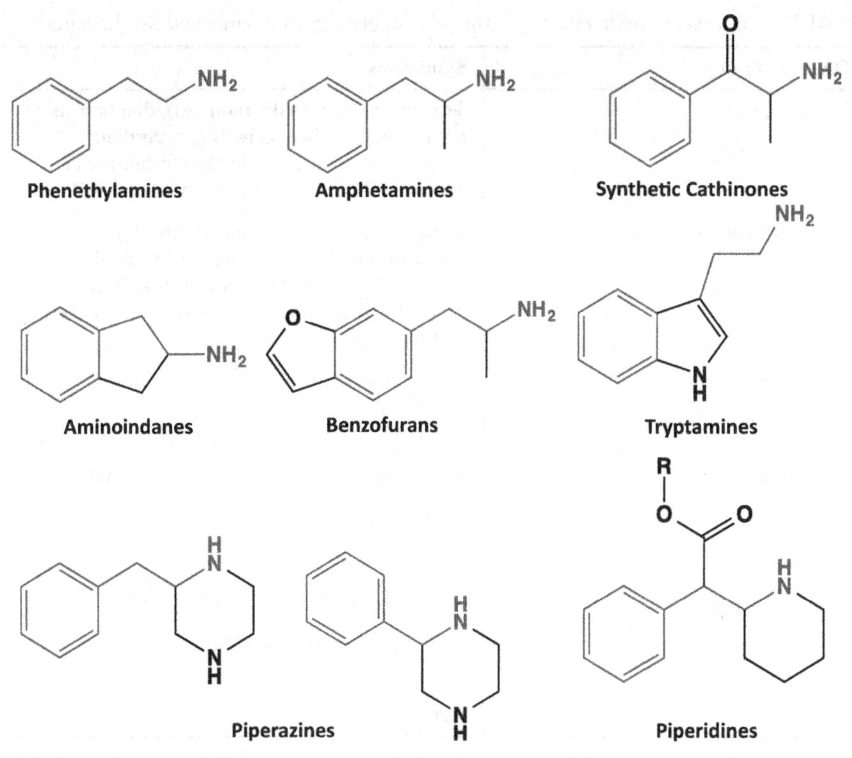

FIGURE 3.2 Structural relationships of various NPS stimulants and psychedelics.

have stimulant and empathogenic properties. As stimulants, they are reported to be less potent than their amphetamine analogs. Methylenedioxy-substituted aminoindanes such as MDAI (5,6-methylenedioxy-2-aminoindane) and some ring-substituted amino-indanes such as MMAI (5-methyl-6-methoxy-2-aminoindane), and 5-IAI (5-iodo-2-aminoindane) are empathogens (Simmler et al., 2014).

Pharmacologically, aminoindanes are weak inhibitors of monoamine reuptake but potent stimulants of nonvesicular serotonin release. These drugs substitute for MDMA but not LSD and methamphetamine in animal discrimination assays. Initial animal studies suggested that they are less neurotoxic than MDMA, as they do not affect the levels of serotonin and SERT after a one-time high dose, in contrast with MDMA. They became popular when the initial synthetic cathinones being used as substitutes for MDMA, such as methylone, were regulated (Sainsbury et al., 2011; Pinterova et al., 2017; Corkery et al., 2013). Later studies, however, showed that high doses cause serotonin syndrome, presenting as excessive salivation, diaphoresis, seizures, disseminated intravascular coagulopathy, and brain edema. Hyperthermia was also observed,

TABLE 3.6 Acute toxic effects of other classes of NPS stimulants and psychedelics[a].

Organ system	Symptoms
Neurological	Seizure, excessive salivation (AI), diaphoresis, mydriasis (TA), headache (P), hyperthermia, bruxism, nausea, amnesia (TA), catalepsy (TA), insomnia(BF), tremor (P), brain edema (AI)
Psychological	Anxiety, agitation (TA), irritability (P), disorientation (TA), confusion (TA), panic, paranoia, clouding of consciousness (TA), psychosis, delirium (TA), depression, visual and auditory hallucinations
Cardiovascular	Disseminated intravascular coagulopathy (AI, P), tachycardia, hypertension, palpitations, chest pain, QT prolongation (P)
Respiratory	Acute respiratory distress syndrome (AI)
Renal	Acute renal failure (TA, P)
Hepatic	Hepatic failure (AI)
Musculoskeletal	Rhabdomyolysis (TA)
Metabolic	Dehydration (AI), metabolic acidosis (P), respiratory acidosis (P), hyponatremia (P)
Multiorgan system	Serotonin syndrome

[a]*Symptoms unique to a class:* AI, *aminoindane;* BF, *benzofuran;* P, *piperazine;* TA, *tryptamine.*

especially when animals were housed together. This has bad implications for the drug, as most users are rave attendees where use occurs in a crowded setting. Moreover, most rave goers are also after stimulation. Because these drugs have weaker stimulant activity, redosing frequently happens. Hence, reports of intoxication usually show serotonin syndrome and sometimes renal failure, acute respiratory distress syndrome, hepatic failure, and increased risk of valvular heart disease. Three fatal intoxications from MDAI have been reported, with serotonin syndrome likely contributing to death (Corkery et al., 2013). Milder effects reported online by users include dehydration, tachycardia, diaphoresis, anxiety, panic attacks, and depression (Coppola and Mondola, 2011).

Benzofurans

Like aminoindanes, the structures of benzofurans bear close resemblance to amphetamine, where the aromatic ring is simply replaced by a benzofuran or benzodifuran ring. The class is exemplified by 5- and 6-APB, both of which inhibit monoamine reuptake and facilitate

MAT-mediated release at the DAT, NET, and SERT, similar to MDMA (Iversen et al., 2013). They also have high affinity for 5-HT_{2B}, 5-HT_{2C}, and adrenoreceptors alpha-1 and alpha-2C. Therefore they have MDMA-like and hallucinogenic properties. Acute toxic effects are similar to those of amphetamines and include tachycardia, hypertension, palpitations, hyperthermia, visual and auditory hallucinations, anxiety, panic attacks, paranoia, depression, insomnia, psychosis, bruxism, and nausea (Greene, 2013; Liechti, 2015).

The archetypal benzodifuran is Bromo-DragonFLY (Fig. 3.3), named for the resemblance of the chemical structure to the insect. It has high potency at 5-HT_1 and 5-HT_{2A}, and its adverse effects are proportionately strong. It is used for its hallucinogenic properties. Onset of action can be within 20–90 min, but very slow onset (up to 6 h) has also been reported. Come-down has been reported from 4 h to about 3 days. The hydrophobicity of the structure most likely allows greater tissue distribution, contributing to the long-lasting effects. Because of their potency, benzodifurans have been implicated in a number of fatalities (Corazza et al., 2011; Greene, 2013). Deaths in Scandinavian countries have been linked to prolong arteriolar vasoconstriction facilitated by potent agonist action on 5-HT and alpha-adrenoreceptors.

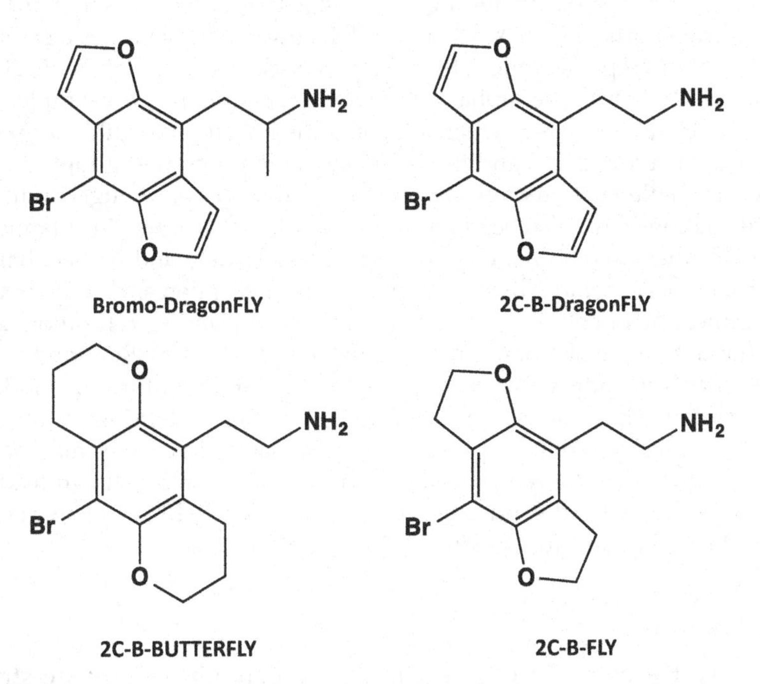

FIGURE 3.3 Representative drugs of the FLY series. Bromo-DragonFLY and 2C-B-DragonFLY are dibenzofurans.

Tryptamines

These are analogs of phenethylamine where the benzene ring is replaced by an indole. Their psychedelic hallucinogenic action is similar to phenethylamines. Naturally occurring tryptamines include serotonin (5-hydroxytrptamine), melatonin (*N*-acetyl-5-methoxytryptamine), bufotenin (*N,N*-dimethylserotonin) from the skin of toads, DMT (*N,N*-dimethyltryptamine) in ayahuasca, and psilocin (4-hydroxy-DMT) and psilocybin (O-phosphoryl-4-hydroxy-DMT) in magic mushrooms (Araújo et al., 2015). Naturally occurring tryptamines became popular in the United States in the 1950s, and recently, their therapeutic properties have been intensely explored (Roseman et al., 2017; Muttoni et al., 2019; Palhano-Fontes et al., 2019). Their synthetic counterparts started to appear in the illicit drug market in the 1990s. Recent ones are sold as legal highs. Examples are 5-MeO-DALT (*N,N*-diallyl-5-methoxytryptamine), AMT (alpha-methyltryptamine), and 4-OH-DALT (*N,N*-diallyl-4-hydroxytryptamine). Pharmacologically, they are agonists of 5-HT_{1A}, 5-HT_{2A}, and 5-HT_{2C}, with slightly lower selectivity for 5-HT_{2A} compared with phenethylamines. Activities toward 5-HT_{1A} and 5-HT_{2A} are associated with their hallucinogenic property (Fantegrossi et al., 2008; Rickli et al., 2016; Madsen et al., 2019), but other receptors and transporters have also been implicated, including the sigma-1 receptor, vesicular monoamine transporter 2 (VMAT 2), and SERT, so some tryptamines also have stimulant or empathogenic activity (Cozzi et al., 2009; Blough et al., 2014; Rickli et al., 2016). The alpha-methylated tryptamines, for example, are stimulants. In most cases, when a tryptamine is a stimulant, this property dominates at low doses and the hallucinogenic properties at high doses.

Visual hallucinations are the most common effects, though auditory hallucinations have also been reported for DiPT (diisopropyltryptamine). As with phenethylamine, the effects can vary widely and be peculiar to the user. These include distortions in sensory perception and body image, intensification of colors, mood lability, depersonalization, relaxation, and euphoria (Shulgin and Shulgin, 1997; Nichols, 2004, 2018). Reported acute toxic effects include agitation, disorientation, panic, confusion, clouding of consciousness, intense hallucinations, psychosis, delirium, amnesia, catalepsy, tachycardia, hypertension, mydriasis, hyperthermia, acute renal failure, rhabdomyolysis, and serotonin syndrome that can lead to death (Muller, 2004; Boland et al., 2005; Alatrash et al., 2006; Itokawa et al., 2007; Taljemark and Johansson, 2012; Jovel et al., 2014).

Piperazines

Unlike the other three classes in this section, piperazines are structurally related to amphetamines or phenethylamines but are not directly derived from either class. They are also entirely synthetic, not based on naturally occurring structures. They are primarily stimulants, with some also empathogenic and/or hallucinogenic. By the turn of the century, they were commonly found in Ecstasy pills as substitutes for MDMA (Baumann et al., 2005; Wood et al., 2008). The two subclasses are based on the main structural scaffolds of benzylpiperazines and phenylpiperazines. They induce dopamine, norepinephrine, and serotonin release and inhibit their reuptake. Some have nonselective affinity for 5-HT_2. They have both stimulant and hallucinogenic activities, typically the former at low doses and the latter at high doses.

Benzylpiperazine (BZP), one of the most popular, was initially investigated as an antihelminthic for livestock in 1994 and abandoned because of its high propensity to induce seizures. It inhibits dopamine reuptake and at the same time is an agonist of dopaminergic and adrenoreceptors. It is also a nonselective serotonin receptor agonist (Arbo et al., 2012; De Boer et al., 2001). Consequently, it has both stimulant and hallucinogenic activities. In contrast, the phenylpiperazine TFMPP does not have dopaminergic and noradrenergic activities but binds the serotonin receptors and prevents serotonin reuptake. Thus, it is empathogenic. Users coingest BZP and TFMPP to produce MDMA-like effects; this combination also produces dissociative effects (Wood et al., 2008; Thompson et al., 2010; Schep et al., 2011).

Most acute intoxication effects of piperazines are mild: agitation, anxiety, confusion, irritability, insomnia, headache, and nausea. Severe intoxication may involve paranoia, auditory hallucinations, and seizures, which can sometimes be repetitive, mixed metabolic and respiratory acidosis, tachycardia, hypertension, chest pain, QT interval prolongation (prolongation of ventricular repolarization), hyperthermia, mydriasis, diaphoresis, and tremors. In very rare cases, hyponatremia, serotonin toxicity, nephrotoxicity, and disseminated intravascular coagulopathy were followed by death (Arbo et al., 2012; Gee et al., 2005, 2008, 2010; Kovaleva et al., 2008; Schep et al., 2011).

Piperidines

Like piperazines, drugs in this class are not direct derivatives of amphetamines and are entirely synthetic. Two common subclasses are the pipradrol and methylphenidate derivatives. Pipradrol, its derivatives, and their closely related analogs containing the pyrrolidine ring (e.g., dihydroprolinol or D2PM) are lipophilic stimulants. Their structures are related to beta-phenylmethamphetamine (Fig. 3.4), a long-acting potent stimulant. The aromatic and cycloalkyl rings in their structures impart a strong hydrophobic property that translates to long half-life. In animal studies, pipradrol and its desoxy form, 2-DPMP, inhibit the reuptake and

β-Phenylmethamphetamine Pipradrol Diphenylprolinol (D2PM)

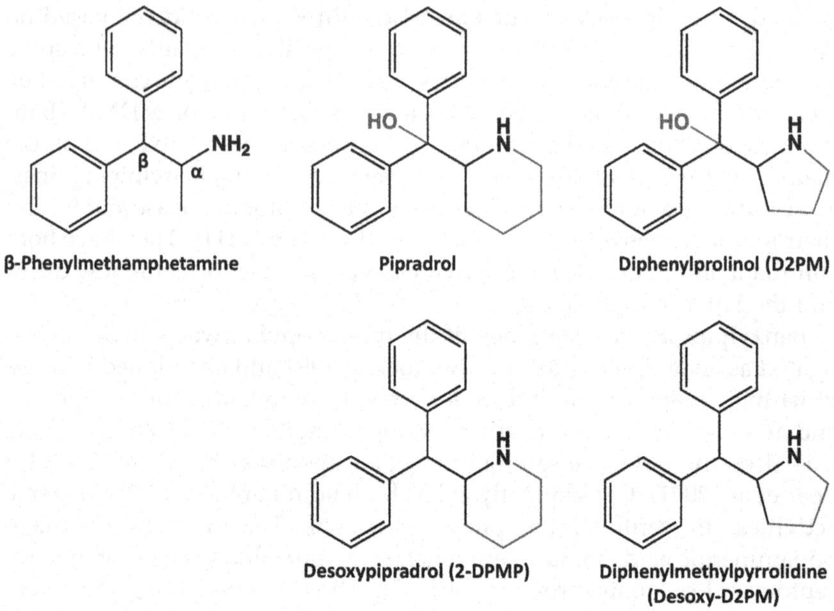

Desoxypipradrol (2-DPMP) Diphenylmethylpyrrolidine (Desoxy-D2PM)

FIGURE 3.4 Representative structures of pipradrol and its derivatives.

stimulate the release of dopamine and norepinephrine. Pipradrol was shown to be less potent than D-amphetamine (Robbins et al., 1983), while desoxypipradrol was shown to be more potent than cocaine at dopamine terminals in rat brain slices (Ferris and Tang, 1979). No pharmacological studies on D2PM and its derivatives have been conducted, but they are assumed to follow pipradrol's pharmacology.

There are no formal toxicological studies on pipradrol and its derivatives. However, in a number of case reports involving desoxypipradrol and D2PM, the clinical symptoms observed were very similar to amphetamine toxicity (Lidder et al., 2008; Wood et al., 2012a; Murray et al., 2012a; Murray et al., 2012b). These include severe agitation, aggression, anxiety, restlessness, paranoia, hallucinations, insomnia, tachycardia, hypertension, chest pain, and myocardial damage. Notably, in some of these cases, the symptoms lasted for 3–7 days, suggestive of this drug's long half-life.

Ethylphenidate is the archetype of methylphenidate derivatives. Fifty years ago, it was first used as an internal standard for the chromatographic analysis of methylphenidate (Ritalin) (Iden and Hungund, 1979; Chan et al., 1980); it was later discovered to be a metabolite formed when methylphenidate is taken with ethanol (Markowitz et al., 1999). As an NPS, it first appeared in Internet drug forums in 2010 and was first reported to the EMCDDA Early Warning System in 2011. It was popular in

Europe in the first half of the past decade, sold under the street name "Nopaine" (Markowitz and Patrick, 2013). However, its detection in samples has been reported as late as 2019. Being a very close structural homolog, its pharmacology is expected to follow methylphenidate; that is, it inhibits dopamine and norepinephrine reuptake more potently than serotonin reuptake. It does not also elicit transporter-mediated efflux of monoamines. This pharmacology is shared by other methylphenidate derivatives such as *N*-benzylethylphenidate, 4-fluoromethylphenidate, and isopropylphenidate, among others (Luethi et al., 2018). Notably, ethylphenidate selectively targets reuptake inhibition through the dopamine transporter, whereas methylphenidate equally targets the dopamine and norepinephrine transporters.

Information on ethylphenidate's toxicity is derived from case reports and case series (Bailey et al., 2015; Ho et al., 2015). Similar to pipradrol derivatives, it causes stimulant toxicity, which includes anxiety, paranoia, insomnia, bruxism, diaphoresis, chest pain, tachycardia, and hypertension. It was also detected in some fatalities with or without other drugs (Krueger et al., 2014; Maskell et al., 2016).

Dissociatives

Arylcyclohexylamines, which include analogs of phencyclidine and ketamine, exhibit a different type of hallucinogenic activity, one that involves out-of-body experience. Pharmacologically, these drugs are agonists of the *N*-methyl-D-aspartate (NMDA) receptor instead of the serotonin receptor (Anis et al., 1983; Morris and Wallach, 2014; Wallach et al., 2016). Popular examples are 3-methoxyphencyclidine (3-MeO-PCP), methoxphenidine (MXP), and methoxetamine (MXE) (Wallach and Brandt, 2018a, 2018b).

Although similar to ketamine in effects, MXE has slower onset and is longer lasting. Acute toxicity is similar to that of stimulants and ketamine (Table 3.7). Stimulant-like toxicity includes cardiovascular, neurological, and psychological effects such as tachycardia, hypertension, palpitation, mydriasis, diaphoresis, agitation, confusion, aggression, hysteria, mania, paranoia, hallucinations, psychosis, stupor, somnolence, seizures, coma, and cerebellar dysfunctions such as ataxia and nystagmus. Cerebellar dysfunction is unique to this class. Dissociative-like toxicity includes catatonia, dystonia, hypertonia, and tetany (Ward et al., 2011; Hofer et al., 2012; Shields et al., 2012; Wood et al., 2012b; Imbert et al., 2014; Zawilska, 2014).

Methoxphenidine and methoxy-PCP have several positional isomers and exert PCP-like effects. Toxicity is similar to that of MXE, but in laboratory-confirmed fatal cases, pulmonary congestion and edema, heart enlargement, and severe hypertensions were reported (Hofer et al., 2014; Backberg et al., 2015b; Helander et al., 2015; Johansson et al., 2017; Thornton et al., 2017).

TABLE 3.7 Acute toxic effects of dissociatives.

Organ system	Symptoms
Neurological	Seizure, mydriasis, diaphoresis, stupor, somnolence, coma, ataxia, nystagmus, catatonia
Psychological	Agitation, confusion, aggression, hysteria, mania, paranoia, hallucination, psychosis
Cardiovascular	Tachycardia, hypertension, palpitations, heart enlargement
Respiratory	Pulmonary congestion, pulmonary edema
Musculoskeletal	Dystonia, hypertonia, tetany

New synthetic opioids

Fentanyl analogs are the largest subclass of NSO. Since 2012, however, nonfentanyl and non–phenanthrene-based opioids have emerged, including benzamide analogs, piperazine analogs, piperazine amides, and benzimidazole opioids. Moreover, since restrictions and regulations on all fentalogs in most parts of the world have become tighter starting in 2019, some of these alternative scaffolds have supplanted fentalogs as the leading NSO. Currently, the nitazenes or benzimidazole opioids are the most popular.

Regardless of subclass, the pharmacology of all NSO is the same. They target the mu opioid receptor (Maguire et al., 1992; Baumann et al., 2018), and acute adverse effects are the opioid toxidrome: CNS depression, coma, miosis (pinpoint pupils), bradypnea or apnea, hypothermia, hypotension, bradycardia, hyporeflexia, and pulmonary edema (Table 3.8) (Helander et al., 2014, 2016, 2017; Backberg et al., 2015c; Domanski et al., 2017; Schneir et al., 2017; Muller et al., 2019). Because of CNS and respiratory depression, many fatalities arise from NSO overdose. In a few reported fentalog intoxications, the opioid toxidrome was accompanied by

TABLE 3.8 Acute toxic effects of new synthetic opioids.

Organ system	Symptoms
Neurological	CNS depression, coma, miosis, hypothermia
Cardiovascular	Hypotension, bradycardia
Respiratory	Respiratory depression, bradypnea, pulmonary edema
Musculoskeletal	Hyporeflexia

tachycardia and hypertension. However, some of these cases are polydrug use, so it is difficult to ascertain whether the other effects can be accurately ascribed to NSO.

Opioids are one of the very few classes for which an antidote is available. Naloxone reverses the toxidrome. However, because of the wide range of potencies, determining the correct naloxone dose can be quite challenging. Fentanyl itself is 100 times more potent than morphine. A sampling of others versus morphine: carfentanil 10,000×, sufentanil 1000×, furanylfentanyl 700×, U-47700 7.5×, metonitazene 100×, protonitazene 200×. The rescue approach is to administer small boluses of naloxone until revival occurs (Kim and Nelson, 2015).

Designer benzodiazepines

These first became available in 2007. Most members are proposed pharmaceuticals synthesized in the 1960 and 1970s that did not get market authorization. A few are regulated prescription drugs in some countries (Manchester et al., 2018; Backberg et al., 2019). Phenazepam is prescribed in Russia, etizolam in Japan, Korea, and Italy. Other examples are flubromazolam, flubromazepam, flualprazolam, and clonazolam.

Pharmacologically, they behave like prescription benzodiazepines and bind the $GABA_A$ receptor. Although some are significantly more potent than prescription benzodiazepines (e.g., phenazepam and etizolam are 5–10 times more potent than diazepam), adverse effects from acute intoxication are generally milder than for most NPS classes. The patient usually presents with sedation, drowsiness, dizziness, ataxia, blurred vision, headache, depression, confusion, amnesia, muscle weakness, and incoordination. Fatality is extremely rare or nonexistent. Similar to prescription benzodiazepines, however, combination with alcohol or opioids can be deadly (Jones et al., 2012; Backberg et al., 2019; Carpenter et al., 2019; Zawilska and Wojcieszak, 2019) (Table 3.9).

TABLE 3.9 Acute toxic effects of designer benzodiazepines.

Organ system	Symptoms
Neurological	Sedation, drowsiness, dizziness, ataxia, blurred vision, headache, amnesia
Psychological	Depression, confusion
Musculoskeletal	Muscle weakness, incoordination

Closeup: An early, laboratory-confirmed outbreak of synthetic cannabinoid-associated toxicity

Synthetic cannabinoid (SC)-containing herbal products first appeared in the United States in 2009. SCs and other new psychoactive substances are now recognized as a significant source of morbidity and mortality. However, at that time, little to nothing was known about their health effects and toxicity. Analytical laboratory capability and capacity was extremely limited, and clinician and medical provider awareness of the looming public health problem was just beginning. Despite these challenges, in 2012, public health, law enforcement, toxicologists, and in particular analytical toxicology laboratorians collaborated to detect an outbreak of acute kidney injury (AKI), confirm a previously unknown SC compound in case patients and determine the scope and severity of the public health threat (Centers for Disease Control and Prevention, 2013).

In March 2012, the Wyoming Department of Health was notified by Natrona County officials of three hospitalized patients with unexplained AKI epidemiologically linked to each other and to the recent use of an SC-containing herbal substance. Public health officials notified law enforcement, developed a case definition, instituted active surveillance, and contacted the Centers for Disease Control and Prevention. Three days later, a fourth epidemiologically linked Wyoming case was admitted with AKI after smoking either a blueberry or bubblegum-flavored SC-containing herbal incense product (Fig. 3.5). The Casper (WY) Police Department sent the seized product from Case 4 to the Arkansas K2

FIGURE 3.5 Samples of the SC-containing product used by a Wyoming case, which tested positive for the previously unknown SC XLR-11. *Photo courtesy of the Casper (WY) Police Department and Dr. Jeff Moran.*

Closeup: An early, laboratory-confirmed outbreak of synthetic cannabinoid-associated toxicity (*cont'd*)

Research Consortium—one of the few analytical laboratories capable of identifying and confirming specific SC compounds in drug products. Clinical specimens from Case 4 and subsequent AKI cases were sent to the University of California San Francisco Clinical Toxicology and Environmental Biomonitoring Laboratory. In both the drug product and clinical specimens, a previously unknown fluorinated SC (1-(5-fluoropentyl)-1H-indole-3-yl) (2,2,3,3-tetramethylcyclopropyl) methanone, now known as XLR-11 (Fig. 3.6), was analytically confirmed.

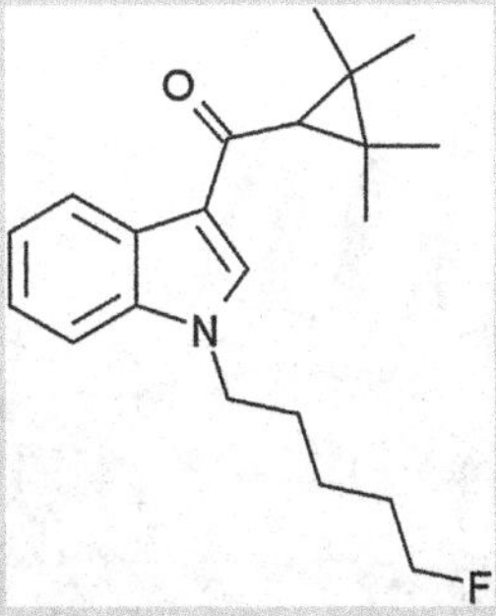

FIGURE 3.6 XLR-11, a novel fluorinated SC implicated in an outbreak of AKI in young, previously healthy SC-containing product users. *Courtesy of Dr. Roy Gerona.*

With a novel SC identified, and its use temporally associated with cases of AKI (a previously unrecognized toxicity after SC use), surveillance was expanded nationally, and medical toxicologists, public health departments, and poison control centers were notified of the association of a novel SC compound with nephrotoxicity. In total, 16 cases of AKI meeting the case definition were reported from six states, including the four Wyoming cases and six in Oregon, between March and December 2012 (Centers for Disease Control and Prevention 2013; Buser et al., 2014; Thornton et al., 2013). No specific named herbal incense product or batch of product was common to all exposures, but XLR-11 was identified in clinical and/or product samples from seven patients. Clinically, AKI resolved in all cases, though five patients required hemodialysis; fortunately, there were no fatalities.

Between early 2012 and mid-2013, a significant proportion of SC-containing products seized by law enforcement nationally contained

continued

Closeup: An early, laboratory-confirmed outbreak of synthetic cannabinoid-associated toxicity (*cont'd*)

XLR-11 (Fig. 3.6, which was named after an early rocket engine, Fig. 3.7) or other tetramethylcyclopropyl ketone indoles, which were common enough to have "street" names. In April 2013, following a trial in US District Court, two women pleaded no contest to supplying the SC-product to four Casper, WY teenagers and were sentenced to 4 years, 2 months in prison. On April 13, 2013, the US Drug Enforcement Administration—citing the AKI outbreak, subsequent investigation, and *MMWR* publication—added XLR-11 to Schedule I of the Controlled Substances Act (Fed. Reg. Volume 78, Number 71), the second ever scheduling action for an SC compound.

FIGURE 3.7 A Reaction motors (later morton Thiokol) XLR-11 liquid-fueled rocket engine from the 1970s. (NASA public photo).

In spite of the challenges faced by public health, clinicians, law enforcement, toxicologists, and forensic laboratory scientists, these stakeholders collaborated in the early days of the new psychoactive substance epidemic to identify the outbreak, investigate the scope of the threat, and ultimately protect the public from a novel synthetic cannabinoid and its severe and previously unknown toxicity.

Michael D. Schwartz

Biomedical Advanced Research and Development Authority (BARDA)/Administration for Preparedness and Response (ASPR)/

Closeup: An early, laboratory-confirmed outbreak of synthetic cannabinoid-associated toxicity (*cont'd*)

Department of Health and Human Services, Washington, DC, United States

Disclaimer: The findings and conclusions in this essay are those of the author and do not represent the views of the US Department of Health and Human Services or its components. It has not been formally disseminated by the US Department of Health and Human Services. It does not represent and should not be construed to represent any agency determination or policy.

References

Adamowicz P, Meissner E, Maślanka M. Fatal intoxication with new synthetic cannabinoids AMB-FUBINACA and EMB-FUBINACA. Clin Toxicol 2019 Nov;57(11):1103–8. https://doi.org/10.1080/15563650.2019.1580371. Epub 2019 Feb 26. PMID: 30806094.

Adams AJ, Banister SD, Irizarry L, Trecki J, Schwartz M, Gerona R. Zombie" outbreak caused by the synthetic cannabinoid AMB-FUBINACA in New York. N Engl J Med 2017 Jan 19; 376(3):235–42. https://doi.org/10.1056/NEJMoa1610300. Epub 2016 Dec 14. PMID: 27973993.

Alatrash G, Majhail NS, Pile JC. Rhabdomyolysis after ingestion of "foxy," a hallucinogenic tryptamine derivative. Mayo Clin Proc 2006;81(4):550–1. https://doi.org/10.4065/81.4.550.

Anis NA, Berry SC, Burton NR, Lodge D. The dissociative anaesthetics, ketamine and phencyclidine, selectively reduce excitation of central mammalian neurones by N-methyl-aspartate. Br J Pharmacol 1983;79(2):565–75. https://doi.org/10.1111/j.1476-5381.1983.tb11031.x. PMID: 6317114; PMCID: PMC2044888.

Araújo AM, Carvalho F, Bastos Mde L, Guedes de Pinho P, Carvalho M. The hallucinogenic world of tryptamines: an updated review. Arch Toxicol 2015;89(8):1151–73. https://doi.org/10.1007/s00204-015-1513-x. Epub 2015 Apr 16. PMID: 25877327.

Arbo MD, Bastos ML, Carmo HF. Piperazine compounds as drugs of abuse. Drug Alcohol Depend 2012;122(3):174–85. https://doi.org/10.1016/j.drugalcdep.2011.10.007.

Backberg M, Beck O, Helander A. Phencyclidine analog use in Sweden—intoxication cases involving 3-MeO-PCP and 4-MeO-PCP from the STRIDA project. Clin Toxicol 2015b; 53(9):856–64. https://doi.org/10.3109/15563650.2015.1079325.

Backberg M, Beck O, Jonsson KH, Helander A. Opioid intoxications involving butyrfentanyl, 4-fluorobutyrfentanyl, and fentanyl from the Swedish STRIDA project. Clin Toxicol 2015c;53(7):609–17. https://doi.org/10.3109/15563650.2015.1054505.

Backberg M, Lindeman E, Beck O, Helander A. Character- istics of analytically confirmed 3-MMC-related intoxications from the Swedish STRIDA project. Clin Toxicol 2015a;53(1): 46–53. https://doi.org/10.3109/15563650.2014.981823.

Backberg M, Pettersson Bergstrand M, Beck O, Helander A. Occurrence and time course of NPS benzodiazepines in Sweden—results from intoxication cases in the STRIDA project. Clin Toxicol 2019;57(3):203–12. https://doi.org/10.1080/15563650.2018.1506130.

Bailey GP, Ho JH, Hudson S, Dines A, Archer JR, Dargan PI, et al. Nopaine no gain: recreational ethylphenidate toxicity. Clin Toxicol 2015 Jun;53(5):498–9. https://doi.org/10.3109/15563650.2015.1033062. Epub 2015 Apr 14.

Baumann MH, Clark RD, Budzynski AG, Partilla JS, Blough BE, Rothman RB. N-substituted piperazines abused by humans mimic the molecular mechanism of 3,4-methylenedioxymethamphetamine (MDMA, or 'Ecstasy'). Neuropsychopharmacology 2005;30(3):550–60. https://doi.org/10.1038/sj.npp.1300585.

Baumann MH, Clark RD, Woolverton WL, Wee S, Blough BE, Rothman RB. In vivo effects of amphetamine analogs reveal evidence for serotonergic inhibition of mesolimbic dopamine transmission in the rat. J Pharmacol Exp Therapeut 2011 Apr;337(1):218–25. https://doi.org/10.1124/jpet.110.176271.

Baumann MH, Partilla JS, Lehner KR, Thorndike EB, Hoffman AF, Holy M. Powerful cocaine-like actions of 3,4-methylenedioxypyrovalerone (MDPV), a principal constituent of psychoactive "bath salts" products. Neuropsychopharmacology 2013;38(4):552–62.

Baumann MH, Majumdar S, Le Rouzic V, Hunkele A, Uprety R, Huang XP, Xu J, Roth BL, Pan YX, Pasternak GW. Pharmacological characterization of novel synthetic opioids (NSO) found in the recreational drug marketplace. Neuropharmacology 2018;134(Pt A):101–7. https://doi.org/10.1016/j.neuropharm.2017.08.016.

Beck O, Franzen L, Bäckberg M, Signell P, Helander A. Intoxications involving MDPV in Sweden during 2010-2014: results from the STRIDA project. Clin Toxicol 2015 Nov; 53(9):865–73. https://doi.org/10.3109/15563650.2015.1089576. Epub 2015 Oct 14. PMID: 26462932.

Beck O, Franzén L, Bäckberg M, Signell P, Helander A. Toxicity evaluation of α-pyrrolidinovalerophenone (α-PVP): results from intoxication cases within the STRIDA project. Clin Toxicol 2016 Aug;54(7):568–75. https://doi.org/10.1080/15563650.2016.1190979. PMID: 27412885.

Blough BE, Landavazo A, Decker AM, Partilla JS, Baumann MH, Rothman RB. Interaction of psychoactive tryptamines with biogenic amine transporters and serotonin receptor subtypes. Psychopharmacology 2014;231(21):4135–44. https://doi.org/10.1007/s00213-014-3557-7.

Boland DM, Andollo W, Hime GW, Hearn WL. Fatality due to acute a-methyltryptamine intoxication. J Anal Toxicol 2005;29(5):394–7. https://doi.org/10.1093/jat/29.5.394.

Bosak A, LoVecchio F, Levine M. Recurrent seizures and serotonin syndrome following "2C-I" ingestion. J Med Toxicol 2013 Jun;9(2):196–8. https://doi.org/10.1007/s13181-013-0287-x.

Boulanger-Gobeil C, St-Onge M, Laliberte M, Auger PL. Seizures and hyponatremia related to ethcathinone and methy- lone poisoning. J Med Toxicol 2012;8(1):59–61. https://doi.org/10.1007/s13181-011-0159-1.

Brandt SD, Freeman S, Sumnall HR, Measham F, Cole J. Analysis of NRG 'legal highs' in the UK: identification and formation of novel cathinones. Drug Test Anal 2011;3:569–75.

Brunt TM, Atkinson AM, Nefau T, Martinez M, Lahaie E, Malzcewski A, et al. Online test purchased new psychoactive substances in 5 different European countries: a snapshot study of chemical composition and price. Int J Drug Pol 2017;44:105–14.

Buser GL, Gerona RR, Horowitz BZ, Vian KP, Troxell ML, Hendrickson RG, et al. Acute kidney injury associated with smoking synthetic cannabinoid. Clin Toxicol 2014 Aug;52(7):664–73. https://doi.org/10.3109/15563650.2014.932365. PMID: 25089722.

Carpenter JE, Murray BP, Dunkley C, Kazzi ZN, Gittinger MH. Designer benzodiazepines: a report of exposures recorded in the national poison data system, 2014–2017. Clin Toxicol 2019;57(4):282–6. https://doi.org/10.1080/15563650.2018.1510502.

Chan YM, Soldin SJ, Swanson JM, Deber CM, Thiessen JJ, Macleod S. Gas chromatographic/mass spectrometric analysis of methylphenidate (ritalin) in serum. Clin Biochem 1980 Dec;13(6):266–72. https://doi.org/10.1016/s0009-9120(80)80007-5.

Centers for Disease Control and Prevention. MMWR Morb Mortal Wkly Rep 2013 Feb 15; 62(6):93–8.

Coppola M, Mondola R. 5-Iodo-2-aminoindan (5-IAI): chemistry, pharmacology, and toxicology of a research chemical producing MDMA-like effects. Toxicol Lett 2013;218(1): 24–9. https://doi.org/10.1016/j.toxlet.2013.01.008.

Corazza O, Schifano F, Farre M, Deluca P, Davey Z, Torrens M, et al. Designer drugs on the internet: a phenomenon out-of-control? The emergence of hallucinogenic drug Bromo-Dragonfly. Curr Clin Pharmacol 2011;6:125–9.

Corkery JM, Elliott S, Schifano F, Corazza O, Ghodse AH. MDAI (5,6-methylenedioxy-2-aminoindane; 6,7-dihydro-5H-cyclopenta- [f][1,3]benzodioxol-6-amine; 'sparkle'; 'mindy') toxicity: a brief overview and update. Hum Psychopharmacol 2013;28:345–55.

Cozzi NV, Sievert MK, Shulgin AT, Jacob 3rd P, Ruoho AE. Inhibition of plasma membrane monoamine transporters by beta-ketoamphetamines. Eur J Pharmacol 1999 Sep 17;381(1): 63–9. https://doi.org/10.1016/s0014-2999(99)00538-5. PMID: 10528135.

Cozzi NV, Gopalakrishnan A, Anderson LL, Feih JT, Shulgin AT, Daley PF, et al. Dimethyltryptamine and other hallucinogenic tryptamines exhibit substrate behavior at the serotonin uptake transporter and the vesicle monoamine transporter. J Neural Transm 2009; 116(12):1591–9. https://doi.org/10.1007/s00702-009-0308-8.

De Boer D, Bosman IJ, Hidvegi E, Manzoni C, Benkö AA, dos Reys LJ, et al. Piperazine-like compounds: a new group of designer drugs-of-abuse on the European market. For Sci Int 2001;121:47–56.

De Letter EA, Coopman VA, Cordonnier JA, Piette MH. One fatal and seven non-fatal cases of 4–methylthioamphetamine (4–MTA) intoxication: clinico-pathological findings. Int J Leg Med 2001;114:352–6.

Dean BV, Stellpflug SJ, Burnett AM, et al. 2C or not 2C: phenethylamine designer drug review. J Med Toxicol 2013;9:172–8. https://doi.org/10.1007/s13181-013-0295-x.

Dolder PC, Strajhar P, Vizeli P, Hammann F, Odermatt A, Liechti ME. Pharmacokinetics and pharmacodynamics of lisdexamfetamine compared with D-amphetamine in healthy subjects. Front Pharmacol 2017;8:617. https://doi.org/10.3389/fphar.2017.00617.

Domanski K, Kleinschmidt KC, Schulte JM, Fleming S, Frazee C, Menendez A, et al. Two cases of intoxication with new synthetic opioid, U-47700. Clin Toxicol 2017;55(1):46–50. https://doi.org/10.1080/15563650.2016.1209763.

Fantegrossi WE, Murnane KS, Reissig CJ. The behavioral pharmacology of hallucinogens. Biochem Pharmacol 2008;75:17–33.

Fattore L. Synthetic cannabinoids-further evidence supporting the relationship between cannabinoids and psychosis. Biol Psychiatr 2016 Apr 1;79(7):539–48. https://doi.org/10.1016/j.biopsych.2016.02.001. Epub 2016 Feb 4. PMID: 26970364.

Ferris RM, Tang FL. Comparison of the effects of the isomers of amphetamine, methylphenidate and deoxypipradrol on the uptake of l-[3H]norepinephrine and [3H]dopamine by synaptic vesicles from rat whole brain, striatum and hypothalamus. J Pharmacol Exp Therapeut 1979 Sep;210(3):422–8. PMID: 39160.

Fleckenstein AE, Haughey HM, Metzger RR, Kokoshka JM, Riddle EL, Hanson JE, et al. Differential effects of psychostimulants and related agents on dopaminergic and serotonergic transporter function. Eur J Pharmacol 1999 Oct 1;382(1):45–9. https://doi.org/10.1016/s0014-2999(99)00588-9. PMID: 10556503.

Franzen L, Backberg M, Beck O, Helander A. Acute intoxications involving a-pyrrolidinobutiophenone (a-PBP): results from the Swedish STRIDA project. J Med Toxicol 2018;14(4):265–71. https://doi.org/10.1007/s13181-018-0668-2.

Gee P, Richardson S, Woltersdorf W, Moore G. Toxic effects of BZP-based herbal party pills in humans: a prospective study in Christchurch. New Zealand N Z Med J 2005;118(1227): U1784.

Gee P, Gilbert M, Richardson S, Moore G, Paterson S, Graham P. Toxicity from the recreational use of 1-benzylpiperazine. Clin Toxicol 2008;46(9):802–7. https://doi.org/10.1080/15563650802307602.

Gee P, Jerram T, Bowie D. Multiorgan failure from 1-benzylpiperazine ingestion—legal high or lethal high? Clin Toxicol 2010;48(3):230–3. https://doi.org/10.3109/15563651003592948.

Greene SL. Benzofurans and benzodifurans. In: Dargan PI, Wood DM, editors. Novel Psychoactive Substances:classification, pharmacology and toxicology. Amsterdam: Elsevier; 2013.

Hadlock GC, Webb KM, McFadden LM, Chu PW, Ellis JD, Allen SC, et al. 4-Methylmethcathinone (mephedrone): neuropharmacological effects of a designer stimulant of abuse. J Pharmacol Exp Therapeut 2011 Nov;339(2):530–6. https://doi.org/10.1124/jpet.111.184119.

Helander A, Bäckberg M, Beck O. MT-45, a new psychoactive substance associated with hearing loss and unconsciousness. Clin Toxicol 2014;52(8):901–4. https://doi.org/10.3109/15563650.2014.943908.

Helander A, Beck O, Bäckberg M. Intoxications by the dissociative new psychoactive substances diphenidine and methoxphenidine. Clin Toxicol 2015;53(5):446–53. https://doi.org/10.3109/15563650.2015.1033630.

Helander A, Bäckberg M, Beck O. Intoxications involving the fentanyl analogs acetylfentanyl, 4-methoxybutyrfentanyl and furanylfentanyl: results from the Swedish STRIDA project. Clin Toxicol 2016;54(4):324–32. https://doi.org/10.3109/15563650.2016.1139715.

Helander A, Bäckberg M, Signell P, Beck O. Intoxications involving acrylfentanyl and other novel designer fentanyls – results from the Swedish STRIDA project. Clin Toxicol 2017;55(6):589–99. https://doi.org/10.1080/15563650.2017.1303141.

Hermanns-Clausen M, Kneisel S, Szabo B, Auwärter V. Acute toxicity due to the confirmed consumption of synthetic cannabinoids: clinical and laboratory findings. Addiction 2013 Mar;108(3):534–44. https://doi.org/10.1111/j.1360-0443.2012.04078.x. Epub 2012 Nov 1.

Hill SL, Dargan PI. Patterns of acute toxicity associated with new psychoactive substances. Handb Exp Pharmacol 2018;252:475–94. https://doi.org/10.1007/164_2018_135. PMID: 29896654.

Hill SL, Najafi J, Dunn M, Acheampong P, Kamour A, Grundlingh J, et al. Clinical toxicity following analytically confirmed use of the synthetic cannabinoid receptor agonist MDMB-CHMICA. A report from the Identification of Novel psychoActive substances (IONA) study. Clin Toxicol 2016 Sep;54(8):638–43. https://doi.org/10.1080/15563650.2016.1190980. Epub 2016 Jun 2.

Ho JH, Bailey GP, Archer JR, Dargan PI, Wood DM. Ethylphenidate: availability, patterns of use, and acute effects of this novel psychoactive substance. Eur J Clin Pharmacol 2015 Oct;71(10):1185–96. https://doi.org/10.1007/s00228-015-1906-z. Epub 2015 Jul 22.

Hofer KE, Grager B, Müller DM, Rauber-Lüthy C, Kupferschmidt H, Rentsch KM, et al. Ketamine-like effects after recreational use of methoxetamine. Ann Emerg Med 2012;60(1):97–9. https://doi.org/10.1016/j.annemergmed.2011.11.018.

Hofer KE, Degrandi C, Müller DM, Zürrer-Härdi U, Wahl S, Rauber- Lüthy C, et al. Acute toxicity associated with the recreational use of the novel dissociative psychoactive substance methoxphenidine. Clin Toxicol 2014;52(10):1288–91. https://doi.org/10.3109/15563650.2014.974264.

Hondebrink L, Nugteren-van Lonkhuyzen JJ, Rietjens SJ, Brunt TM, Venhuis B, Soerdjbalie-Maikoe V, et al. Fatalities, cerebral hemorrhage, and severe cardiovascular toxicity after exposure to the new psychoactive substance 4-fluoroamphetamine: a prospective cohort study. Ann Emerg Med 2017;71:294–305. https://doi.org/10.1016/j.annemergmed.2017.07.482.

Huffman JW, Szklennik PV, Almond A, Bushell K, Selley DE, He H, et al. 1-Pentyl-3-phenylacetylindoles, a new class of cannabimimetic indoles. Bioorg Med Chem Lett 2005;15:4110–3.

Hysek CM, Simmler LD, Nicola V, Vischer N, Donzelli M, Krähenbühl S, et al. Duloxetine inhibits effects of MDMA ("ecstasy") in vitro and in humans in a randomized placebo-controlled laboratory study. PLoS One 2012;7:e36476.

Iden CR, Hungund BL. A chemical ionization selected ion monitoring assay for methylphenidate and ritalinic acid. Biomed Mass Spectrom 1979 Oct;6(10):422–6. https://doi.org/10.1002/bms.1200061003. PMID: 526558.

Imbert L, Boucher A, Delhome G, Cueto T, Boudinaud M, Maublanc J, Dulaurent S, Descotes J, Lachâtre G, Gaulier JM. Analytical findings of an acute intoxication after inhalation of methoxetamine. J Anal Toxicol 2014;38(7):410–5. https://doi.org/10.1093/jat/bku052. Epub 2014 Jun 5. PMID: 24904014.

Itokawa M, Iwata K, Takahashi M, Sugihara G, Sasaki T, Abe Y, et al. Acute confusional state after designer tryptamine abuse. Psychiatr Clin Neurosci 2007;61(2):196–9. https://doi.org/10.1111/j.14401819.2007.01638.x.

Iversen L, Gibbons S, Treble R, Setola V, Huang XP, Roth BL. Neurochemical profiles of some novel psychoactive substances. Eur J Pharmacol 2013;700:147–51.

Johansson A, Lindstedt D, Roman M, Thelander G, Nielsen EI, Lennborn U, et al. A non-fatal intoxication and seven deaths involving the dissociative drug 3-MeO-PCP. Forensic Sci Int 2017;275:76–82. https://doi.org/10.1016/j.forsciint.2017.02.034.

Jones JD, Mogali S, Comer SD. Polydrug abuse: a review of opioid and benzodiazepine combination use. Drug Alcohol Depend 2012;125(1–2):8–18. https://doi.org/10.1016/j.drugalcdep.2012.07.004.

Jovel A, Felthous A, Bhattacharyya A. Delirium due to intoxication from the novel synthetic tryptamine 5-MeO-DALT. J Forensic Sci 2014;59(3):844–6. https://doi.org/10.1111/1556-4029.12367.

Kasper AM, Ridpath AD, Arnold JK, Chatham-Stephens K, Morrison M, Olayinka O, et al. Severe illness associated with reported use of synthetic cannabinoids - Mississippi, April 2015. MMWR Morb Mortal Wkly Rep 2015 Oct 9;64(39):1121–2. https://doi.org/10.15585/mmwr.mm6439a7. PMID: 26447715.

Kim HK, Nelson LS. Reducing the harm of opioid overdose with the safe use of naloxone: a pharmacologic review. Expet Opin Drug Saf 2015;14(7):1137–46. https://doi.org/10.1517/14740338.2015.1037274.

Kovaleva J, Devuyst E, De Paepe P, Verstraete A. Acute chlorophenylpiperazine overdose: a case report and review of the literature. Ther Drug Monit 2008;30(3):394–8. https://doi.org/10.1097/FTD.0b013e318170a879.

Kraemer M, Boehmer A, Madea B, Maas A. Death cases involving certain new psychoactive substances: a review of the literature. Forensic Sci Int 2019 May;298:186–267. https://doi.org/10.1016/j.forsciint.2019.02.021. Epub 2019 Feb 25. PMID: 30925344.

Kronstrand R, Roman M, Andersson M, Eklund A. Toxicological findings of synthetic cannabinoids in recreational users. J Anal Toxicol 2013 Oct;37(8):534–41. https://doi.org/10.1093/jat/bkt068. Epub 2013 Aug 22. PMID: 23970540.

Krueger J, Sachs H, Musshoff F, Dame T, Schaeper J, Schwerer M, et al. First detection of ethylphenidate in human fatalities after ethylphenidate intake. Forensic Sci Int 2014 Oct;243:126–9. https://doi.org/10.1016/j.forsciint.2014.07.017.

Lidder S, Dargan P, Sexton M, Button J, Ramsey J, Holt D, et al. Cardiovascular toxicity associated with recreational use of diphenylprolinol (diphenyl-2-pyrrolidinemethanol [D2PM]). J Med Toxicol 2008 Sep;4(3):167–9. https://doi.org/10.1007/BF03161195.

Liechti M. Novel psychoactive substances: overview and pharmacology of modulators of monoamine signalling. Swiss Med Wkly 2015;145:w14043. https://doi.org/10.4414/smw.2015.14043.

Liechti ME, Baumann C, Gamma A, Vollenweider FX. Acute psychological effects of 3,4-methylenedioxymethamphetamine (MDMA, "Ecstasy") are attenuated by the serotonin uptake inhibitor citalopram. Neuropsychopharmacology 2000 May;22(5):513–21. https://doi.org/10.1016/S0893-133X(99)00148-7. PMID: 10731626.

López-Arnau R, Martínez-Clemente J, Pubill D, Escubedo E, Camarasa J. Comparative neuropharmacology of three psychostimulant cathinone derivatives: butylone, mephedrone and methylone. Br J Pharmacol 2012 Sep;167(2):407–20. https://doi.org/10.1111/j.1476-5381.2012.01998.x.

Luethi D, Kaeser PJ, Brandt SD, Krähenbühl S, Hoener MC, Liechti ME. Pharmacological profile of methylphenidate-based designer drugs. Neuropharmacology 2018;134:133–40.

Lurie Y, Gopher A, Lavon O, Almog S, Sulimani L, Bentur Y. Severe paramethoxymethamphetamine (PMMA) and paramethoxyamphetam- ine (PMA) outbreak in Israel. Clin Toxicol 2012;50:39–43.

Madsen MK, Fisher PM, Burmester D, Dyssegaard A, Stenbaek DS, Kristiansen S, et al. Psychedelic effects of psilocybin correlate with serotonin 2A receptor occupancy and plasma psilocin levels. Neuropsychopharmacology 2019;44(7):1328–34. https://doi.org/10.1038/s41386-019-0324-9.

Maguire P, Tsai N, Kamal J, Cometta-Morini C, Upton C, Loew G. Pharmacological profiles of fentanyl analogs at l, d and j opiate receptors. Eur J Pharmacol 1992;213(2):219–25. https://doi.org/10.1016/0014-2999(92)90685-w.

Manchester KR, Lomas EC, Waters L, Dempsey FC, Maskell PD. The emergence of new psychoactive substance (NPS) benzodiazepines: a review. Drug Test Anal 2018;10(1):37–53. https://doi.org/10.1002/dta.2211.

Markowitz JS, Logan BK, Diamond F, Patrick KS. Detection of the novel metabolite ethylphenidate after methylphenidate overdose with alcohol coingestion. J Clin Psychopharmacol 1999 Aug;19(4):362–6. https://doi.org/10.1097/00004714-199908000-00013.

Markowitz JS, Patrick KS. Ethylphenidate: from biomarker to designer drug. Mental Health Clinician 2013;3(6):318–20. https://doi.org/10.9740/mhc.n183949.

Maskell PD, Smith PR, Cole R, Hikin L, Morley SR. Seven fatalities associated with ethylphenidate. Forensic Sci Int 2016 Aug;265:70–4. https://doi.org/10.1016/j.forsciint.2015.12.045.

Meltzer PC, Butler D, Deschamps JR, Madras BK. 1-(4-Methylphenyl)-2-pyrrolidin-1-yl-pentan-1-one (Pyrovalerone) analogues: a promising class of monoamine uptake inhibitors. J Med Chem 2006 Feb 23;49(4):1420–32. https://doi.org/10.1021/jm050797a.

Meyyappan C, Ford L, Vale A. Poisoning due to MDMB-CHMICA, a synthetic cannabinoid receptor agonist. Clin Toxicol 2017 Feb;55(2):151–2. https://doi.org/10.1080/15563650.2016.1227832. Epub 2016 Sep 16. PMID: 27635694.

Morris H, Wallach J. From PCP to MXE: a comprehensive review of the non-medical use of dissociative drugs. Drug Test Anal 2014;6(7–8):614–32. https://doi.org/10.1002/dta.1620. Epub 2014 Mar 26. PMID: 24678061.

Muller AA. New drugs of abuse update: foxy Methoxy. J Emerg Nurs 2004;30(5):507–8. https://doi.org/10.1016/j.jen.2004.07.037.

Muller D, Neurath H, Neukamm MA, Wilde M, Despicht C, Blaschke S, et al. New synthetic opioid cyclopropylfentanyl together with other novel synthetic opioids in respiratory insufficient comatose patients detected by toxicological analysis. Clin Toxicol 2019; 57(9):806–12. https://doi.org/10.1080/15563650.2018.1554187.

Muttoni S, Ardissino M, John C. Classical psychedelics for the treatment of depression and anxiety: a systematic review. J Affect Disord 2019;258:11–24. https://doi.org/10.1016/j.jad.2019.07.076.

Murphy T, Van Houten C, Gerona RR, Moran J, Kirschner R, Marraffa J, et al. Acute kidney injury associated with synthetic cannabinoids use, multiple states, 2012. MMWR (Morb Mortal Wkly Rep) 2013;62(6):93–8.

Murray BL, Murphy CM, Beuhler MC. Death following recreational use of designer drug "bath salts" containing 3,4-Methylenedioxypyrovalerone (MDPV). J Med Toxicol 2012a; 8(1):69–75. https://doi.org/10.1007/s13181-011-0196-9.

Murray DB, Potts S, Haxton C, Jackson G, Sandilands EA, Ramsey J, et al. Ivory wave' toxicity in recreational drug users; integration of clinical and poisons information services to manage legal high poisoning. Clin Toxicol 2012b;50(2):108–13. https://doi.org/10.3109/15563650.2011.647992. Epub 2012 Jan 6. PMID: 22224933.

Nelson ME, Bryant SM, Aks SE. Emerging drugs of abuse. Emerg Med Clin 2014:1–28. https://doi.org/10.1016/j.emc.2013.09.001.

Nichols DE. Differences between the mechanism of action of MDMA, MBDB, and the classic hallucinogens. Identification of a new therapeutic class: entactogens. J Psychoact Drugs 1986;18(4):305–13. https://doi.org/10.1080/02791072.1986.10472362.

Nichols DE. Hallucinogens. Pharmacol Ther 2004;101:131–81.

Nichols DE. Chemistry and structure-activity relationships of psychedelics. Curr Top Behav Neurosci 2018;36:1–43. https://doi.org/10.1007/7854_2017_475. PMID: 28401524.

Ong RS, Kappatos DC, Russell SGG, Poulsen HA, Banister SD, Gerona RR, et al. Simultaneous analysis of 29 synthetic cannabinoids and metabolites, amphetamines, and cannabinoids in human whole blood by liquid chromatography-tandem mass spectrometry - a New Zealand perspective of use in 2018. Drug Test Anal 2020 Feb;12(2):195–214. https://doi.org/10.1002/dta.2697. Epub 2019 Nov 28. PMID: 31595682.

Palhano-Fontes F, Barreto D, Onias H, Andrade KC, Novaes MM, Pessoa JA, et al. Rapid antidepressant effects of the psychedelic ayahuasca in treatment-resistant depression: a randomized placebo-controlled trial. Psychol Med 2019;49(4):655–63. https://doi.org/10.1017/s0033291718001356.

Pinterova N, Horsley RR, Palenicek T. Synthetic aminoin- danes: a summary of existing knowledge. Front Psychiatr 2017;8:236. https://doi.org/10.3389/fpsyt.2017.00236.

Reith ME, Blough BE, Hong WC, Jones KT, Schmitt KC, Baumann MH, et al. Behavioral, biological, and chemical perspectives on atypical agents targeting the dopamine transporter. Drug Alcohol Depend 2015;147:1–19.

Rickli A, Moning OD, Hoener MC, Liechti ME. Receptor interaction profiles of novel psychoactive tryptamines compared with classic hallucinogens. Eur Neuropsychopharmacol 2016;26(8):1327–37. https://doi.org/10.1016/j.euroneuro.2016.05.001.

Robbins TW, Watson BA, Gaskin M, Ennis C. Contrasting interactions of pipradrol, d-amphetamine, cocaine, cocaine analogues, apomorphine and other drugs with conditioned reinforcement. Psychopharmacology (Berl) 1983;80(2):113–9. https://doi.org/10.1007/BF00427952. PMID: 6136060.

Roseman L, Nutt DJ, Carhart-Harris RL. Quality of acute psychedelic experience predicts therapeutic efficacy of psilocybin for treatment-resistant depression. Front Pharmacol 2017;8:974. https://doi.org/10.3389/fphar.2017.00974.

Rothman RB, Baumann MH, Dersch CM, Romero DV, Rice KC, Carroll FI, et al. Amphetamine-type central nervous system stimulants release norepinephrine more potently than they release dopamine and serotonin. Synapse 2001 Jan;39(1):32–41. https://doi.org/10.1002/1098-2396(20010101)39:1<32::AID-SYN5>3.0.CO;2-3. PMID: 11071707.

Rothman RB, Baumann MH. Monoamine transporters and psychostimulant drugs. Eur J Pharmacol 2003;479(1–3):23–40. https://doi.org/10.1016/j.ejphar.2003.08.054.

Rothman RB, Baumann MH. Balance between dopamine and serotonin release modulates behavioral effects of amphetamine-type drugs. Ann N Y Acad Sci 2006 Aug;1074: 245–60. https://doi.org/10.1196/annals.1369.064. PMID: 17105921.

Sainsbury PD, Kicman AT, Archer RP, King LA, Braithwaite RA. Aminoindanes-the next wave of 'legal highs'. Drug Test Anal 2011;3:479–82.

Sanders B, Lankenau SE, Bloom JJ, Hathazi D. "Research chemicals": tryptamine and phenethylamine use among high-risk youth. Subst Use Misuse 2008;43(3–4):389–402. https://doi.org/10.1080/00952990701202970.

Schep LJ, Slaughter RJ, Vale JA, Beasley DM, Gee P. The clinical toxicology of the designer "party pills" benzylpiperazine and trifluoromethylphenylpiperazine. Clin Toxicol 2011; 49(3):131–41. https://doi.org/10.3109/15563650.2011.57207.

Schifano F, Orsolini L, Papanti GD, Corkery J. Novel psychoactive substances of interest for psychiatry. World Psychiatr 2015;14:15–26.

Schneir A, Ly BT, Casagrande K, Darracq M, Offerman SR, Thornton S, et al. Comprehensive analysis of "bath salts" purchased from California stores and the internet. Clin Toxicol 2014 Aug;52(7):651–8. https://doi.org/10.3109/15563650.2014.933231. PMID: 25089721.

Schneir A, Metushi IG, Sloane C, Benaron DJ, Fitzgerald RL. Near death from a novel synthetic opioid labeled U-47700: emergence of a new opioid class. Clin Toxicol 2017;55(1): 51–4. https://doi.org/10.1080/15563650.2016.1209764.

Schwartz MD, Trecki J, Edison LA, Steck AR, Arnold JK, Gerona RR. A common source outbreak of severe delirium associated with exposure to the novel synthetic cannabinoid ADB-PINACA. J Emerg Med 2015 May;48(5):573–80. https://doi.org/10.1016/j.jemermed.2014.12.038.

Shields JE, Dargan PI, Wood DM, Puchnarewicz M, Davies S, Waring WS. Methoxetamine associated reversible cerebellar toxicity: three cases with analytical confirmation. Clin Toxicol 2012;50(5):438–40. https://doi.org/10.3109/15563650.2012.683437.

Shulgin A, Shulgin A. Tryptamines that i have known and loved (TiHKAL): the continuation. 1st ed. Berkeley, CA: Transform Press; 1997. p. 433–557.

Silva JP, Carmo H, Carvalho F. The synthetic cannabinoid XLR-11 induces in vitro nephrotoxicity by impairment of endocannabinoid-mediated regulation of mitochondrial function homeostasis and triggering of apoptosis. Toxicol Lett 2018 May 1;287:59–69. https://doi.org/10.1016/j.toxlet.2018.01.023. Epub 2018 Feb 3. PMID: 29410032.

Simmler LD, Buser TA, Donzelli M, Schramm Y, Dieu LH, Huwyler J, et al. Pharmacological characterization of designer cathinones in vitro. Br J Pharmacol 2013;168(2):458–70. https://doi.org/10.1111/j.1476-5381.2012.02145.x.

Simmler LD, Liechti ME. Pharmacology of MDMA- and amphetamine-like new psychoactive substances. Handb Exp Pharmacol 2018;252:143–64. https://doi.org/10.1007/164_2018_113. PMID: 29633178.

Simmler LD, Rickli A, Hoener MC, Liechti ME. Monoamine transporter and receptor interaction profiles of a new series of designer cathinones. Neuropharmacology 2014;79: 152–60.

Sitte HH, Freissmuth M. Amphetamines, new psychoactive drugs and the monoamine transporter cycle. Trends Pharmacol Sci 2015;36(1):41–50.

Spiller HA, Ryan ML, Weston RG, Jansen J. Clinical experience with and analytical confirmation of "bath salts" and "legal highs" (synthetic cathinones) in the United States. Clin Toxicol 2011;49:499–505.

Springer YP, Gerona R, Scheunemann E, Shafer SL, Lin T, Banister SD, et al. Increase in adverse reactions associated with use of synthetic cannabinoids - Anchorage, Alaska, 2015–2016. MMWR Morb Mortal Wkly Rep 2016 Oct 14;65(40):1108–11. https://doi.org/10.15585/mmwr.mm6540a4. PMID: 27736839.

Suzuki J, Dekker MA, Valenti ES, Arbelo Cruz FA, Correa AM, Poklis JL, et al. Toxicities associated with NBOMe ingestion-a novel class of potent hallucinogens: a review of the literature. Psychosomatics 2015 ;56(2):129–39. https://doi.org/10.1016/j.psym.2014.11.002.

Tai S, Fantegrossi WE. Pharmacological and toxicological effects of synthetic cannabinoids and their metabolites. Curr. Top. Behav. Neurosci. 2016;32:249–62.

Taljemark J, Johansson BA. Drug-induced acute psychosis in an adolescent first-time user of 4-HO-MET. Eur Child Adolesc Psychiatr 2012;21(9):527–8. https://doi.org/10.1007/s00787-012-0282-9.

Thompson I, Williams G, Caldwell B, Aldington S, Dickson S, Lucas N, et al. Randomised double-blind, placebo-controlled trial of the effects of the "party pills" BZP/TFMPP alone and in combination with alcohol. J Psychopharmacol 2010;24:1299–308.

Thornton SL, Gerona RR, Tomaszewski CA. Psychosis from a bath salt product containing flephedrone and MDPV with serum, urine, and product quantification. J Med Toxicol 2012 Sep;8(3):310–3. https://doi.org/10.1007/s13181-012-0232-4.

Thornton SL, Wood C, Friesen MW, Gerona RR. Synthetic cannabinoid use associated with acute kidney injury. Clin Toxicol 2013 Mar;51(3):189–90. https://doi.org/10.3109/15563650.2013.770870. PMID: 23473465.

Thornton S, Lisbon D, Lin T, Gerona R. Beyond ketamine and phencyclidine: analytically confirmed use of multiple novel arylcyclohexylamines. J Psychoact Drugs 2017;49(4):289–93. https://doi.org/10.1080/02791072.2017.1333660.

Topeff JM, Ellsworth H, Willhite LA, Cole JB, Edwards EM. A case series of symptomatic patients, including one fatality, following 2C-E exposure. Clin Toxicol 2011;49:526.

Trecki J, Gerona RR, Schwartz MD. Synthetic cannabinoid-related illnesses and deaths. N Engl J Med 2015 Jul 9;373(2):103–7. https://doi.org/10.1056/NEJMp1505328. PMID: 26154784.

Tyndall JA, Gerona R, De Portu G, Trecki J, Elie MC, Lucas J, et al. An outbreak of acute delirium from exposure to the synthetic cannabinoid AB-CHMINACA. Clin Toxicol 2015;53(10):950–6. https://doi.org/10.3109/15563650.2015.1100306.

Umebachi R, Aoki H, Sugita M, Taira T, Wakai S, Saito T, et al. Clinical characteristics of a-pyrrolidinovalerophenone (a-PVP) poisoning. Clin Toxicol 2016;54(7):563–7. https://doi.org/10.3109/15563650.2016.1166508.

Vevelstad M, Oiestad EL, Middelkoop G, Hasvold I, Lilleng P, Delaver GJ, et al. The PMMA epidemic in Norway: comparison of fatal and non-fatal intoxications. Forensic Sci Int 2012;219:151–7.

Wallach J, Brandt SD. 1,2-diarylethylamine- and ketamine-based new psychoactive substances. Handb Exp Pharmacol 2018a;252:305–52. https://doi.org/10.1007/164_2018_148.

Wallach J, Brandt SD. Phencyclidine-based new psychoactive substances. Handb Exp Pharmacol 2018b;252:261–303. https://doi.org/10.1007/164_2018_124.

Wallach J, Colestock T, Cicali B, Elliott SP, Kavanagh PV, Adejare A, Dempster NM, Brandt SD. Syntheses and analytical characterizations of N-alkyl-arylcyclohexylamines. Drug Test Anal 2016;8(8):801–15. https://doi.org/10.1002/dta.1861. Epub 2015 Sep 11. PMID: 26360516.

Ward J, Rhyee S, Plansky J, Boyer E. Methoxetamine: a novel ketamine analog and growing health-care concern. Clin Toxicol 2011;49(9):874–5. https://doi.org/10.3109/15563650.2011.617310.

Wijers CH, van Litsenburg RT, Hondebrink L, Niesink RJ, Croes EA. Acute toxic effects related to 4-fluoroamphetamine. Lancet 2017 Feb 11;389(10069):600. https://doi.org/10.1016/S0140-6736(17)30281-7. Erratum in: Lancet. 2017 Mar 4;389(10072):908. PMID: 28195054.

Wiley JL, Marusich JA, Huffman JW, Balster RL, Thomas BF. Hijacking of basic research: the case of synthetic cannabinoids. Methods report. RTI Press; 2011. https://doi.org/10.3768/rtipress.2011.op.0007.1111.

Wiley JL, Marusich JA, Huffmann JW. Moving around the molecule: relationship between chemical structure and in vivo activity of synthetic cannabinoids. Life Sci 2013;97:55–63.

Wiley JL, Marusich JA, Lefever TW, Antonazzo KR, Wallgren MT, Cortes RA, et al. AB-CHMINACA, AB-PINACA, and FUBIMINA: affinity and potency of novel synthetic cannabinoids in producing d9-tetrahydrocannabinol-like effects in mice. J Pharmacol Exp Therapeut 2015;354(3):328–39.

Winstock A, Lynskey M, Borschmann R, Waldron J. Risk of emergency medical treatment following consumption of cannabis or synthetic cannabinoids in a large global sample. J Psychopharmacol 2015 Jun;29(6):698–703. https://doi.org/10.1177/0269881115574493. Epub 2015 Mar 10. PMID: 25759401.

Wood DM, Button J, Lidder S, Ramsey J, Holt DW, Dargan PI. Dissociative and sympathomimetic toxicity associated with recreational use of 1–(3–trifluoromethylphenyl) piperazine (TFMPP) and 1–benzylpiperzine (BZP). J Med Toxicol 2008;4:254–7.

Wood DM, Puchnarewicz M, Johnston A, Dargan PI. A case series of individuals with analytically confirmed acute diphenyl-2-pyrrolidinemethanol (D2PM) toxicity. Eur J Clin Pharmacol 2012a;68(4):349–53. https://doi.org/10.1007/s00228-011-1142-0.

Wood DM, Davies S, Puchnarewicz M, Johnston A, Dargan PI. Acute toxicity associated with the recreational use of the ketamine derivative methoxetamine. Eur J Clin Pharmacol 2012b;68(5):853–6. https://doi.org/10.1007/s00228-011-1199-9.

Wood DM, Sedefov R, Cunningham A, Dargan PI. Prevalence of use and acute toxicity associated with the use of NBOMe drugs. Clin Toxicol 2015;53:85–92. https://doi.org/10.3109/15563650.2015.1004179.

Zawilska JB. Methoxetamine—a novel recreational drug with potent hallucinogenic properties. Toxicol Lett 2014;230(3):402–7. https://doi.org/10.1016/j.toxlet.2014.08.011.

Zawilska JB, Wojcieszak J. An expanding world of new psychoactive substances-designer benzodiazepines. Neurotoxicology 2019;73:8–16. https://doi.org/10.1016/j.neuro.2019.02.015.

CHAPTER

4

NPS analysis

Workflow

Clinical and forensic drug testing is routinely done according to a defined workflow, and this applies to new psychoactive substance (NPS) analysis as well. It consists of (1) sample documentation, (2) sample processing and preparation, (3) sample testing, (4) data analysis, and (5) results reporting. Testing is done in two steps: screening and confirmation. Screening allows the qualitative detection of a drug (e.g., fentanyl) or class (e.g., opiates), and the sample is declared to be either presumptive positive or negative. If the result is positive, another assay is done to confirm and quantify the specific drug (CLSI, 2014). In a hospital emergency department, the screening results might be reported immediately, before confirmation.

The typical screening methods are immunoassays. Advanced laboratories have begun to use high-resolution mass spectrometry (HRMS). For confirmation, gas chromatography–mass spectrometry (GC-MS) and liquid chromatography-tandem mass spectrometry (LC-MS/MS) are common, the latter being the gold standard in clinical labs. Liquid chromatography–high-resolution mass spectrometry (LC-HRMS), exemplified by liquid chromatography–quadrupole time-of-flight mass spectrometry (LC-QTOF/MS), is also becoming a method of choice for confirmation (Meyer and Maurer, 2016; Yuan et al., 2015).

The expenditure of time and care in building and validating the processes for sample preparation and testing protocols (steps 1–3) is necessary to ensure that the method is accurate and reproducible. Method characteristics determined and optimized during validation include accuracy, precision, linear dynamic range, sensitivity, selectivity, matrix effect, recovery, and stability. Each of these parameters is optimized for each drug in an assay (Kruve et al., 2015a, 2015b). It is a laborious and tedious process. For a panel of less than 10 drug analytes, method development

Designer Drugs
https://doi.org/10.1016/B978-0-12-811764-4.00003-3

and validation in LC-MS/MS or GC-MS take about 3–6 months barring any difficult hurdle in the compatibility of drugs included in the panel. With a larger number of target drugs in the panel, the required time proportionately lengthens. Federal agencies and other organizations issue guidelines for validation and acceptable values for parameters, adherence to which ensure comparability of results within and between laboratories.

Further checks are applied once a validated method is implemented in the laboratory. Quality control samples are included in each batch run. Periodically, the laboratory also participates in proficiency testing, where it analyzes blind samples sent by an organization that administers the program. Results for proficiency samples are reported to the organization, which in turn evaluates and reports the method's current accuracy and precision according to how close it comes to the true values of the drugs in the proficiency testing sample. All of these checks are well established for traditional recreational drugs (TRD), including cocaine, methamphetamine, amphetamine, MDMA, MDA, THC and its metabolites, opiates, fentanyl, benzodiazepines, phencyclidine, and tricyclic antidepressants. The same is not true, however, for NPS because of their constantly evolving composition, the sheer number available, and the scarcity of well-equipped laboratories with analytical capabilities for NPS testing.

Chain of custody and documentation

The veracity and traceability of data obtained from drug and NPS analysis are absolute requirements in clinical and forensic settings. Different countries have different approaches to ensure that the detection, identification, and quantification of controlled drugs and NPS in drug products and biological samples related to their unlawful manufacture, distribution, diversion, or misuse are fully vetted and strictly monitored. The policies and protocols that govern implementation depend on available laboratory resources, access to training and expertise in analysis, and the perceived public health impact of controlled drug misuse. In the United States, the Drug Enforcement Administration (DEA) is mandated to oversee the formulation of rules and guidelines governing controlled substance regulation and their implementation in accordance with international agreements.

Because reported drug levels can be used in determining the cause of a toxidrome or death, and analytical reports can be entered as evidence in legal proceedings, the data obtained from an analysis should provide the scientific basis for rational interpretation and judicious conclusions on drug use and misuse. Hence, the standard procedure should have verifiable accuracy, precision, and reproducibility, and the facility at which the analysis is done as well as the activities of its personnel should be well regulated and monitored to avoid any legal doubt about the integrity of the sample and the analysis performed to obtain the data. To safeguard sample integrity, judicious and meticulous documentation of the proper chain of custody is paramount. Chain of custody refers to the location and movement of physical evidence from the time it is obtained until the time it and data obtained from its analysis are presented in court. Guidelines governing chain of custody along with the requirements for laboratory facilities, and the handling of and working with controlled substances including their testing, are described in the amended Controlled Substances Act.

The DEA determines, implements and monitors the requirements for the facilities where drug testing is done, handling of controlled substances, and handling and testing drug exhibits for legal proceedings. For any laboratory to engage in testing controlled substances, it has to secure the DEA license to acquire, store, handle, and test them. The DEA requirements on handling include (1) the provision and use of a secure area or vault for storing controlled substances; (2) control of keys and designated personnel who can access the controlled substances; (3) detailed standard operating procedure and documentation on the handling, storage, access, and use of controlled substances, and (4) a designated custodian who monitors personnel's access to controlled substances, enforces procedures for their handling, and ensures that proper documentation is maintained. Furthermore, the DEA guidance on handling a drug recommends documentation of all procedures required for its receipt and inventory, sampling of the material for analysis, analytical testing, data storage, archiving of residual materials, and destruction once authorized.

All laboratory staff handling the material are required to record, retain, and archive all specimens and samples, documentation of the chain of custody, testing procedures, raw data obtained from the analysis, and summary reports. Laboratory access records, laboratory notebooks, drug inventory activities, and chain of custody should be archived and readily available on demand should they be requested. The facilities, inventory, data management system, and equipment used for analysis should be registered, licensed if required, and validated. Finally, all of the laboratory facility, equipment, personnel, procedures and processes, inventory and data management systems, records, and documentation are subject to both internal and external inspection, review, and audit to ensure that the entire laboratory system can withstand all legal and civil challenges and objections. Although meant for DEA licensed laboratories, these recommendations ensure verifiable and traceable data that any laboratory engaged in drug and NPS testing can follow.

Sample matrices

A sample matrix is the vehicle in which an analyte (NPS in this case) is contained. The nature and composition of the matrix significantly affect the analysis. Biological samples such as blood and urine contain thousands of endogenous metabolites and other substances that can suppress or enhance the physical or chemical property of the analyte. In immunoassays, some of these substances can cross-react with the antibody that was raised to detect the analyte. In colorimetric or spectrophotometric assays, they may absorb light at the same maximum wavelength at which the absorbance of the analyte is being measured. In techniques involving MS, they may interfere with the ionization of the analyte.

As with TRD, NPS analysis can be performed on different matrices, including drug products, drug paraphernalia, and biological samples (Kim et al., 2021; Ambach et al., 2015; Lung et al., 2016). The most commonly analyzed biological matrices are blood and urine. Others are useful, especially hair, oral fluid, and dried blood spot, depending on the purpose of the analysis. Vitreous humor is popular in forensic analysis, and cerebrospinal fluid can be helpful in clinical investigation.

Because NPS come in different forms, the samples for analysis can be powder, capsule, tablet, liquid, herbal mixture, vaping fluid, blotter paper, and edible materials. Generally, powder, capsule, tablet, liquid, and blotter paper contain much less other substance than do biological samples, which simplifies analysis. When the substance taken by the subject is no longer available, residues in paraphernalia such as pipes, bongs, syringes, foils, and vaping delivery systems are analyzed (Fig. 4.1).

FIGURE 4.1 Types of samples used in the analysis of NPS products and paraphernalia.

The type of biological matrix to use depends on the purpose of the analysis: forensic investigation, clinical testing, workplace drug screening, driving under the influence (DUI) testing, abstinence monitoring, therapeutic adherence monitoring, and sports doping testing. These end goals vary in their requirement for relevant drug detection window, the time frame in which drug levels are measurable. Abstinence monitoring, therapeutic adherence monitoring, and forensic testing may require determining the pattern of drug use, so a longer detection window is paramount. Oral fluid, which reflects the most immediate drug use, is not relevant; hair is. On the other extreme, DUI testing requires immediate detection, while the subject is still under the influence, so oral fluid is good. Detectability in a biological matrix depends on elimination half-life and biological compartment distribution; the detection window for each drug and NPS varies for each matrix (Khadehjian, 2005).

Urine

This is the most common matrix in emergency departments for drugs of abuse screening and confirmation, and in most types of testing. The relative ease of collection in large volume without sticking the patient makes it preferable, especially if the patient is mentally altered (Maurer, 2004). The United Nations Office on Drugs and Crime in its general guidelines for submission of specimens for NPS detection recommends urine as the sample of choice primarily for this reason. Also, drugs and their metabolites concentrate in urine, allowing detection even by relatively insensitive assays. Most drugs remain detectable in urine for a long period, typically 48–72 h. Generally, total elimination from the body takes about five half-lives. Table 4.1 presents the half-lives of common TRD and some NPS.

The long detection window in urine can be a disadvantage if the immediate cause of a patient's toxidrome is being sought. Detection indicates only exposure. If that happened 24–48 h earlier, testing will still detect the drug or its metabolite, and that might not be the cause of current intoxication. For detection of drugs that are extensively metabolized, it is their metabolites and not the parent that are relevant. For example, over 90% of synthetic cathinones administered orally undergo metabolism prior to excretion. In vitro metabolic studies of several synthetic cannabinoids likewise suggest that their metabolites are better targets for detection and identification. This can be problematic for some NPS because their primary metabolite might not be known; or even if known, the reference standard to confirm detection might not be available. Another concern is that some NPS share the same metabolite. Because some NPS are designed by simple changes in the structure of a previous iteration, this issue is encountered more often than in TRD. JWH-018 and

TABLE 4.1 Half-lives of common traditional recreational drugs and some new psychoactive substances (in bold font).

Drug	Half-life	Drug	Half-life
Amphetamine[a]	9–11 h (D) 11–14 h (L)	LSD	3.6 h
Carfentanil	7.7 h	MDMA[a]	3.8–7.8 h (R) 2.5–4.5 h (S)
Cocaine	0.7–1.5 h	**MDPV**	1–1.5 h
Diazepam	20–100 h	Methamphetamine	9–12 h
Fentanyl	3.5 h (terminal)	Morphine	2–3 h
Heroin	2–3 min	Oxycodone	2–4.5 h
Hydrocodone	3.3–4.4 h	Phencyclidine	7–46 h
JWH-018	1.7 h	THC	1.6–59 h

[a]*The D/L and R/S notations refer to the stereoisomers of the drugs.*

AM-2201, for example, have the same primary metabolites: 5-hydroxypentyl JWH-018 and JWH-018 N-pentanoic acid. Because urine sampling is usually not visually observed, it is also prone to adulteration and manipulation. And a drug user who anticipates testing can drink a large volume of water to induce dilution. In such cases, urine creatinine can be measured to normalize urine concentration. Despite these challenges, most NPS analyses reported so far were developed in urine (Peters, 2014; Ambach et al., 2015).

Blood

This is the second most common matrix. It comes as whole blood, plasma, or serum. Plasma and serum are generally antemortem, whole blood postmortem. Blood level is closer than urine level to the effective concentration of drug that binds receptors to effect its action, and so is more useful in the investigation of toxidrome causality. For this reason, reference ranges have been established for TRD in blood (Table 4.2). In the absence of NPS ranges, TRD ranges are used to assess NPS levels in clinical and forensic investigation, especially if the relative potency of the NPS with respect to a reference TRD is known. Because they have been cleared of cellular components, plasma and serum samples generally pose lesser matrix effects and are simpler to analyze than whole blood. Of course, the sample preparation can also affect this comparison.

Aside from requiring venipuncture, which may not always be available (e.g., roadside sampling for DUI), blood also contains active enzymes that

TABLE 4.2 Reference ranges of common traditional recreational drugs.

Drug	Reference range (ng/mL)		
	Recreational	Toxic	Lethal
Amphetamine	20–150	200	500–1000
Cocaine	50–300	250–5000	1,000–20,000
Diazepam	125–500	1500	5000
Fentanyl	1–2	2–20	>20
Hydrocodone	2–50	100	100–200
LSD	0.5–5	1	2–5
MDMA	100–350	350–500	400–800
Methamphetamine	10–50	200–2500	10,000–40,000
Morphine	10–120	150–500	50–4000
Oxycodone	5–50	200	600
Phencyclidine	10–200	7–240	300–5000

may continue to metabolize the drug past sampling, so samples should be refrigerated after initial processing. Long-term storage of serum and plasma requires freezing. Moreover, the blood level of some drugs may change precipitously from the time of intake, especially those that are rapidly metabolized, so detection requires higher sensitivity of the analytical platform. Nevertheless, validated methods for various NPS classes are as common for blood as for urine (Adamowicz and Tokarczyk, 2016; Ambach et al., 2015; Vaiano et al., 2016; Lehmann et al., 2017; Giorgetti et al., 2022).

Hair

This is a unique matrix in that drugs and any substances that are embedded in its cortex as it grows stay for a very long time. It is ideal for monitoring chronic use. Hair grows at slightly varying rates on different parts of the head (Nakahara, 1999; Boumba et al., 2006). The occipital area is ideal because it has the fastest growth and remains available in male pattern baldness. Occipital hair growth rate is approximately 1 cm per month; if sampling is done carefully and the end closest to the scalp can be identified, analysis allows monitoring of the temporal drug use pattern. There are several advantages to using hair: (1) sampling is noninvasive, (2) drugs and metabolites are stable once embedded in the hair cortex, (3) sampled hair is stable for a long period, (4) storage can be done at room temperature and does not require much space, (5) transport does not require cold chain custody, and (6) hair has no biohazard risk.

The primary challenge is interpretation. Detection certainly indicates use, but quantitation is problematic. Because blood levels have been correlated to the intensity of effect and toxicity of most drugs, correlative studies between blood and hair levels would be useful. Not many have been done for TRD, and very few for NPS. Furthermore, hair color affects the amount of some drugs that can be deposited. Eumelanin, the pigment that gives black hair its hue, contains a lot of negatively charged 5,6-dihydroxyindole carboxylic acid residues, which serve as binding sites for drugs that contain amine or basic groups (Fig. 4.2). Most drugs, including cocaine, amphetamines, phenethylamines, tryptamines, and piperazines, contain such groups. This means that the darker the hair, the greater its ability to bind basic or amine-containing drugs. Thus, a light-haired person can show a lower methamphetamine hair level than a dark-haired person when both have consumed the same amount.

Other factors can affect hair levels. Drug can be deposited outside the hair shaft when a person either smokes it or is present where it is being smoked, so extensive washing of the sample is required. Chemical

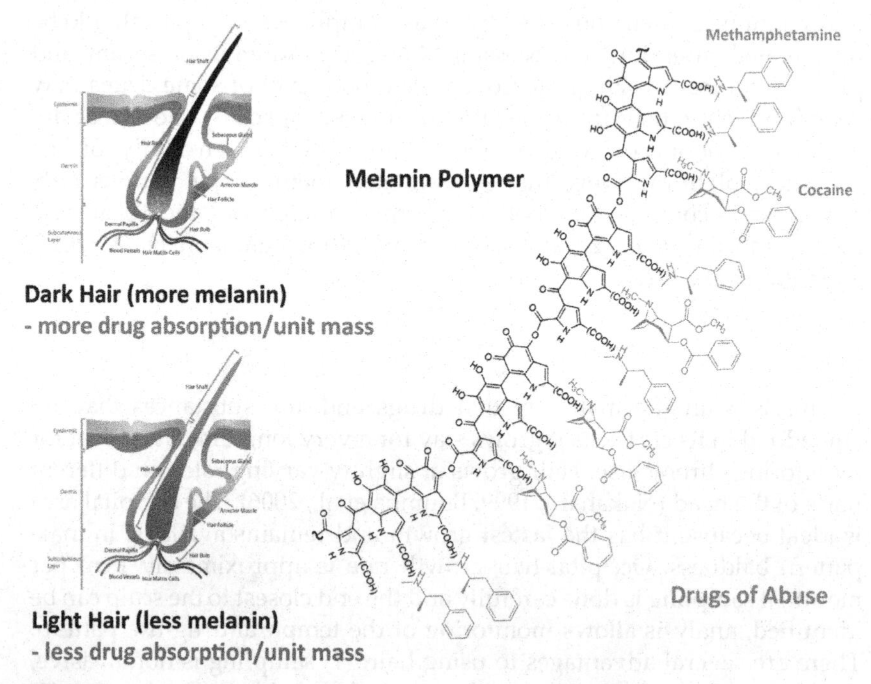

FIGURE 4.2 Functional groups in eumelanin that interacts with basic and amine-containing drugs.

treatment during perming or dyeing can extract or degrade drugs. Harsh environmental condition such as excessive sunlight can affect the stability of drugs in the cortex. Some drugs undergo lateral diffusion as the hair grows; this becomes critical when temporal use is important. Despite these disadvantages, hair is a useful matrix for forensic investigation, abstinence monitoring, therapeutic adherence monitoring, and establishing a use pattern over time. Validated methods have been published for synthetic cannabinoids, synthetic cathinones, amphetamines, phenethylamines, and piperazines (Salomone et al., 2016; Lendoiro et al., 2017; Montesano et al., 2017; Florou and Boumba, 2021).

Oral fluid

Recently, this matrix has gained popularity in specific situations such as rapid on-site screening. Its utility is proven for DUI testing, primarily because of its noninvasiveness, ease of sampling, and the reduced risk of manipulation or adulteration, as collection can be directly observed. As with blood, oral fluid drug levels reflect recent intake; the detection window is at the phase when the subject is still under the influence. Because drug in the oral cavity has not undergone first-pass metabolism, the parent is more salient than metabolites. Over the past decade, interpretation of oral fluid levels has become easier as more studies have been conducted on the oral fluid-to-blood ratios of common TRD (Bosker and Huestis, 2009; Desrosiers and Huestis, 2019). Synthetic cannabinoids, synthetic cathinones, amphetamines, phenethylamines, and piperazines have been measured in oral fluid (Amaratunga et al., 2013; Amaratunga et al., 2014; Rocchi et al., 2018). The primary disadvantage is its inherent heterogeneity, which can impose differing matrix effects on a drug. This in turn affects quantitation of drug levels.

Dried blood spot

This matrix, DBS, is desired when only a small sample can be obtained, such as in neonates. The method has attracted recent interest in situations when rapid onsite sample collection outside the hospital setting is required, and in problematic populations such as mentally altered drug abusers, psychiatric patients, children, and the elderly. The sample is taken by a finger prick (also heel or toe for neonates) using a lancet and collecting the drop on absorbent filter paper. The blood is allowed to air-dry on the paper, and then a disc of the spot (~6 mm) is punched out for analyte extraction. The volume in a punched disc can vary between 6 and 12 μL. The composition of the spot from a finger prick is similar to that of whole blood. However, it is capillary blood, as opposed to that obtained from venipuncture. The difference is minimal, but capillary blood has higher hemoglobin and hematocrit (percentage of red blood cells) than venous blood. Water leaves as blood passes through capillaries, but it gets reabsorbed in the venules.

DBS has several advantages: (1) less invasiveness than venipuncture, (2) minimal postcollection processing, (3) simple storage at room temperature, (4) longer stability in storage, (5) no cold chain custody for transport, and (6) low biohazard risk (Sadones et al., 2014). The primary disadvantages are its small volume and the heterogeneity of hematocrit across samples. The small volume of sample requires accuracy in the sampling procedure, because a unit difference in sampling has significant impact on the total sample volume. Likewise, the limited amount of sample requires high sensitivity in the analytical method; most DBS samples are analyzed by MS. The relative level of hematocrit significantly contributes to the matrix effect and the quantitative measurement of drug. Hence, a hematocrit measurement concurrent with drug analysis using DBS can be useful in normalizing for the variability of hematocrit across samples. This may not always be available in on-site and field sampling. Nevertheless, NPS analysis using DBS has been reported for synthetic cannabinoids, synthetic cathinones, amphetamines, phenethylamines, piperazines, tryptamines, and opioids (Ambach et al., 2014).

Others

Aside from the five discussed before, other types of biological samples have been used in TRD and NPS analysis. Most have specialized applications. Meconium, for example, allows assessment of fetal drug exposure. It is the thick, sticky, dark green to black substance that fills out the fetal intestines before birth. For about 3–5 days after birth, it is obtained when the baby defecates. Because it is roughly equivalent to the fetus' stool made up of dead cells, proteins, fats, intestinal secretions, and other substances that the fetus was exposed to in the womb, it captures drugs in the mother's circulation that pass through the placenta. A method to simultaneously detect synthetic cannabinoids, synthetic cathinones, amphetamines, phenethylamines, piperazines, tryptamines, benzofurans, piperidines, and new synthetic opioids in meconium has been reported (Lopez-Rabuñal et al., 2021). The inherent heterogeneity of meconium and the scarcity of validated methods for comparison make interpreting quantitative levels of drugs challenging, but it is a good way to confirm fetal exposure.

Vitreous humor, the gelatinous substance that fills the eyeball between the lens and retina, is commonly sampled in postmortem forensics. About 99% of it is water, the rest is collagen, proteins, sugar, and salts. Drugs partition to the vitreous humor. This matrix is valuable because its isolation from blood and other body fluids spares it from postmortem changes such as drug redistribution and hemoconcentration.

Furthermore, it resists decay longer than other body fluids. Several NPS have been reported in it, including synthetic cannabinoids, synthetic cathinones, amphetamines, phenethylamines, new synthetic opioids, and designer benzodiazepines (Andreasen et al., 2009; Fels et al., 2019; Cartiser et al., 2021).

Cerebrospinal fluid, bile, gastric contents, and various human tissues have also been analyzed for NPS, usually postmortem (Thornton et al., 2012). Fingernails and sweat collected in patches have also been reported (Pichini et al., 1996; Kim et al., 2010). Like hair, nails have a long detection window.

Sample preparation strategies

Because of the innate complexity of sample matrices, they need to be processed and prepared before they undergo testing in an analytical platform. The type of preparation depends on the matrix, analytical platform, and required sensitivity. Samples can be directly analyzed in an instrument, but this is more often an exception than the rule, especially for biological samples. In most cases, the sample undergoes extraction protocols to isolate the drug or NPS of interest from most components of the matrix. In blood, for example, drugs need to be separated from cellular, protein, and lipid components.

Drug products often require minimal processing, especially if they are in the powder, capsule, tablet, or liquid form. These have either simple matrices or insoluble excipients in which the drug is incorporated. Drugs are extracted by organic solvents, typically methanol. The extract is then separated from the insoluble components and evaporated so that it can be reconstituted in a solvent that is compatible with the analysis. Other solvents are also used for extraction, depending on the polarity of the target compounds. Ethyl acetate and methyl tert-butyl ether (MTBE) are common extracting solvents for nonpolar drugs. Most NPS, however, have enough polarity to dissolve in methanol.

Among biological samples, urine is the most amenable to direct analysis. In clinical testing, the most common preparation method is "dilute and shoot," which is simply dilution with either the solvent system used for analysis or water. If the assay is chromatography, the initial composition of the mobile phase is used for dilution. The sample is then injected directly to the analytical platform. In some instances, the diluted sample is centrifuged or filtered before injection to make sure that it is free of particulates that could clog the instrument. Because conjugated metabolites (phase II) of NPS are significantly present in urine, deconjugation using acid hydrolysis or enzymes is often applied prior to dilution. Commercially available enzymes such as *Helix pomatia* glucuronidase that has both deglucuronidation and desulfation activities are usually applied to facilitate deconjugation. Dilute and shoot has also been applied in oral fluid. Several methods for the analysis of synthetic cannabinoids, synthetic cathinones, phenethylamines, amphetamines, benzofurans, arylcyclohexylamines, new synthetic opioids, and designer benzodiazepines using dilute and shoot in urine and oral fluid have been published (Beck and Ericsson, 2014; Peters, 2014; Malaca et al., 2019; Fan et al., 2020, 2021). Direct analysis offers a rapid and easy approach to sample preparation.

For all other biological samples, NPS or drug analysis is done indirectly following more elaborate sample preparation. Preparation strategies aim to extract and isolate NPS or any drug of interest from other components of the matrix that could interfere with the analysis. Complete isolation of a drug from all components of the matrix is impossible, but extraction methods simplify the matrix, thereby minimizing matrix effect. This has the added benefit of improving the sensitivity of the assay. The three most common approaches in extracting NPS from biological fluids are protein precipitation, liquid–liquid extraction (LLE), and solid–phase extraction (SPE). Recently, microextraction techniques were also introduced.

Protein precipitation

The protein components of blood samples (whole blood, plasma, serum) can interfere with analysis. A simple method of eliminating these components is the salting-out effect afforded by adding 3× to 5× volume of water-miscible organic solvent such as methanol or acetonitrile. Solvent can be used in pure form or various combination ratios. Addition of the organic solvent strips the water of solvation of proteins that maintains the repulsive forces keeping them soluble in aqueous solution. With lowered repulsive forces, they crash out and are easily removed by centrifugation. The resulting extract is evaporated and reconstituted in an appropriate solvent for analysis. This method is popularly used in LC-MS/MS and LC-HRMS. Evaporation of the extract allows concentration or dilution of the analytes, depending on the volume of solvent used in reconstitution. Not all components of blood are proteinaceous, so concentration during reconstitution can also *increase* the matrix effect. Some drugs are protein bound in plasma or serum and so can be lost by precipitation. Notwithstanding these disadvantages, protein precipitation still provides a simple, fast, cost-effective preparation compatible with high-throughput analysis. In the past decade, precipitation methods in 96-well plates have been developed. They come with filtration systems that enable separation of the crashed protein and at the same time eliminate lipid components. Most published analyses for NPS in blood use protein precipitation, especially those aimed at a broad spectrum of NPS classes (Adamowicz and Tokarczyk, 2016; Lung et al., 2016; Vaiano et al., 2016).

Liquid–liquid extraction

LLE is the most commonly used method for biological fluids. Unlike protein precipitation, application is not limited to blood. It is a simple preparation method that allows extraction of a wide spectrum of drugs. It uses immiscible solvent and takes advantage of the differential solubility of drugs in two immiscible liquid phases. To increase the efficiency of drug extraction, rigorous shaking before phase separation is typically done. Because biological fluids are aqueous, hydrophobic solvents are usually employed; the most common ones are ethyl acetate, MTBE, and chlorinated ones such as dichloromethane and 1-chlorobutane. These are efficient at extracting neutral NPS; therefore, to increase the efficiency of extracting basic NPS (most amine-containing ones such as amphetamines, synthetic cathinones, and tryptamines), basic conditions are induced by NaOH or ammonium and carbonate buffers. LLE has been applied in the analysis of all NPS classes in blood, urine, and oral fluid (Mortier et al., 2002; Alexandridou et al., 2020).

Solid-phase extraction

SPE is an alternative to LLE that affords more selectivity. It uses a sorbent material, packed in a column, which is selective for the class to be extracted. Columns can employ nonpolar (C8 or C18), anion exchange, and cation exchange retention mechanisms. To cover most drugs of abuse and NPS, a combination of nonpolar and strong cation exchange retention mechanisms are commercially available.

The procedure has four steps: conditioning, loading, washing, and elution. After a column is conditioned with aqueous solution at an appropriate pH, target analytes are retained in the sorbent from the biological fluid applied. The column is washed with an appropriate solvent to remove any retained matrix component before elution of the target analyte. The type of solvent used to elute the analyte depends on the type of sorbent: for C8 or C18, methanol; for anion or cation exchange, acidic or basic solvents. High selectivity and enrichment of NPS can be achieved; this facilitates better sensitivity. SPE is typically combined with LC-MS/MS or LC-HRMS. A few GC-MS methods have also been reported. Like LLE, SPE is used for all NPS classes in blood, urine, and oral fluid, and for alternative matrices such as vitreous humor, bile, and tissue homogenate (Lehmann et al., 2017; Montesano et al., 2017; Morini et al., 2017; Rojek et al., 2020).

Automated SPE systems with 96-well plates have been introduced over the past decade. Some can be incorporated in-line on LC-MS/MS and LC-HRMS platforms, allowing high throughput.

Presumptive and confirmatory analyses

NPS analysis, like TRD analysis, is done in two steps in the clinical and forensic setting: screening (presumptive) and confirmation (CLSI, 2014). Screening assays qualitatively detect the presence or absence of a drug. The goal is to cast a wide net and catch as many drugs and classes as possible. Sensitivity is more important than selectivity. A presumptive positive result must be verified by a confirmatory assay anyway, so false positives can still be minimized.

Immunoassay is the most popular screening method. An antibody for a drug, or more often a class, is raised and provides the necessary antigen (drug)–antibody selectivity. The assay is administered through a kit or more commonly incorporated in a laboratory autoanalyzer. The antibody–drug interaction is measured by spectrophotometry, chemiluminescence, or electrochemiluminescence. Although the interaction is quantified, immunoassays are qualitative. A response cutoff is established above which the result is reported as presumptive for a drug or class. The urine drug screen given at an emergency department is just a collection of immunoassays for different drugs and classes. The most common ones are for amphetamines, cocaine, cannabinoids (THC and its metabolites), opiates, fentanyl, benzodiazepines, and tricyclic antidepressants. Others are for oxycodone/oxymorphone, methadone, buprenorphine, phencyclidine, and barbiturates. Because antibodies are raised on functional groups that can be shared by a target drug with other drugs, cross-reactivity is a problem (Saitman et al., 2014). Some amphetamine assays, for example, cross-react with bupropion and metformin; the cannabinoid assay cross-reacts with efavirenz (an HIV medication), ibuprofen, and naproxen. So the result of an immunoassay is at best presumptive positive and requires confirmation. Table 4.3 presents the detection cutoff and cross-reacting drugs of common drugs of abuse immunoassays.

Although most NPS do not cross-react in immunoassays for common TRD, some that have close structural relationship can: synthetic cathinones with amphetamine, fentalogs with fentanyl, and designer benzodiazepines with benzodiazepines. Assays have been developed for a few NPS classes, notably synthetic cannabinoids and synthetic cathinones. The challenge for immunoassays is the rapid, continuous transformation of scaffolds for some NPS classes; synthetic cannabinoids and new synthetic opioids are notorious. Most of the time, the "half-life" of a given NPS class scaffold in the recreational drug market is shorter than the time required to raise a new antibody against it. Hence, if false positive due to cross-reactivity is a challenge for TRD immunoassays, false negative is a challenge for NPS immunoassays.

There are two alternatives to immunoassay for screening. Activity-based assays, exemplified by one for synthetic cannabinoids (Cannaert

TABLE 4.3 Cutoff ranges and cross-reacting drugs to common drugs of abuse in immunoassays.

Assay	Typical cut-off (ng/mL)	Common drugs causing false positive
Amphetamines	1000	Bupropion, desipramine, ephedrine, methylphenidate, pseudoephedrine, trazodone
Barbiturates	200	Fenoprofen, ibuprofen, naproxen
Benzodiazepines	200	Oxaprozine, sertraline
Cannabinoids	50	Efavirenz, fenoprofen, ibuprofen, naproxen
Opiates	300	Dextomethorphan, diphenhydramine, verapamil
Methadone	300	Clomipramine, diphenhydramine, doxylamine, quetiapine, thioridazine, verapamil
Phencyclidine	35	Dextomethorphan, diphenhydramine, doxylamine, ibuprofen, ketamine, meperidine, venlafaxine
Tricyclic antidepressant	1000	Diphenhydramine, quetiapine, cyclobenzaprine

et al., 2016, 2017), are promising. They use a cellular system that stably expresses CB1 and CB2 receptors that are connected to the luciferase reporter system. Binding of a synthetic cannabinoid to either receptor recruits beta-arrestin, which facilitates luciferase activity and produces measurable bioluminescence (Cannaert et al., 2018). This approach depends on receptor binding instead of a specific structural scaffold, which is susceptible to structural changes. The other approach is the use of high-resolution MS (Sundstrom et al., 2013, 2015). Unlike LC-MS/MS, LC-HRMS facilitates nontargeted data acquisition, which allows collection of all ions generated from a biological sample. The resulting total ion chromatogram can then be queried in multiple ways using drug databases that can include even those that the laboratory does not have reference standards for. Detection of ions obtained from the sample requires matching with drugs contained in the library facilitated through accurate mass and isotope cluster matching. Therefore, the analysis is not limited to known drugs and can be used to discover previously unreported NPS (Wu et al., 2012; Wu and Colby, 2016). The primary challenges are the initial acquisition cost for the instrument and the high level of technical expertise needed in data analysis.

Results from a screening assay require confirmation from a separate assay that has high selectivity. To produce unequivocal identification and quantitation for a final report, a confirmatory assay is mandatory (CLSI, 2014). Recent literature in analytical toxicology indicates that MS-based assays are suitable to fulfill this requirement (Maurer, 2010; Meyer and Maurer, 2016). They are linked to a gas chromatography or more commonly liquid chromatography platform. The most common are GC-MS, LC-MS/MS, and LC-HRMS. LC-MS/MS is the gold standard for quantitative analysis. The routine use of monitoring for at least two product ions through multiple reaction monitoring allows high specificity for this assay. Recent innovations in the quadrupole and detectors also allow high sensitivity and wide linear dynamic range—four or five orders of magnitude. Confirmatory assays for all types of NPS using GC-MS, LC-MS/MS, and LC-HRMS have been reported; Table 4.4 compares them.

TABLE 4.4 Comparison of commonly used mass spectrometry platforms in confirmatory drug testing.

Platform	Strengths	Limitations	Applications
GC-MS	High sensitivity Wide dynamic range of detection Lower cost Compatibility of MS/MS spectral data across platform brands	Low mass resolution Not applicable to thermolabile drugs Requires derivatization for nonvolatile drugs	Targeted analysis Quantitative analysis MS/MS spectral collection Structural elucidation Nonpolar and slightly polar drugs and derivatives
LC-MS/MS	Highest sensitivity Wide linear dynamic range Moderate cost	Low mass resolution	Targeted analysis Quantitative analysis Slight polar to polar drugs
LC-(Q)TOF/MS	High mass resolution High sensitivity Medium linear dynamic range of detection Wide mass range	Lower sensitivity than LC-MS/MS for some analytes More costly than LC-MS/MS Incompatibility of MS/MS spectral data across platform brands Requires more technical expertise	Targeted analysis Nontargeted analysis Quantitative analysis MS/MS spectral collection Structural elucidation Slightly polar to polar drugs

Closeup: NPS in the trenches: the crazy monkey

Humanity faces constant, recurring epidemics of drug abuse, an ever-changing quagmire of addiction, misery, and death. An apt comparison here is to the Lernaean Hydra, the mythological beast whose many heads regrow when severed: as one substance of abuse is brought under control through legislation, education, and treatment efforts, another takes its place.

The human toll associated with this phenomenon is staggering, and nowhere is this more strongly felt than in emergency departments across the United States. From 1990 to 2019, while the number of deaths per 100,000 people attributed to drug use disorders increased from 1.15 to 1.57 globally, this number increased more than ninefold in the United States, from 2 to 19 (Fig. 4.3) (Our World in Data).

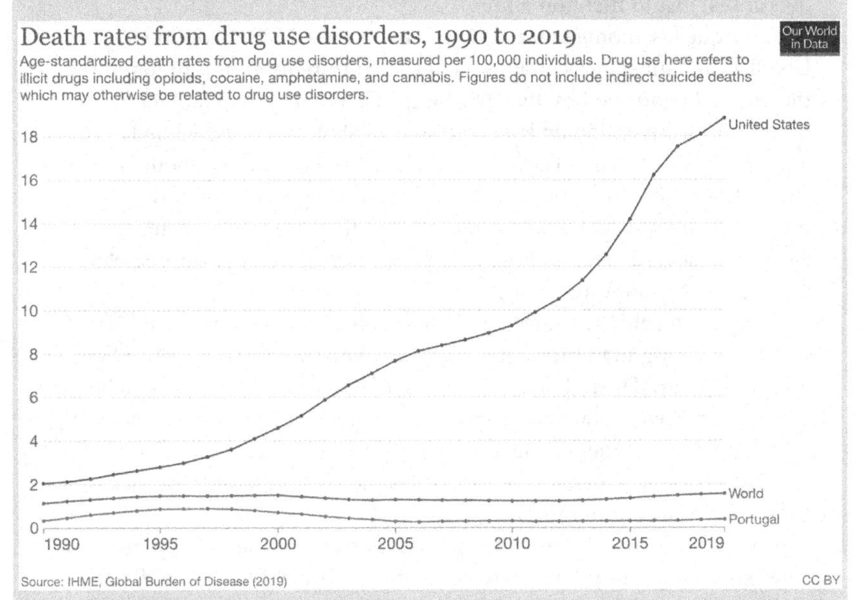

FIGURE 4.3 Death rates from drug use disorders, 1990 to 2019.

The concurrent drug-related disability-adjusted life years in the United States—a measure that reflects those years of life lost due to premature death and disability—increased more than threefold in the 15- to 49-year-old cohort. The astute reader will realize that neither the deaths nor the disability-adjusted life years in that period include the COVID-19 pandemic. Preliminary data from the United Nations Office on Drugs

continued

Closeup: NPS in the trenches: the crazy monkey (*cont'd*)

and Crime (UNODC) show a 22% increase in drug use worldwide from 2010 to 2021 (United Nations, 2021).

As an emergency department physician and medical toxicology fellow-in-training in San Francisco, I found these phenomena fascinating, heartbreaking, and seemingly insurmountable. As a field of medicine dedicated to the evaluation and treatment of poisoned and envenomated patients, medical toxicology has a unique front-row seat to the waves of intoxicants that periodically sweep through this country. The range of situations encountered can be huge: snake envenomations; prescription medication interactions and overdoses; carbon monoxide and cyanide poisonings from house fires; plant ingestions; glue sniffing; and a seemingly never-ending stream of Tylenol (paracetamol, acetaminophen) overdoses, just to mention a few.

A unique fascinoma in the deluge of exposures, NPS are defined by the UNODC as "substances of abuse, either in a pure form or a preparation, that are not controlled by the 1961 Single Convention on Narcotic Drugs or the 1971 Convention on Psychotropic Substances, but which may pose a public health threat." (UNODC, 2021). Of note, nothing about NPS is straightforward: they belong to a broad range of categories, they frequently are not actually new xenobiotics, they are inconsistently packaged and labeled, and their clinical effects are extremely unpredictable and frequently life-threatening.

One case nicely illustrates all of these points (Gugelmann et al., 2014). In 2013, a young man brought his dog to a veterinary clinic in San Francisco's Mission District, where he complained that his pet had been twitching all over. The veterinarian was just suspecting seizure activity when the dog's owner collapsed on the ground and proceeded to shake uncontrollably in a generalized tonic–clonic seizure (stiffening and uncontrollable jerking of muscles).

An ambulance was immediately summoned, the emergency medical crew stabilized the patient with a short-acting antiseizure medication, and he was brought to the emergency department. Emergency department staff first noted that the patient had with him a small jar labeled "Crazy Monkey." Although initially sleepy to the point of being unarousable, he quickly transitioned to a state of significant agitation and violence, requiring physical restraint, sedation, paralysis, and the insertion of a breathing tube, to ensure his and the staff's safety. He was put on a breathing machine and sent to the intensive care unit, where he spent the afternoon and night in a chemically induced coma under close observation. The next day, the sedating medications were weaned, the

Closeup: NPS in the trenches: the crazy monkey *(cont'd)*

breathing tube was removed, and the patient was discharged from the hospital. He was reunited shortly afterward with his dog. Fortunately, neither he nor his dog seemed to have suffered any obvious permanent damage. Three months later, however, he was back in the emergency department, again with seizures, another batch of Crazy Monkey clutched in his hand.

Typically, a patient presenting with seizures in an emergency department is stabilized with medications while an immediate concurrent laboratory workup is begun, starting with a blood glucose test. Hypoglycemia (low blood sugar) can cause profound, life-threatening seizures that will progress to coma, and brain death if not immediately corrected.

This patient's blood sugar was normal, so the concurrent workup continued with the next line of investigation, a urine drug screen. Rapid, cheap, and relatively accurate (with some notable exceptions), these tests are able to detect several common substances of abuse, which can help in pinpointing the cause of several life-threatening clinical situations. In this case, however, the urine drug screen brought the clinical team no closer to clarity; the only substance identified was a benzodiazepine, probably among the medications given by the emergency medical crew to control the seizures.

As the patient became increasingly agitated and was chemically sedated and sent to the intensive care unit, a variety of other tests remained unremarkable. A head CT revealed no traumatic injuries, and blood tests showed no electrolyte imbalance or infection that might explain the seizures. In parallel to the excitement of the clinical conundrum, a scientific investigation only rarely undertaken in the United States unfolded. Contrary to what is depicted in TV shows, the time, cost, and resources required to pinpoint the exact cause of seizures are frequently unavailable. At the hospital where he was treated, however, the samples were subjected to a more sophisticated battery of tests designed to detect the presence of 12 different seizure-inducing prescription and illicit substances. This was also unrevealing.

Luckily, there was one final option: liquid chromatography quadrupole time-of-flight mass spectrometry (LC-QTOF/MS), a complex and highly sophisticated approach that can identify rare and novel drugs of abuse. The patient's blood and urine, along with his dog's blood and the Crazy Monkey substance itself, were screened for nearly 300 different high-likelihood substances. This finally identified the culprit, a novel synthetic cannabinoid called PB-22 (QUPIC), designed to bind the same

continued

Closeup: NPS in the trenches: the crazy monkey (*cont'd*)

receptors as the THC in marijuana but with unpredictable and sometimes devastating effects.

One further element of this case serves to illustrate further how quickly the field of NPS is evolving. The patient's samples also contained metabolites of UR-144, a second synthetic cannabinoid. In a pattern of abuse that is familiar to those of us who have worked in the trenches of clinical emergency medicine, this patient was using multiple NPS to titrate a desired clinical effect.

With the perspective of my years in clinical practice, I have found these ever-evolving novel substances of abuse both fascinating and extremely disheartening. But one fact, reflected in Fig. 4.3, should provide a glimmer of hope. While the death rate from drug use disorders over the past 30 years has skyrocketed in the United States, it has been remarkably stable in Portugal.

In their 2021 commentary, Rêgo et al. provide an excellent review of the context for Portugal's successes in controlling the drug epidemic (Rego et al., 2021). In 2000, Portugal's problematic heroin use was the second highest in Europe. Although the country had bought into the War on Drugs in the 1970s, political debate leading up to the year 2000 focused on the vulnerability rather than the criminality of high-risk users, and the Portuguese Drug Policy Model (PDPM) was created. In one perspective, the PDPM represents the transition from a "tough on drugs" approach to a recognition that although dealers are delinquents in need of legal intervention, drug use itself should be seen as a medical condition requiring treatment. Since 2000, Portuguese drug use prevalence has remained low compared with other European countries, and rates of infectious disease and incarceration for drug-related offenses also have remained remarkably stable.

Although the stigma associated with drug use has not been eradicated in Portugal, the PDPM seems to indicate that legislative changes recognizing the fundamental rights and medical needs of drug users could ultimately have a tremendous positive effect for patients and society as a whole.

Hallam Melville Gugelmann

Senior Medical Director, Mirum Pharmaceuticals, Inc., Foster City, CA, United States

Lecturer, Emergency Physician, and Medical Toxicologist, University Hospital of Basel, Basel, Switzerland

References

Adamowicz P, Tokarczyk B. Simple and rapid screening procedure for 143 new psychoactive substances by liquid chromatography-tandem mass spectrometry. Epub 2015 May 14 Drug Test Anal 2016 Jul;8(7):652–67. https://doi.org/10.1002/dta.1815. PMID: 25976069.

Alexandridou A, Mouskeftara T, Raikos N, Gika HG. GC-MS analysis of underivatised new psychoactive substances in whole blood and urine. Epub 2020 Aug 11 J Chromatogr B Anal Technol Biomed Life Sci 2020 Nov 1;1156:122308. https://doi.org/10.1016/j.jchromb.2020.122308. PMID: 33038866.

Amaratunga P, Lorenz Lemberg B, Lemberg D. Quantitative measurement of synthetic cathinones in oral fluid. J Anal Toxicol 2013 ;37(9):622–8. https://doi.org/10.1093/jat/bkt080. PMID: 24123620.

Amaratunga P, Thomas C, Lemberg BL, Lemberg D. Quantitative measurement of XLR11 and UR-144 in oral fluid by LC-MS-MS. Epub 2014 May 7 J Anal Toxicol 2014 ;38(6): 315–21. https://doi.org/10.1093/jat/bku040. PMID: 24812645.

Ambach L, Hernández Redondo A, König S, Weinmann W. Rapid and simple LC-MS/MS screening of 64 novel psychoactive substances using dried blood spots. Epub 2013 Jul 19 Drug Test Anal 2014;6(4):367–75. https://doi.org/10.1002/dta.1505. PMID: 23868723.

Ambach L, Redondo AH, König S, Angerer V, Schürch S, Weinmann W. Detection and quantification of 56 new psychoactive substances in whole blood and urine by LC-MS/MS. Bioanalysis 2015;7(9):1119–36. https://doi.org/10.4155/bio.15.48. PMID: 26039809.

Andreasen MF, Telving R, Birkler RI, Schumacher B, Johannsen M. A fatal poisoning involving Bromo-Dragonfly. Forensic Sci Int 2009 Jan 10;183(1–3):91–6. https://doi.org/10.1016/j.forsciint.2008.11.001. PMID: 19091499.

Beck O, Ericsson M. Methods for urine drug testing using one-step dilution and direct injection in combination with LC-MS/MS and LC-HRMS. Bioanalysis 2014 Aug;6(17): 2229–44. https://doi.org/10.4155/bio.14.192. PMID: 25383734.

Bosker WM, Huestis MA. Oral fluid testing for drugs of abuse. Clin Chem 2009 Nov;55(11): 1910–31. https://doi.org/10.1373/clinchem.2008.108670. PMID: 19745062.

Boumba VA, Ziavrou KS, Vougiouklakis T. Hair as a biological indicator of drug use, drug abuse or chronic exposure to environmental toxicants. Int J Toxicol 2006 ;25(3):143–63. https://doi.org/10.1080/10915810600683028. PMID: 16717031.

Cannaert A, Storme J, Franz F, Auwärter V, Stove CP. Detection and activity profiling of synthetic cannabinoids and their metabolites with a newly developed bioassay. Epub 2016 Nov 7 Anal Chem 2016 Dec 6;88(23):11476–85. https://doi.org/10.1021/acs.analchem.6b02600. PMID: 27779402.

Cannaert A, Franz F, Auwärter V, Stove CP. Activity-based detection of consumption of synthetic cannabinoids in authentic urine samples using a stable cannabinoid reporter system. Epub 2017 Aug 18 Anal Chem 2017 Sep 5;89(17):9527–36. https://doi.org/10.1021/acs.analchem.7b02552. PMID: 28771321.

Cannaert A, Storme J, Hess C, Auwärter V, Wille SMR, Stove CP. Activity-based detection of cannabinoids in serum and plasma samples. Epub 2018 Mar 20 Clin Chem 2018 Jun;64(6): 918–26. https://doi.org/10.1373/clinchem.2017.285361. PMID: 29559524.

Cartiser N, Sahy A, Advenier AS, Franchi A, Revelut K, Bottinelli C, et al. Fatal intoxication involving 4-methylpentedrone (4-MPD) in a context of chemsex. Epub 2020 Dec 11 Forensic Sci Int 2021 Feb;319:110659. https://doi.org/10.1016/j.forsciint.2020.110659. PMID: 33370656.

CLSI. Liquid chromatography-mass spectrometry methods; approved guideline. CLSI document C62-A. Wayne, PA: Clinical and Laboratory Standards Institute; 2014.

Desrosiers NA, Huestis MA. Oral fluid drug testing: analytical approaches, issues and interpretation of results. J Anal Toxicol 2019 Jul 24;43(6):415–43. https://doi.org/10.1093/jat/bkz048. PMID: 31263897.

Fan SY, Zang CZ, Shih PH, Ko YC, Hsu YH, Lin MC, et al. A LC-MS/MS method for determination of 73 synthetic cathinones and related metabolites in urine. Epub 2020 Jul 31 Forensic Sci Int 2020 Oct;315:110429. https://doi.org/10.1016/j.forsciint.2020.110429. PMID: 32784041.

Fan SY, Zang CZ, Shih PH, Ko YC, Hsu YH, Lin MC, Tseng SH, Wang DY. Simultaneous LC-MS/MS screening for multiple phenethylamine-type conventional drugs and new psychoactive substances in urine. Forensic Sci Int. 2021 Aug;325:110884. https://doi.org/10.1016/j.forsciint.2021.110884. Epub 2021 Jun 26. PMID: 34245937.

Fels H, Lottner-Nau S, Sax T, Roider G, Graw M, Auwärter V, et al. Postmortem concentrations of the synthetic opioid U-47700 in 26 fatalities associated with the drug. Forensic Sci Int 2019 Aug;301:e20–8. https://doi.org/10.1016/j.forsciint.2019.04.010. PMID: 31097357.

Florou D, Boumba VA. Hair analysis for New Psychoactive Substances (NPS): still far from becoming the tool to study NPS spread in the community? Toxicol Rep 2021 Sep 28;8: 1699–720. https://doi.org/10.1016/j.toxrep.2021.09.003. PMID: 34646750.

Giorgetti A, Barone R, Pelletti G, Garagnani M, Pascali J, Haschimi B, et al. Development and validation of a rapid LC-MS/MS method for the detection of 182 novel psychoactive substances in whole blood. Drug Test Anal 2022 Feb;14(2):202–23. https://doi.org/10.1002/dta.3170. PMID: 34599648.

Gugelmann H, Gerona R, Li C, Tsutaoka B, Olson KR, Lung D. 'Crazy Monkey' poisons man and dog: human and canine seizures due to PB-22, a novel synthetic cannabinoid. Clin Toxicol 2014 Jul;52(6):635–8.

Khadehjian LJ. Specimens for drugs of abuse testing. In: Wong RC, Tse HY, editors. Drugs of abuse- body fluid testing. Totowa, NJ: Humana Press; 2005. p. 11–28.

Kim JY, Shin SH, In MK. Determination of amphetamine-type stimulants, ketamine and metabolites in fingernails by gas chromatography-mass spectrometry. Epub 2009 Nov 18 Forensic Sci Int 2010 Jan 30;194(1–3):108–14. https://doi.org/10.1016/j.forsciint.2009.10.023. PMID: 19926235.

Kim NS, Lim NY, Choi HS, Lee JH, Kim H, Baek SY. Application of a simultaneous screening method for the detection of new psychoactive substances in various matrix samples using liquid chromatography/electrospray ionization tandem mass spectrometry and liquid chromatography/quadrupole time-of-flight mass spectrometry. Rapid Commun Mass Spectrom 2021 May 30;35(10):e9067. https://doi.org/10.1002/rcm.9067. PMID: 33656207.

Kruve A, Rebane R, Kipper K, Oldekop ML, Evard H, Herodes K, et al. Tutorial review on validation of liquid chromatography-mass spectrometry methods: part I. Epub 2015 Feb 13 Anal Chim Acta 2015a;870:29–44. https://doi.org/10.1016/j.aca.2015.02.017. PMID: 25819785.

Kruve A, Rebane R, Kipper K, Oldekop ML, Evard H, Herodes K, et al. Tutorial review on validation of liquid chromatography-mass spectrometry methods: part II. Epub 2015 Feb 13 Anal Chim Acta 2015b;870:8–28. https://doi.org/10.1016/j.aca.2015.02.016. PMID: 25819784.

Lehmann S, Kieliba T, Beike J, Thevis M, Mercer-Chalmers-Bender K. Determination of 74 new psychoactive substances in serum using automated in-line solid-phase extraction-liquid chromatography-tandem mass spectrometry. Epub 2017 Sep 7 J Chromatogr, B: Anal Technol Biomed Life Sci 2017 Oct 1;1064:124–38. https://doi.org/10.1016/j.jchromb.2017.09.003. PMID: 28922649.

Lendoiro E, Jiménez-Morigosa C, Cruz A, Páramo M, López-Rivadulla M, de Castro A. An LC-MS/MS methodological approach to the analysis of hair for amphetamine-type-stimulant (ATS) drugs, including selected synthetic cathinones and piperazines. Drug Test Anal 2017 Jan;9(1):96–105. https://doi.org/10.1002/dta.1948. PMID: 26914712.

López-Rabuñal Á, Di Corcia D, Amante E, Massano M, Cruz-Landeira A, de-Castro-Ríos A, et al. Simultaneous determination of 137 drugs of abuse, new psychoactive substances, and novel synthetic opioids in meconium by UHPLC-QTOF. Anal Bioanal Chem 2021 Sep;413(21):5493–507. https://doi.org/10.1007/s00216-021-03533-y. PMID: 34286357.

Lung D, Wilson N, Chatenet FT, LaCroix C, Gerona R. Non-targeted screening for novel psychoactive substances among agitated emergency department patients. Epub 2016 Feb 5 Clin Toxicol 2016;54(4):319–23. https://doi.org/10.3109/15563650.2016.1139714. PMID: 26846684.

Malaca S, Busardò FP, Gottardi M, Pichini S, Marchei E. Dilute and shoot ultra-high performance liquid chromatography tandem mass spectrometry (UHPLC-MS/MS) analysis of psychoactive drugs in oral fluid. Epub 2019 Mar 16 J Pharm Biomed Anal 2019 Jun 5;170: 63–7. https://doi.org/10.1016/j.jpba.2019.02.039. PMID: 30904741.

Maurer HH. Position of chromatographic techniques in screening for detection of drugs or poisons in clinical and forensic toxicology and/or doping control. Clin Chem Lab Med 2004;42(11):1310–24. https://doi.org/10.1515/CCLM.2004.250. PMID: 15576292.

Maurer HH. Perspectives of liquid chromatography coupled to low- and high-resolution mass spectrometry for screening, identification, and quantification of drugs in clinical and forensic toxicology. Ther Drug Monit 2010 Jun;32(3):324–7. https://doi.org/10.1097/FTD.0b013e3181dca295. PMID: 20418802.

Meyer MR, Maurer HH. Review: LC coupled to low- and high-resolution mass spectrometry for new psychoactive substance screening in biological matrices - where do we stand today? Anal Chim Acta 2016 Jul 13;927:13–20. https://doi.org/10.1016/j.aca.2016.04.046. PMID: 27237833.

Montesano C, Vannutelli G, Massa M, Simeoni MC, Gregori A, Ripani L, et al. Multi-class analysis of new psychoactive substances and metabolites in hair by pressurized liquid extraction coupled to HPLC-HRMS. Epub 2016 Aug 8 Drug Test Anal 2017 May;9(5): 798–807. https://doi.org/10.1002/dta.2043. PMID: 27448433.

Morini L, Bernini M, Vezzoli S, Restori M, Moretti M, Crenna S, et al. Death after 25C-NBOMe and 25H-NBOMe consumption. Epub 2017 Sep 5 Forensic Sci Int 2017 Oct;279:e1–6. https://doi.org/10.1016/j.forsciint.2017.08.028. PMID: 28893436.

Mortier KA, Dams R, Lambert WE, De Letter EA, Van Calenbergh S, De Leenheer AP. Determination of paramethoxyamphetamine and other amphetamine-related designer drugs by liquid chromatography/sonic spray ionization mass spectrometry. Rapid Commun Mass Spectrom 2002;16(9):865–70. https://doi.org/10.1002/rcm.657. PMID: 11948818.

Nakahara Y. Hair analysis for abused and therapeutic drugs. J Chromatogr B Biomed Sci Appl 1999 Oct 15;733(1–2):161–80. https://doi.org/10.1016/s0378-4347(99)00059-6. PMID: 10572981.

Our World in Data. https://ourworldindata.org/grapher/death-rates-substance-disorders?tab=chart&country=USA~OWID_WRL~PRT.

Peters FT. Recent developments in urinalysis of metabolites of new psychoactive substances using LC-MS. Bioanalysis 2014 Aug;6(15):2083–107. https://doi.org/10.4155/bio.14.168. PMID: 25322784.

Pichini S, Altieri I, Zuccaro P, Pacifici R. Drug monitoring in nonconventional biological fluids and matrices. Clin Pharmacokinet 1996 Mar;30(3):211–28. https://doi.org/10.2165/00003088-199630030-00003. PMID: 8882302.

Rego X, Oliveira MJ, Lameira C, Cruz OS. 20 years of Portuguese drug policy - developments, challenges and the quest for human rights. Subst Abuse Treat Prev Pol 2021;16:59.

Rocchi R, Simeoni MC, Montesano C, Vannutelli G, Curini R, Sergi M, et al. Analysis of new psychoactive substances in oral fluids by means of microextraction by packed sorbent followed by ultra-high-performance liquid chromatography-tandem mass spectrometry. Epub 2017 Nov 27 Drug Test Anal 2018 May;10(5):865–73. https://doi.org/10.1002/dta.2330. PMID: 29078252.

Rojek S, Maciów-Głąb M, Kula K, Romańczuk A, Synowiec K. Forensic-psychiatric relationships in the context of forensic medical examination of new psychoactive substance-related deaths. English Arch Med Sadowej Kryminol 2020;70(4):202–21. https://doi.org/10.5114/amsik.2020.105019. PMID: 34431645.

Sadones N, Capiau S, De Kesel PM, Lambert WE, Stove CP. Spot them in the spot: analysis of abused substances using dried blood spots. Bioanalysis 2014 Aug;6(17):2211–27. https://doi.org/10.4155/bio.14.156. PMID: 25383733.

Saitman A, Park HD, Fitzgerald RL. False-positive interferences of common urine drug screen immunoassays: a review. Epub 2014 Jul 1 J Anal Toxicol 2014 Sep;38(7):387–96. https://doi.org/10.1093/jat/bku075. PMID: 24986836.

Salomone A, Gazzilli G, Di Corcia D, Gerace E, Vincenti M. Determination of cathinones and other stimulant, psychedelic, and dissociative designer drugs in real hair samples. Epub 2015 Dec 17 Anal Bioanal Chem 2016 Mar;408(8):2035–42. https://doi.org/10.1007/s00216-015-9247-4. PMID: 26680593.

Sundström M, Pelander A, Angerer V, Hutter M, Kneisel S, Ojanperä I. A high-sensitivity ultra-high performance liquid chromatography/high-resolution time-of-flight mass spectrometry (UHPLC-HR-TOFMS) method for screening synthetic cannabinoids and other drugs of abuse in urine. Epub 2013 Aug 18 Anal Bioanal Chem 2013 Oct;405(26):8463–74. https://doi.org/10.1007/s00216-013-7272-8. PMID: 23954996.

Sundström M, Pelander A, Ojanperä I. Comparison between drug screening by immunoassay and ultra-high performance liquid chromatography/high-resolution time-of-flight mass spectrometry in post-mortem urine. Epub 2014 Jun 20 Drug Test Anal 2015 May;7(5):420–7. https://doi.org/10.1002/dta.1683. PMID: 24953563.

Thornton SL, Lo J, Clark RF, Wu AH, Gerona RR. Simultaneous detection of multiple designer drugs in serum, urine, and CSF in a patient with prolonged psychosis. Epub 2012 Nov 20 Clin Toxicol 2012 Dec;50(10):1165–8. https://doi.org/10.3109/15563650.2012.744996. PMID: 23163617.

United Nations. COVID pandemic fuelling major increase in drug use worldwide: UN report. UN News; June 24, 2021. https://news.un.org/en/story/2021/06/1094672.

UNODC. Early warning advisory on new psychoactive substances. Home page; 2021. Accessed 2021 July 4, https://www.unodc.org/LSS/Home/NPS.

Vaiano F, Busardò FP, Palumbo D, Kyriakou C, Fioravanti A, Catalani V, et al. A novel screening method for 64 new psychoactive substances and 5 amphetamines in blood by LC-MS/MS and application to real cases. Epub 2016 Jul 7 J Pharm Biomed Anal 2016 Sep 10;129:441–9. https://doi.org/10.1016/j.jpba.2016.07.009. PMID: 27490334.

Wu AH, Gerona R, Armenian P, French D, Petrie M, Lynch KL. Role of liquid chromatography-high-resolution mass spectrometry (LC-HR/MS) in clinical toxicology. Epub 2012 Aug 13 Clin Toxicol 2012 Sep;50(8):733–42. https://doi.org/10.3109/15563650.2012.713108. PMID: 22888997.

Wu AH, Colby J. High-resolution mass spectrometry for untargeted drug screening. Methods Mol Biol 2016;1383:153–66. https://doi.org/10.1007/978-1-4939-3252-8_17. PMID: 26660184.

Yuan C, Chen D, Wang S. Drug confirmation by mass spectrometry: identification criteria and complicating factors. Epub 2014 Aug 27 Clin Chim Acta 2015 Jan 1;438:119–25. https://doi.org/10.1016/j.cca.2014.08.021. PMID: 25182671.

CHAPTER

5

NPS screening

Screening platforms

Presumptive screening in drug or new psychoactive substance (NPS) analysis primarily aims to maximize diagnostic sensitivity to catch all presumptive drugs or drug classes present in the sample. Screening assays are generally qualitative or semiquantitative at best. They can produce false positives. Therefore, the succeeding confirmatory step consists of analytical tests that optimize diagnostic specificity to sort true positives from false positives. Confirmatory analysis also allows the quantification of the specific drug identified in a sample.

The current screening platforms are colorimetric assays (spot tests), immunoassays, activity-based assays, and mass spectrometry (MS)–based assays (Graziano et al., 2019). Immunoassays are the most popular, especially in clinical setting. Activity-based assays show promise of overcoming the huge challenge imposed by the rapidly evolving molecular structures of NPS; however, these are currently rare. An activity-based assay is described in Chapter 4. MS-based assays are more commonly used for confirmation and will be discussed in the next two chapters.

Colorimetric assays (spot tests)

These are based on chemical reactions that target specific functional groups of a drug or class and produce a colored end product (usually transition metal complexes or charged organic species) that can be distinguished from the test reagent and the sample. Because a functional group can be shared by many compounds, including drugs or metabolites that may or may not be related to the target drug, colorimetric assays have limited specificity and are at best useful in ruling out the presence of a

Designer Drugs
https://doi.org/10.1016/B978-0-12-811764-4.00008-2

drug. However, the simplicity of most of these tests offers portability, rapidity, minimal or no sample preparation, low cost, and minimal expertise requirement in result interpretation. They are also very useful in some settings such as roadside, law enforcement, workplace, and emergency department where there is a need to quickly rule out intoxication. In the past decade, colorimetric assays have been most commonly used on seized drug materials according to an international survey conducted by the UNODC (International Collaborative Exercises, 2015, 2016).

Commercially available spot test kits target either a single drug, a class, or several classes. Most are developed for common traditional recreational drugs (TRD), and only a few screen for NPS. They are typically used for drug products (capsule, tablet, liquid, etc.) and have been applied in biological samples in very few instances. An example of a general screen is the Marquis test, formulated in 1896 by Edward Marquis. It is the most commonly used to test unknown substances. The reagent consists of formaldehyde with sulfuric acid and was originally developed for detecting morphine and other alkaloids (Marquis, 1896). Methanol is sometimes added to slow the rate of reaction for better observation of color change. The reagent reacts with various classes that contain basic amine groups. Each drug or several related drugs form a distinct color that allows identification. The time taken for the color change is also used for identity. Reaction of target drugs with formaldehyde under acidic conditions forms colored carbenium ion products (Fig. 5.1). Different colored end products are produced depending on the structure of the target drug. The Marquis is the primary spot test for

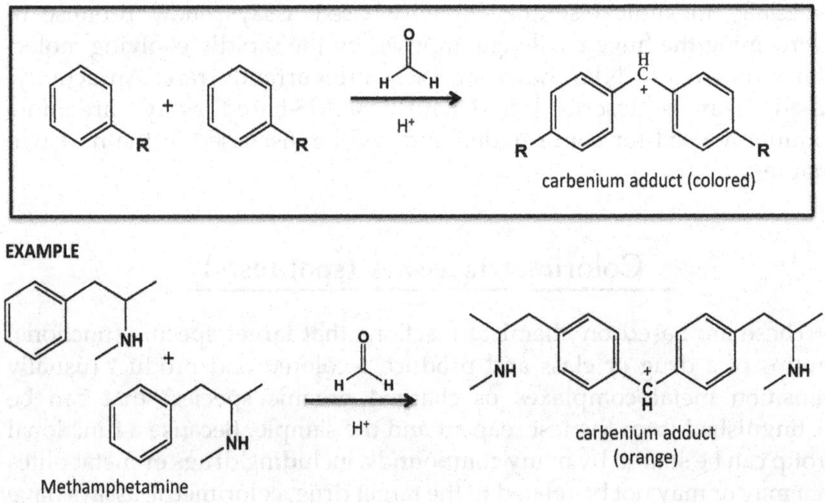

FIGURE 5.1 The Marquis test, a chemical reaction that produces a colored end product.

Ecstasy (3,4-methylenedioxymethamphetamine, MDMA) and is also used for opiates, phenethylamines, LSD, and some tryptamines. Some drugs and their color indicators (O'Neal et al., 2000):

- methamphetamine—deep red orange to dark reddish brown
- MDMA/MDA—deep purple to black
- methylphenidate—yellow orange
- dimethyltryptamine—orange
- heroin—purple
- morphine—intense reddish purple
- codeine—deep purple
- LSD—olive black

The Marquis test can detect some NPS, particularly amphetamines, phenethylamines, synthetic cathinones, tryptamines, and piperazines. Examples are:

- 2C-B/2C-I—greenish yellow in 30 min
- DOB—olive green to yellow
- 25I-NBOMe—orange
- methylone/butylone/MDPV—yellow
- benzphetamine—deep reddish brown
- 5-MeO-MiPY—light brown

Other spot tests can be functional-group-selective or drug-class-selective. Table 5.1 summarizes the most common spot tests and their target drug or drug class. Philip and Fu published a good review on the chemistry (Philp and Fu, 2018).

Reactivity toward spot tests has been tested for a limited number of NPS, including synthetic cannabinoids, synthetic cathinones, amphetamines, phenethylamines, piperazines, tryptamines, and Kratom alkaloids (Tsumura et al., 2005; Toole et al., 2012; Drug Chemistry Section, 2014; UNODC, 2015; Cuypers et al., 2016). Table 5.1 shows some examples.

Because of their lack of specificity, the primary disadvantage of spot tests is the rate of false positives (Binette and Pilon, 2013). Moreover, because color perception is subjective, interpretation of results can vary, especially when the drug concentration in a sample approaches the detection limit or adulterants are present to obscure the color endpoint (Wolchover, 2012).

Immunoassays

These presumptive screens, done primarily in urine, are the most widely used in the clinical setting. They are commercially available as kits incorporated in laboratory autoanalyzers. Most clinical labs offer a drugs-

TABLE 5.1 Common colorimetric tests used in drug screening.

Test reagent	Target drug or drug class	Reacting NPS
Chen–Kao	Ephedrine, pseudoephedrine, norephedrine	Synthetic cathinones
Cobalt thiocyanate	Cocaine, methaqualone, phencyclidine	Synthetic cathinones
Dille–Koppanyi	Barbiturates	Synthetic cathinones
Duquenois–Levine	Cannabinoids in cannabis	Kratom alkaloids
Erlich's	Ergot alkaloids, LSD	Synthetic cathinones
Fast blue B	Cannabinoids in cannabis	
Ferric chloride	Phenols	Synthetic cathinones
Gallic acid	Amphetamines	Synthetic cathinones
Liebermann	Phenols, substituted aromatic rings	Synthetic cathinones and synthetic cannabinoids
Mandelin's	Amphetamines, antidepressants	Amphetamines, phenethylamines, synthetic cathinones, piperazines, synthetic cannabinoids
Marquis	Broad spectrum: mostly opium alkaloids, amphetamines, mescaline, methadone	Amphetamines, phenethylamines, synthetic cathinones, piperazines, synthetic cannabinoids
Mecke	Opium alkaloids and phencyclidine	Amphetamines, phenethylamines, synthetic cathinones, piperazines, synthetic cannabinoids
Scott's	Cocaine and methadone	Amphetamines, phenethylamines, synthetic cathinones, piperazines, synthetic cannabinoids

TABLE 5.1 Common colorimetric tests used in drug screening.—cont'd

Test reagent	Target drug or drug class	Reacting NPS
Simon's	Distinguishes primary and secondary amines; methamphetamine, MDMA, and MDE	Amphetamines, phenethylamines, synthetic cathinones, piperazines, synthetic cannabinoids
Zimmerman	Benzodiazepines	Synthetic cathinones
Zwikker	Barbiturates	Synthetic cathinones

of-abuse screen, a panel that consists of assays for the most common TRD: amphetamines, cocaine, opiates, THC and its metabolites, and benzodiazepines. Depending on a hospital's patient population and resources, other immunoassays that test for fentanyl, oxycodone/oxymorphone, methadone, buprenorphine, phencyclidine (PCP), tricyclic antidepressants, and barbiturates can be added to the panel. In rare instances, assays for synthetic cannabinoids and synthetic cathinones may also be available. These tests require no sample preparation, are easy to perform, and have autoanalyzers that require no specialized training for data interpretation. Although not reliably quantitative, the results are expressed in concentration with respect to a standard drug. Interpretation of whether a sample is presumptive positive for a drug or class is based on a cutoff concentration. Because the antibody used can cross-react with other compounds in the sample, the result reflects all other compounds that bind to the antibody, not just the target analyte. Nevertheless, immunoassays generally provide rapid class-specific results to support emergency toxicology testing, workplace drug testing, detoxification, and rehabilitation (Gerona and French, 2022).

Various designs or formats are used in developing immunoassays. For those used to detect drugs, either the antigen or the antibody is labeled to allow detection and measurement of the antigen–antibody interaction at high sensitivity. The most common labels are enzymatic (e.g., horseradish peroxidase, alkaline phosphatase, and glucose-6-phosphate dehydrogenase), fluorescent (fluorescein), and chemiluminescent (luminol).

Immunoassays (IAs) can be competitive or noncompetitive (Fig. 5.2). If competitive, a labeled antigen competes with the unlabeled antigen (analyte) in a sample for binding sites in a limited, defined amount of the antibody. After separation (usually by centrifugation), the labeled antigen–antibody complex is measured by a technique that depends on the type of label carried by the antigen. The intensity of the signal measured from the complex is inversely related to the concentration of the unlabeled antigen in the sample. If noncompetitive, a labeled antibody

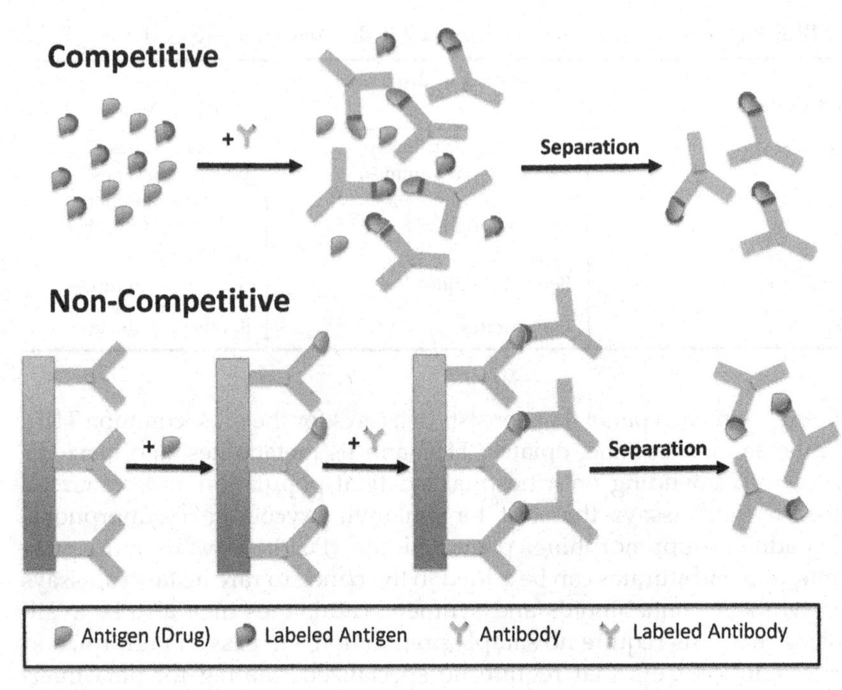

FIGURE 5.2 Differences between competitive and noncompetitive immunoassay.

reagent is used to detect the antigen. After the reaction is allowed to proceed, the labeled antibody–antigen complex is separated and the signal from the label is measured. The concentration of the complex formed is directly proportional to the amount of antigen in the sample.

The sandwich immunoassay is an example of noncompetitive IA (Fig. 5.2). Antigens in the sample are collected by using immobilized unlabeled antibody. After washing unreacted compounds in the mixture, the labeled antibody reagent is added, which then targets another structural motif of the analyte. The amount of labeled antibody that binds the captured antigen by the immobilized unlabeled antibody is then measured to determine the amount of antigen in the sample.

The tests used for drug analysis are variants of the competitive type. The two most common are the cloned enzyme donor immunoassay (CEDIA) and the enzyme-multiplied immunoassay (EMIT). These formats are applied to screening common drugs of abuse and distributed by several commercial vendors as test kits that can be incorporated with autoanalyzers. In the CEDIA (Fig. 5.3), two fragments of an enzyme (e.g., horseradish peroxidase) are used, which are individually inactive. Under the right conditions, the two fragments spontaneously assemble in solution to form the active enzyme. One of the fragments is used to label a target drug molecule. The fragment-labeled drug competes with the free

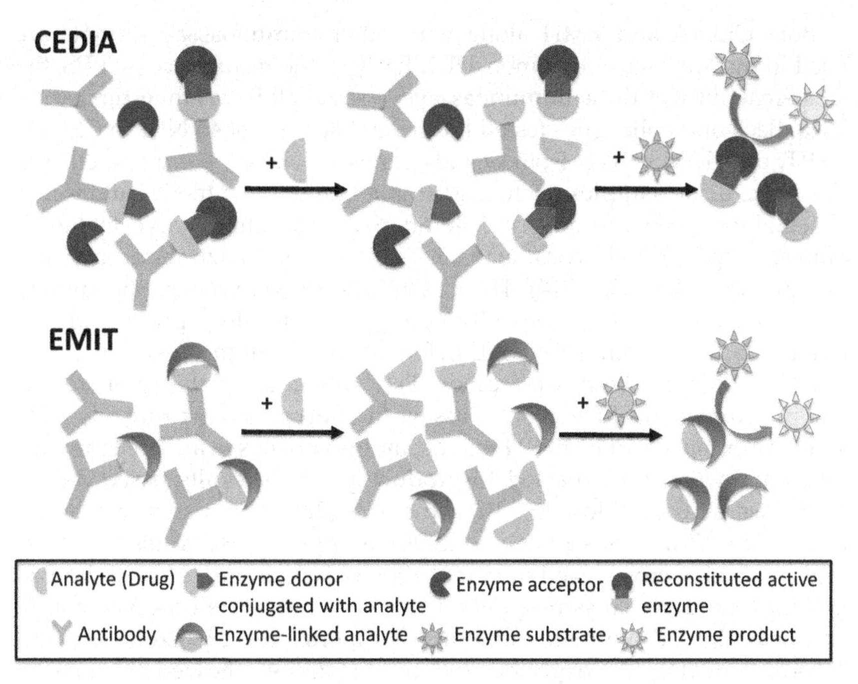

FIGURE 5.3 Common competitive immunoassay formats used in drug screening.

drug in a sample for binding site in the antibody that is added in the reaction mixture. At the end of the analysis, the fragment-labeled drug reagent that was outcompeted combines with the other enzyme fragment to form the active enzyme. Hence, when the appropriate substrate of the enzyme is added to the assay, an enzyme–substrate reaction occurs, forming a product that is then measured. The amount of the product formed is directly proportional to the amount of drug in the sample.

In the EMIT (Fig. 5.3), a known quantity of the target drug is labeled with an enzyme (e.g., glucose-6-phosphoate dehydrogenase) and is used as a reagent. The enzyme–drug complex binds with the antibody, inhibiting enzyme activity. When a sample containing the same drug is introduced, the free drug competes with the enzyme-labeled drug, allowing the release of the enzyme–drug complex. This increases enzyme activity and allows the formation of more product when the appropriate substrate is added. The amount of product formed from enzymatic activity is directly proportional to the amount of target drug in the sample. Other formats of competitive immunoassays have been developed for measuring drugs of abuse, including the kinetic interaction of microparticles in solution (KIMS) and the homogenous enzyme immunoassay (HEIA).

Both CEDIA and EMIT along with other immunoassay formats are used in the analysis of common TRD. Early in the resurgence of NPS, the cross-reactivity of these immunoassays to some NPS was investigated. In 2013, Beck and colleagues tested the cross-reactivity of 45 NPS to CEDIA, EMIT, and KIMS. NPS from a wide range of classes were spiked into drug-free urine samples, including amphetamines, synthetic cathinones, piperazines, benzofurans, aminoindane, piperidine, arycyclohexylamines, tryptamines, cocaine analogs, designer benzodiazepine, and an opioid (Beck et al., 2014). The CEDIA assays were for amphetamine/Ecstasy, cocaine, PCP, benzodiazepines, barbiturates, propoxyphene, opiates, and buprenorphine; EMIT for amphetamines, cocaine, and opiates; KIMS for benzodiazepines, THC, cocaine, opiates, methadone, and propoxyphene. Eleven of the assays did not cross-react with any NPS. With the exception of EMIT-2 Plus cocaine and opiates in all three formats, all assays were not expected to produce positive results based on the NPS classes tested. Close to 50% of the NPS (22) did not cross-react with any of the 17 assays surveyed. CEDIA amphetamine/Ecstasy had the highest level of cross-reactivity. At a cutoff of 500 ng/mL, 24 NPS showed greater than 1.5% cross-reactivity. Cross-reactivity was observed for 10 amphetamines, 7 synthetic cathinones, 4 piperazines, 2 benzofurans, and 1 cocaine analog. As expected, the amphetamines showed the highest levels of cross-reactivity, while the piperazines and synthetic cathinones showed <9% and <6%, respectively.

Only a few of the NPS showed cross-reactivity to the cocaine, PCP, and benzodiazepine assays. Etizolam, a designer benzodiazepine, cross-reacted with both benzodiazepine assays. Alpha-dinopropiophenonepyrroli, but not dimethocaine, cross-reacted with the cocaine assay, and desoxypipradol, but not methoxetamine, cross-reacted with the PCP assay.

The assays were also used in actual urine samples previously tested for NPS by liquid chromatography–tandem MS (LC-MS/MS). With the LC-MS/MS data as the gold standard, the CEDIA amphetamine/Ecstasy assay was in agreement in 62 of the 283 (22%) confirmed NPS. The assay was in agreement in 26 of 69 (38%) amphetamines and 27 of 174 (16%) synthetic cathinones confirmed. These results confirmed that a large number of NPS cannot be detected by common immunoassays, a high false-negative rate. Still, some NPS can be detected. So if a confirmatory panel does not include NPS, false-positive results from some assays can be potentially ascribed to NPS. Other investigations on the cross-reactivity of commercial immunoassays to a broad range of NPS were conducted by other groups (Regester et al., 2015; Mercolini and Protti, 2016). These studies reproduced the observations of Beck and colleagues.

Synthetic cannabinoids

These compounds comprise the largest and most diverse NPS class. The breadth of structural scaffolds has presented a significant challenge to conventional drug screening by immunoassay. Because of their significant structural difference from THC and its metabolites, practically no synthetic cannabinoids were detectable in the regular cannabinoid assay. This served as impetus for the development of commercial assays specifically targeting synthetic cannabinoids. The rapid evolution of structures, however, limited the temporal utility of these assays, so different iterations were needed. Keeping the relevance of a synthetic cannabinoids assay is quite a challenge, because development can be significantly slower than the evolution of structures. IA development can take 1–2 years, while the release of new structures can happen in a couple of months.

Immunalysis and Randox were the first companies to develop synthetic cannabinoid assays. The former has the K2 kit as part of their HEIA series, and the latter has the enzyme-linked immunosorbent assay (ELISA). The first generation targeted common naphthoylindoles such as JWH-018, JWH-073, and their metabolites. They found utility early on when naphthoylindoles were the most popular drugs in herbal incense products (2012–13). In a year or two, the market shifted to other scaffolds. Immunalysis responded with a second-generation HEIA K2 kit targeting the cyclopropylindoles UR-144 and XLR-11. In another couple of years, the third generation targeted the indazoles AB-PINACA and ADB-PINACA (Fig. 5.4).

Barnes and colleagues evaluated the sensitivity, specificity, and efficiency of the first Immunalysis HEIA K2 kit for JWH-018-N-pentanoic acid in urine samples previously analyzed by LC-MS/MS (Barnes et al., 2014). At a cutoff of 10 ng/mL, the sensitivity (ability to detect true positives), specificity (ability to detect true negatives), and diagnostic efficiency were 92.2%, 98.1%, and 97.4%, respectively. These results speak well for the assay. However, when cross-reactivity was tested with 74 other synthetic cannabinoids, 57 showed little (<10%) or none. Only 13 showed ≥50% cross-reactivity, and these were mostly naphthoylindoles. As expected, the more current synthetic cannabinoids at the time of testing were missed. This high false-negative rate for the more relevant compounds is quite problematic and clearly demonstrates the problem of using immunoassays in screening a rapidly evolving NPS class.

Another group developed and validated 96-well ELISA assays designed to detect JWH-018 and JWH-250 separately (Arntson et al., 2013). Both compounds are aminoalkylindoles, but the former is a naphthoylindole and the latter is a phenylacetylindole (Fig. 5.4). Both assays were designed to identify metabolites, using 5-hydroxy-JHW-018

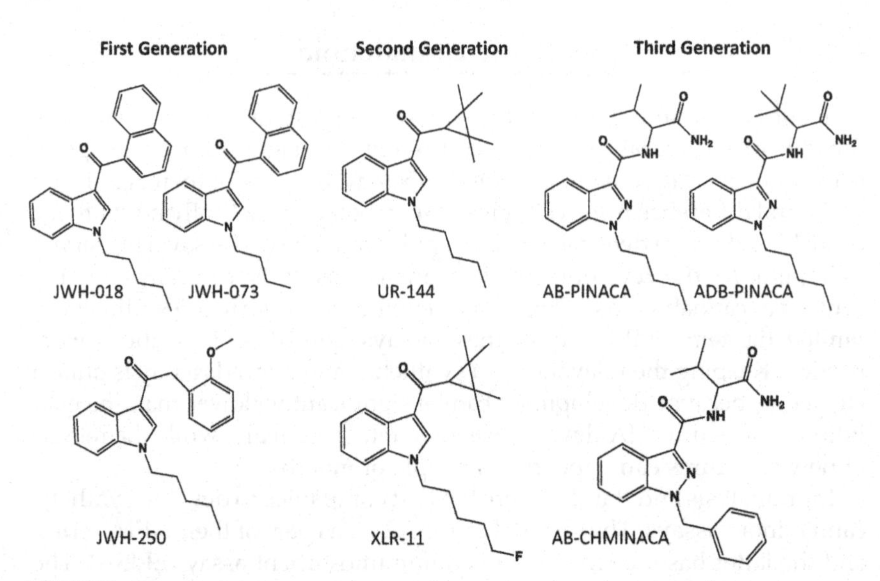

FIGURE 5.4 Molecular evolution of target synthetic cannabinoids in designing immunoassays.

and 4-hydroxy-JWH-250 as calibrators. The assays were applied to 114 urine samples for JWH-018 and 84 for JWH-250, using an LC-MS/MS confirmatory assay for the metabolites of JWH-018, JWH-019, JWH-073, JWH-250, and AM-2201 as gold standards. Accuracy was 98% and sensitivity was >95% for both assays. Similar to the K2 HEIA kit, however, both assays showed limited cross-reactivity to other synthetic cannabinoids, even among other aminoalkylindoles with slightly different scaffolds. The JWH-018 assay showed cross-reactivity of ≥10% to metabolites of some naphthoylindoles, including JWH-018, 19, 22, 73, 81, 122, 200, and 398 and AM-2201, but no or <1% cross-reactivity to 33 other synthetic cannabinoids, including 10 other JWH compounds, HU-210, UR-144, and XLR-11, among others. It registered only 5% cross-reactivity to the JWH-018 parent relative to 5 ng/mL 5-hydroxy-JWH-018. The assay for JWH-250 cross-reacted with only hydroxylated and carboxypentyl metabolites of JWH-250. The parent itself showed <1% cross-reactivity relative to that of 4-hydroxy-JWH-250. These observations amplify the problematic nature of assays developed to a specific scaffold of synthetic cannabinoids.

The second generation of assays, designed to detect the cyclopropylindoles UR-144 and XLR-11, had the same characteristics as the first. Mohr and colleagues demonstrated very good accuracy, specificity, and sensitivity for an ELISA-based assay targeting the N-pentanoic acid metabolite of UR-144, which it shares with XLR-11 (Mohr et al., 2014).

However, the assay reacts only to the metabolites of UR-144 and XLR-11. The parent compounds themselves showed <10% cross-reactivity, and none was observed for aminoalkylindoles or any other synthetic cannabinoids.

Franz and colleagues applied both the first- and second-generation Immunalysis K2 HEIA kits to 547 and 200 urine samples, respectively, from inpatients of forensic-psychiatric clinics to evaluate cross-reactivity to newer generation synthetic cannabinoids (Franz et al., 2017). None of the 547 urine samples tested positive in the first-generation IA even though a separate LC-MS/MS analysis confirmed synthetic cannabinoids in 8% of the samples. With LC-MS/MS as the gold standard, an overall sensitivity of 2%, specificity of 99%, and diagnostic efficiency of 51% were demonstrated for the second-generation IA kit. The lack of synthetic cannabinoid detection by the first-generation IA and the wide disparity in sensitivity and specificity of the second-generation IA can be explained by the fact that the prevalent synthetic cannabinoids in the population at the time, 2015 to 2016, were indazoles. The assays were developed in 2012 and 2013. Samples containing only metabolites of AB-CHMINACA, AB-FUBINACA, ADB-CHMINACA, MDMB-CHMICA, and AM-2201 were not detected as positives by either. This further shows the very limited temporal utility of the assays.

With the inherent limitations on the applicability of synthetic cannabinoid immunoassays, laboratories were compelled to develop MS-based assays. Several LC-MS/MS and LC-HRMS methods have been published. Although a few targeted LC-MS/MS methods were developed, this approach runs into the same limitation as immunoassays: new synthetic cannabinoids appear too rapidly on the recreational market. Each time a method is updated, it must be revalidated, which can take longer than the elapsed time before a new synthetic cannabinoid is released.

What has been effective in responding to the rapid, continuous introduction of new synthetic cannabinoids and their molecular evolution are methods based on HRMS. This platform allows nontargeted data acquisition in which data on all organic compounds that are ionizable are collected during a sample run. The total ion chromatogram obtained from the run can then be analyzed in a targeted manner using databases of known drugs or NPS (targeted analysis and suspect screening) or in a nontargeted manner. The ability to collect data on all ionizable molecules and the opportunity to query the resulting data in a nontargeted manner facilitate discovery of previously unreported NPS. Addition of new targets in a drug database for screening can also be done faster, allowing the method to react and adapt in almost real time to the introduction of new NPS in the market. Details on LC-MS assays including LC-HRMS will be discussed in the next two chapters.

Stimulants, psychedelics, and dissociatives

These three standard pharmacological classes are intertwined in various NPS classes. Some NPS have both stimulant and psychedelic or dissociative properties; hence, their pharmacology and analysis are often discussed together. The classes that fall into these three categories are diverse, but they are similar in having either an alkyl amine or a heterocyclic amine in their structure. These include amphetamines, phenethylamines, synthetic cathinones, aminoindanes, benzofurans, piperazines, piperidines, cocaine analogs, tryptamines, and arylcyclohexylamines (Fig. 5.5).

Because the NPS in these pharmacological classes are structurally related to TRD such as methamphetamine, their cross-reactivity to relevant commercial drugs of abuse was investigated by Regester and colleagues in 2015 (Regester et al., 2015). The study probed 94 NPS comprised of phenethylamines, amphetamines, synthetic cathinones, aminoindanes, benzofurans, tryptamines, and arylcyclohexylamines. 19% −57% of the NPS cross-reacted: CEDIA DAU amphetamine/Ecstasy (57%), Siemens EMIT II Plus amphetamines (43%), Lin-Zhi methamphetamine enzyme (39%), Microgenics DRI phencyclidine (20%), and Microgenics DRI Ecstasy enzyme (19%). Because the CEDIA assay targets both amphetamines and methylenedioxyamphetamines, it showed the highest cross-reactivity rate. Amphetamines (especially the DOx compounds), benzofurans, aminoindanes, methylenedioxycathinones, piperazines, and phenethylamines without bulky substituents in the 4-position of the aromatic ring showed good cross-reactivity. A few tryptamines and

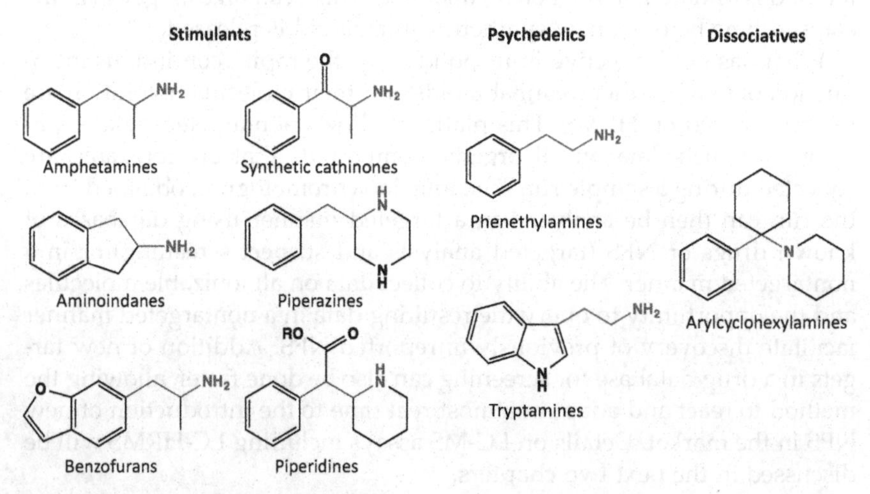

FIGURE 5.5 Molecular comparison of NPS stimulants, psychedelics, and dissociatives.

synthetic cathinones with alkyl substitutions in the aromatic ring also cross-reacted. The EMIT cross-reacted mostly with amphetamines, benzofurans, and aminoindanes. The Lin Zhi cross-reacted with amphetamines and a significant number of synthetic cathinones except for most of the α-pyrrolidinopropiophenones. The Microgenics DRI MDMA cross-reacted mainly with amphetamines, benzofurans, and a few aminoindanes. The Lin Zhi and Microgenics DRI did not cross-react with the DOx compounds, phenethylamines, and tryptamines. The Microgenics DRI PCP cross-reacted with arylcyclohexylamines and α-pyrrolidinopropiophenone. None of the assays designed for methamphetamine and/or MDMA cross-reacted with arylcyclohexylamines.

These results demonstrate that arylcyclohexylamines can be detected by the Microgenics DRI PCP assay. Amphetamines, aminoindanes, and benzofurans exhibit good cross-reactivity with the assays designed for methamphetamine and MDMA. There are wide differences on cross-reactivity to synthetic cathinone and piperazine, while very few phenethylamines and tryptamines cross-react with the assays. Although this may have some use in NPS screening, the challenge in extending these assays to NPS stimulants, psychedelics, and dissociatives is sorting out which NPS cross-react with which assay. The danger of false negatives with respect to drug use is high for relying on the results of immunoassay, especially because samples that produce negative screening results are not typically subjected to confirmatory assays. Other studies bolster this concern. An earlier study on the cross-reactivity of nine commercially available drugs of abuse assays to 2C, 2C-T, and DOx compounds demonstrated only 0.4% cross-reactivity even up to concentrations of 50,000 ng/mL, a concentration that far exceeds what is expected of forensic cases (Kerrigan et al., 2011). Another study, by Swortwood and colleagues, examined the cross-reactivity of 16 commercial immunoassays to 9 synthetic cathinones, 5 phenethylamines, 10 amphetamines, 3 piperazines, and 3 tryptamines (Swortwood et al., 2014). The cross-reactivity of all drugs was $<4\%$ in all assays targeting amphetamine and methamphetamine.

Given either less than satisfactory or defined cross-reactivity of most of the stimulants, psychedelics, and dissociatives to existing assays, new ones were designed to target specific NPS classes, similar to what was done for synthetic cannabinoids. Because synthetic cathinones are the second largest NPS class reported, an assay targeting this class was the first to become commercially available. Ellefsen and colleagues evaluated the Randox Drugs of Abuse V biochip assay (DOA-V), which contains two synthetic cathinone antibodies: Bath Salt I (BSI) targets methcathinone/mephedrone, and Bath Salt II (BSII) targets 3′,4′-methylenedioxy-pyrovalerone (MDPV)/3′,4′-methylenedioxy-α-pyrrolidinobutiophenone (MDPBP) (Ellefsen et al., 2014). They were validated and applied to 20,007 urine samples collected from the military that previously tested negative in screens for amphetamines, benzoylecgonine, opioids, cannabinoids, and PCP. The BSI antibody cross-reacted with 13 of 17 synthetic cathinones, the BSII antibody with 5 of 17, all of which are α-pyrrolidinopropiophenone. Neither antibody cross-reacted with pentylone and pentedrone. Generally, synthetic cathinones with the 4-methylphenyl-based structures cross-reacted with BSI, those with pyrrolidinyl-based structures with BSII. Both assays also showed high negative percent bias; that is, the concentrations of the quality control samples fortified with mephedrone and MDPV were lower than the acceptable criteria of ±20% of the target value.

One hundred six of the urine samples screened positive for synthetic cathinones: 75 in BSI and 40 in BSII. A confirmatory assay in an LC-HRMS method consisting of a panel of 28 synthetic cathinones confirmed only 4 of the 97 presumptive positive samples (4.1%). The large false-positive rate (95.9%) was attributed to the possibility that the synthetic cathinones present in the sample were not covered in the confirmatory assay, the analyte in the sample became unstable during the year elapsed between screening and confirmation, and the low cutoff for a positive screen set by the manufacturer was inappropriate. Because immunoassays do not react with single drugs, the presence of low concentrations of multiple synthetic cathinones in a sample can cumulatively cause a positive response. If the concentrations of the individual synthetic cathinones are below the limit of detection of the confirmatory assay, they will elicit a negative result. Bath salt products in the recreational drug market frequently contain two or more synthetic cathinones.

The Randox BSI kit was also used in a Swortwood et al. study (Swortwood et al., 2014). All 9 synthetic cathinones tested showed cross-reactivity at concentrations as low as 150 ng/mL, while all amphetamine-like compounds tested did not. However, the study did not test the kit with actual urine samples.

Because commercial immunoassays rarely cross-react with phenethylamines, assays designed to detect 2C compounds in lateral flow immunoassay (LFIA) and ELISA formats were developed by Sulakova and colleagues (Šuláková et al., 2019). LFIA is offered in a cassette that can be used for on-site testing, while the ELISA format was designed for clinical testing. Both were designed to detect 2C-B, a prototypical phenethylamine. Both have very good sensitivity (25 ng/mL for LFIA, 6 pg/mL for ELISA). Cross-reactivity with 12 phenethylamines, 10 synthetic cathinones, 8 cannabinoids, 6 piperazines, 1 aminoindane, and 1 tryptamine showed that ELISA is selective for 2C compounds (34.4%–100%). DOx compounds also showed some cross-reactivity (2%–12.7%). The other NPS tested did not. These results are promising for the selective screening of 2C compounds. However, the assays were not applied to clinical or forensic samples.

Aside from those designed for synthetic cathinones and phenethylamines, immunoassays that target piperazines as part of the biochip array technology (BAT) were developed by Randox. BAT is an IA testing platform that allows simultaneous analysis of multiple drugs, including NPS. It combines related IA tests in a single biochip and utilizes a single set of reagents, controls, and calibrators. All assays in the biochip array are based on competitive chemiluminescent immunoassay. Light signals generated from the test regions on the biochip during the assay are detected by a digital imaging technology and are compared with that from a calibration curve, allowing a readout of drug detection.

Castaneto and colleagues assessed the performance of three IA for piperazines included in the BAT panel, two targeting phenylpiperazine at cutoffs of 5 and 7.5 ng/mL (PNP I, II) and one targeting benzylpiperazine at a cutoff of 5 ng/mL (BZP) (Castaneto et al., 2015). Using LC-HRMS as the confirmatory assay, 20,017 randomly collected workplace urine samples were analyzed. Of the 840 that tested presumptive positive in the immunoassays, only 78 were confirmed by LC-HRMS, 72 of which are m-chlorophenylpiperazine (mCPP), a designer drug that is also a metabolite of the antidepressant prescription drug trazodone. Of the 206 randomly selected presumptive negative specimens, one confirmed positive for mCPP at 3.3 ng/mL and another confirmed positive for benzylpiperazine at 3.6 ng/mL. Raising the cutoff of the assays to 25, 42, and 100 ng/mL for PNP I, PNP II, and BZP, respectively, improved the specificity and efficiency of the assays.

Finally, perhaps because most of the amphetamines, benzofurans, and aminoindanes show cross-reactivity to a few conventional immunoassays developed for methamphetamine and MDMA, there is really no strong impetus to develop specialized immunoassays for these NPS classes.

New synthetic opioids and designer benzodiazepines

Fentanyl analogs (a major subclass of new synthetic opioids) and designer benzodiazepines closely resemble their TRD counterparts, and so are expected to have considerable cross-reactivity. Pettersson Bergstrand and colleagues tested CEDIA, HEIA, EMIT II Plus, and KIMS II in 2016 (Pettersson Bergstrand et al., 2017). Thirteen designer benzodiazepines (Fig. 5.6) were spiked into drug-free urine. All showed very good cross-reactivity with all four assays. The only exceptions were flutazolam, etizolam, and deschloroetizolam. Flutazolam, the most divergent in structure, showed 3%–13% with CEDIA, HEIA, and EMIT II Plus. Etizolam and deschloroetizolam showed 2% and 4% with HEIA. CEDIA registered the highest overall degree of cross-reactivity. With authentic

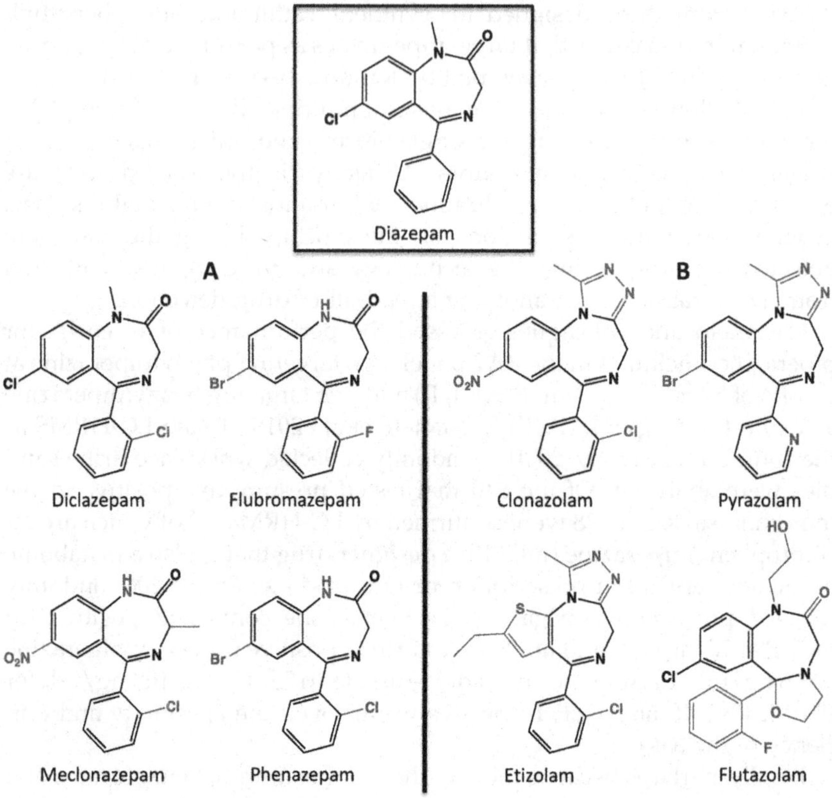

FIGURE 5.6 A sampling of designer benzodiazepines tested for cross-reactivity in commercial benzodiazepine immunoassays, compared with diazepam. Drugs in group A have structures that more closely resemble diazepam, while drugs in group B are structurally less similar.

urine samples previously confirmed to have designer benzodiazepines by UPLC-MS/MS, all assays showed a very high frequency of positive response at 200 ng/mL. CEDIA and KIMS had the best detectability. Both of these assays involve enzyme hydrolysis in their protocol, suggesting that these analytes are conjugated in urine.

Helander and colleagues investigated the cross-reactivity of the Thermo DRI Fentanyl Enzyme Immunoassay, ARK Fentanyl Assay, and Immunalysis Fentanyl Urine SEFRIA Drug Screening Kit with 13 fentanyl analogs (Fig. 5.7) (Helander et al., 2018). All three showed good cross-reactivity (33%–95%) except for 4-methoxybutyrfentanyl (all assays) and 2-fluorofentanyl (DRI). The assays were also applied to 58 authentic urine samples from intoxication cases previously confirmed to have

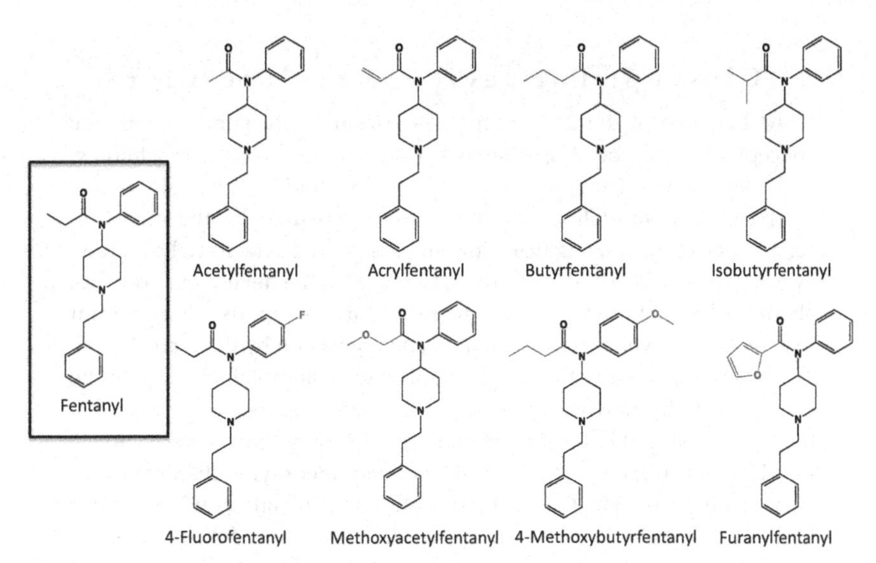

FIGURE 5.7 A sampling of fentalogs tested for cross-reactivity in commercial fentanyl immunoassays, compared with fentanyl.

fentanyl or fentalog by LC-HRMS. All three showed good detectability (79%–100%).

A study in the United States expanded the investigation to 30 fentalogs in nine LFIA, seven heterogenous, and three homogenous kits (Wharton et al., 2021). The selection of analytes was based on surveillance reports from the Drug Enforcement Administration and the National Forensic Laboratory Information Systems between 2015 and 2018. Similar to the findings of Helander and colleagues, the assays showed good detectability except for 4-methoxybutyrfentanyl and 3-methylfentanyl. Group additions to the piperidine ring and bulky groups or long alkyl chain modifications in the N-aryl or alkyl amide regions caused poor detectability.

In stark contrast with synthetic cannabinoids, these investigations convincingly showed that commercial immunoassays for fentanyl and benzodiazepines can be used for fentalogs and designer benzodiazepines and that development of separate assays for these classes of NPS is unnecessary. Thus, the presence of these NPS should always be considered in positive fentanyl or benzodiazepine screens.

No immunoassays or cross-reactivity studies for other subclasses of new synthetic opioids have been reported so far.

Closeup: The devil you can't test for

At the turn of the 21st century, physicians were familiar with the concept of drugs of abuse being heroin, cocaine, methamphetamine, and marijuana. Though these substances could cause significant morbidity and mortality, physicians were reassured by the ability to readily detect them using urine immunoassay drug screens (UDS). However, by the late 2000s, this relative comfort with the devil you know was about to be shattered, as there were increasing reports of significant toxicity from novel substances that evaded detection by the typical UDS.

As it happened, my training in medical toxicology coincided with the beginning of this new world of toxicology. Gone was the old paradigm of the UDS. Now patients were presenting with sometimes severe clinical toxicity, yet routinely available tests were not identifying the novel substances. In particular, the synthetic cathinones ("bath salts" or "plant food") and synthetic cannabinoid receptor agonists ("K2" or "Spice") were undergoing dramatic increases in abuse without being detected on standard screens. New and more powerful testing modalities were needed. Liquid chromatography–quadrupole time-of-flight mass spectrometry (LC-QTOF-MS) was there to answer the challenge.

This was illustrated early in my career when a 23-year-old male presented to the emergency department with severe psychomotor agitation requiring treatment with multiple intravenous sedatives. Though reported to have taken a bath salt, his UDS was negative. Fortunately, blood and urine samples from this patient could be sent to a lab at the University of California San Francisco that ran LC-QTOF/MS. It identified a known synthetic cathinone, 3,4-methylenedioxypyrovalerone (MDPV) and, for the first time clinically, another synthetic cathinone, 4-fluoromethcathinone (flephedrone) (Thornton, Gerona, et al., 2012).

A subsequent case was even more dramatic. An 18-year-old male presented with such agitation and psychosis that, along with multiple intravenous sedatives, he required a lumbar puncture to rule out other etiologies. Again, the UDS was negative, but this time blood, urine, and cerebrospinal fluid were sent for LC-QTOF/MS. The results demonstrated just how much the landscape of drug use had evolved, as four different substances—JWH-072, cannabicyclohexanol, 3′,4′-methylenedioxy-α-pyrrolidinopropiophenone (MDPPP), and methylenedioxyamphetamine (MDA)—were detected (Thornton, Lo, et al., 2012). The synthetic cannabinoid receptor agonist JWH-072 had not been previously reported in the medical literature.

As my career progressed, the use of NPS continued to evolve, but the challenge of detection remained. A case of prolonged encephalopathy (over 10 days) in a 16-year-old who reportedly ingested "Hot Molly"

Closeup: The devil you can't test for *(cont'd)*

highlighted this. The UDS was positive only for marijuana metabolites, but LC-QTOF/MS of the cerebrospinal fluid detected 1-(3,4-methylenedioxybenzyl)piperazine (MDBP), which had not been previously reported in the medical literature (Albadareen et al., 2015).

In my 10+ years as a medical toxicologist, NPS have evolved, but routine laboratory identification remains a common and constant challenge. Fortunately, I have also seen firsthand the utility that advanced detection methods such as LC-QTOF/MS can play in identifying these devils we do not know and allowing appropriate clinical and public health decisions to be made.

Stephen L. Thornton
Medical Director, Kansas Poison Control Center, KU Medical Center, Kansas City, KS 66160, United States

References

Albadareen R, Thornton S, Heshmati A, Gerona R, Lowry J. Unusually prolonged presentation of designer drug encephalopathy responsive to steroids. Pediatrics 2015 Jul;136(1): e246–8. https://doi.org/10.1542/peds.2015-0073. Epub 2015 Jun 8. PMID: 26055852.

Arntson A, Ofsa B, Lancaster D, Simon JR, McMullin M, Logan B. Validation of a novel immunoassay for the detection of synthetic cannabinoids and metabolites in urine specimens. J Anal Toxicol 2013 Jun;37(5):284–90. https://doi.org/10.1093/jat/bkt024. Epub 2013 Apr 26. PMID: 23625703.

Barnes AJ, Young S, Spinelli E, Martin TM, Klette KL, Huestis MA. Evaluation of a homogenous enzyme immunoassay for the detection of synthetic cannabinoids in urine. Forensic Sci Int 2014 Aug;241:27–34. https://doi.org/10.1016/j.forsciint.2014.04.020. Epub 2014 Apr 24. PMID: 24845968; PMCID: PMC4127333.

Beck O, Rausberg L, Al-Saffar Y, Villen T, Karlsson L, Hansson T, et al. Detectability of new psychoactive substances, 'legal highs', in CEDIA, EMIT, and KIMS immunochemical screening assays for drugs of abuse. Drug Test Anal 2014 May;6(5):492–9. https://doi.org/10.1002/dta.1641. Epub 2014 Mar 24. PMID: 24665024.

Binette MJ, Pilon P. Detecting black cocaine using various presumptive drug tests. Microgram J 2013;10:8.

Castaneto MS, Barnes AJ, Concheiro M, Klette KL, Martin TA, Huestis MA. Biochip array technology immunoassay performance and quantitative confirmation of designer piperazines for urine workplace drug testing. Anal Bioanal Chem 2015 Jun;407(16):4639–48. https://doi.org/10.1007/s00216-015-8660-z. Epub 2015 Apr 23. PMID: 25903022.

Cuypers E, Bonneure AJ, Tytgat J. The use of presumptive colour tests for new psychoactive substances. Drug Test Anal 2016;8:136–40. https://doi.org/10.1002/dta.1847.

Drug Chemistry Section. Technical procedure for preliminary color tests. Raleigh, North Carolina: North Carolina Department of Justice (NCDOJ); 2014.

Ellefsen KN, Anizan S, Castaneto MS, Desrosiers NA, Martin TM, Klette KL, et al. Validation of the only commercially available immunoassay for synthetic cathinones in urine:

Randox Drugs of Abuse V Biochip Array Technology. Drug Test Anal 2014 ;6(7–8): 728–38. https://doi.org/10.1002/dta.1633. Epub 2014 Mar 21. PMID: 24659527; PMCID: PMC4107059.

Franz F, Angerer V, Jechle H, Pegoro M, Ertl H, Weinfurtner G, et al. Immunoassay screening in urine for synthetic cannabinoids - an evaluation of the diagnostic efficiency. Clin Chem Lab Med 2017 Aug 28;55(9):1375–84. https://doi.org/10.1515/cclm-2016-0831. PMID: 28130957.

Gerona RR, French D. Drug testing in the era of new psychoactive substances. Adv Clin Chem 2022;111:217–63. https://doi.org/10.1016/bs.acc.2022.08.001. Epub 2022 Sep 23. PMID: 36427911.

Graziano S, Anzillotti L, Mannocchi G, Pichini S, Busardò FP. Screening methods for rapid determination of new psychoactive substances (NPS) in conventional and non-conventional biological matrices. J Pharm Biomed Anal 2019 Jan 30;163:170–9. https://doi.org/10.1016/j.jpba.2018.10.011. Epub 2018 Oct 4. PMID: 30316062.

Helander A, Stojanovic K, Villén T, Beck O. Detectability of fentanyl and designer fentanyls in urine by 3 commercial fentanyl immunoassays. Drug Test Anal 2018 Mar 12. https://doi.org/10.1002/dta.2382. Epub ahead of print. PMID: 29529707.

International Collaborative Exercises (ICE). Summary report seized materials. Vienna, Austria: UNODC; 2015.

International Collaborative Exercises (ICE). Summary report seized materials. Vienna, Austria: UNODC; 2016.

Kerrigan S, Mellon MB, Banuelos S, Arndt C. Evaluation of commercial enzyme-linked immunosorbent assays to identify psychedelic phenethylamines. J Anal Toxicol 2011 Sep;35(7):444–51. https://doi.org/10.1093/anatox/35.7.444. PMID: 21871153.

Marquis E. Über den Verbleib des Morphin im tierischen Organismus. Jurjev (Dorpat), Russia. University of Dorpat; 1896.

Mercolini L, Protti M. Biosampling strategies for emerging drugs of abuse: towards the future of toxicological and forensic analysis. J Pharm Biomed Anal 2016;130:202–19. https://doi.org/10.1016/j.jpba.2016.06.046.

Mohr AL, Ofsa B, Keil AM, Simon JR, McMullin M, Logan BK. Enzyme-linked immunosorbent assay (ELISA) for the detection of use of the synthetic cannabinoid agonists UR-144 and XLR-11 in human urine. J Anal Toxicol 2014 Sep;38(7):427–31. https://doi.org/10.1093/jat/bku049. Epub 2014 Jun 7. PMID: 24908262.

O'Neal CK, Crouch DJ, Fatah AA. Validation of twelve chemical spot tests for the detection of drugs of abuse. Forensic Sci Int 2000;109:189–201.

Pettersson Bergstrand M, Helander A, Hansson T, Beck O. Detectability of designer benzodiazepines in CEDIA, EMIT II Plus, HEIA, and KIMS II immunochemical screening assays. Drug Test Anal 2017 Apr;9(4):640–5. https://doi.org/10.1002/dta.2003. Epub 2016 Jul 1. PMID: 27366870.

Philp M, Fu S. A review of chemical 'spot' tests: a presumptive illicit drug identification technique. Drug Test Anal 2018 Jan;10(1):95–108. https://doi.org/10.1002/dta.2300. Epub 2017 Nov 10. PMID: 28915346.

Regester LE, Chmiel JD, Holler JM, Vorce SP, Levine B, Bosy TZ. Determination of designer drug cross-reactivity on five commercial immunoassay screening kits. J Anal Toxicol 2015;39:144–51. https://doi.org/10.1093/jat/bku133.

Šuláková A, Fojtíková L, Holubová B, Bártová K, Lapčík O, Kuchař M. Two immunoassays for the detection of 2C-B and related hallucinogenic phenethylamines. J Pharmacol Toxicol Methods 2019 ;95:36–46. https://doi.org/10.1016/j.vascn.2018.11.001. Epub 2018 Nov 24. PMID: 30481558.

Swortwood MJ, Hearn WL, DeCaprio AP. Cross-reactivity of designer drugs, including cathinone derivatives, in commercial enzyme-linked immunosorbent assays. Drug Test Anal

2014 ;6(7–8):716–27. https://doi.org/10.1002/dta.1489. Epub 2013 May 15. PMID: 23677923.

Thornton SL, Gerona RR, Tomaszewski CA. Psychosis from a bath salt product containing flephedrone and MDPV with serum, urine, and product quantification. J Med Toxicol 2012;8(3):310–3. https://doi.org/10.1007/s13181-012-0232-4. PMID: 22528592; PMCID: PMC3550171.

Thornton SL, Lo J, Clark RF, Wu AH, Gerona RR. Simultaneous detection of multiple designer drugs in serum, urine, and CSF in a patient with prolonged psychosis. Clin Toxicol 2012;50(10):1165–8. https://doi.org/10.3109/15563650.2012.744996. Epub 2012 Nov 20. PMID: 23163617.

Toole KE, Fu S, Shimmon R, Krayem N, Taflaga S. Colour tests for the preliminary identification of methcathinone and analogues of methcathinone. Microgram 2012;9:27–32.

Tsumura Y, Mitome T, Kimoto S. False positives and false negatives with a cocaine-specific field test and modification of test protocol to reduce false decision. Forensic Sci Int 2005;155:158–64.

United Nations Office on Drugs and Crime (UNODC). Recommended methods for the identification and analysis of synthetic cathinones in seized materials. New York, USA: United Nations; 2015. available at: https://www.unodc.org/unodc/en/scientists/recommended-methods-for-the-identification-and-analysis-of-synthetic-cathinones-in-seized-materials.html.

Wharton RE, Casbohm J, Hoffmaster R, Brewer BN, Finn MG, Johnson RC. Detection of 30 fentanyl analogs by commercial immunoassay kits. J Anal Toxicol 2021 Feb 13;45(2):111–6. https://doi.org/10.1093/jat/bkaa181. PMID: 33580693.

Wolchover N. Your colour red really could be my blue. 2012. Available at: http://www.livescience.com/21275-colour-red-blue-scientists.html.

CHAPTER

6

NPS confirmation using targeted analysis

Confirmatory analysis

Once a sample has a presumptive positive test, a confirmatory assay follows to identify and quantify a specific drug. Confirmatory assays optimize specificity and also have the best sensitivity. Mass spectrometry–based methods are the most popular. In clinical and forensic settings, liquid chromatography–tandem mass spectrometry (LC-MS/MS) is the gold standard for confirmatory and quantitative analysis (Peters, 2011; Viette et al., 2012). Liquid chromatography–high-resolution mass spectrometry (LC-HRMS) has started to gain popularity, but its main utility is in nontargeted testing, which will be explained in the next chapter.

Before LC-MS/MS, gas chromatography–mass spectrometry (GC-MS) was the primary method, and some labs still use it. Like LC-MS/MS and LC-HRMS, it has two separate but linked methods. Most of these techniques combine chromatography with MS because chromatography allows the separation of components based on their polarity, while MS facilitates the identification of components based on their mass to charge (m/z) property. Since most methods primarily identify ions with unit charge, the analyte's m/z is equivalent to the mass of their ionic form.

Because biological samples containing drugs and new psychoactive substances (NPS) are rarely gas in their natural state, analytes are converted to their gas phase prior to chromatographic separation in GC-MS. Hence, analytes need to be volatile and stable at high temperature. Not all analytes, including a significant number of drugs, are compatible with this requirement. Those that are not easily or stably converted to gas phase can be converted to derivatives that are volatile. Esterification is

Designer Drugs
https://doi.org/10.1016/B978-0-12-811764-4.00007-0

often used to derivatize the sample for GC. However, this adds a step that can affect precision and overall recovery of the analyte. If complete derivatization cannot be ensured, variations in the reaction efficiency can introduce imprecision. Losses during derivatization also decrease the analyte's recovery. All these disadvantages are overcome in LC-MS/MS, which is why it is favored by most labs for confirmatory testing.

Liquid chromatography—tandem mass spectrometry

LC-MS/MS (Fig. 6.1) combines the separation capability of chromatography based on the analytes' polarity differences with the sensitivity and specificity of mass analysis through triple quadrupole MS. Although LC-MS was first commercialized in the 1970s, the utility of LC-MS/MS in clinical and forensic testing was not exhaustively explored until the 1990s. By the turn of the century, LC-MS/MS had replaced GC-MS as the gold standard for quantitative drug analysis. LC has one or more mobile phase, a column, and a detector, while MS has an ion source, mass analyzer, and mass detector. In LC-MS/MS, the mass spectrometer is the LC's detector.

Liquid chromatography

In this separation technique, the components of a liquid sample partition between two immiscible phases, mobile and stationary, based on their relative polarity. LC is done by several mechanisms: partition, adsorption, affinity, ion exchange, and size exclusion. In LC-MS/MS, partition is primarily used. The mobile phase consisting of pure solvent or combination of solvents is pumped into a stationary phase, which consists of spherical or irregularly shaped silica microparticles (~2–5 μm) coated with specific functional groups that impart the phase's polarity. The particles are tightly packed in columns with narrow internal diameter, so the column acts as the stationary phase.

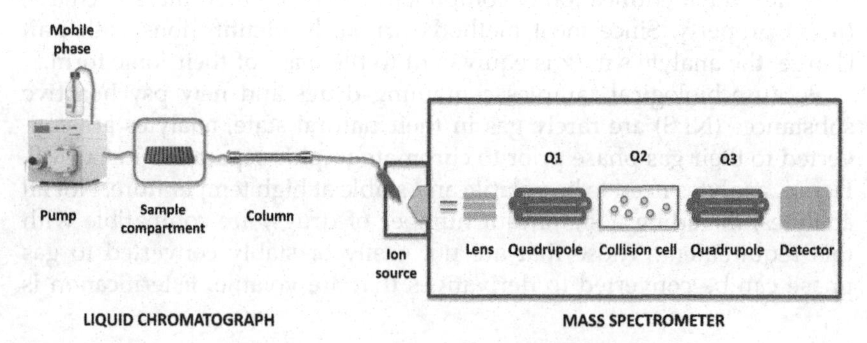

FIGURE 6.1 Schematic of LC-MS/MS.

Partition chromatography can be done in two modes based on the polarity of the stationary phase: normal phase or reverse phase. Reverse phase is most common in LC-MS, where the stationary phase is nonpolar and the mobile phase is polar. Typically, buffered or nonbuffered water, either by itself or mixed with polar solvents such as methanol and acetonitrile, is used as the mobile phase. Using a high-pressure pump, the mobile phase is delivered into the stationary phase, which consists of a column where the silica particles are coated with functionalized long-chain alkyl groups that may be 18 C (n-octadecyl, C18) or 8 C atoms (n-octyl, C8) long. The sample, typically 10–20 μL, is injected into the stream of the mobile phase, which carries it into the column. There the individual components of the liquid sample interact with the stationary phase, undergoing repeated sorption and desorption. The length of time a component is retained inside the column depends on its polarity. The more polar the component, the less affinity it has with the nonpolar or less polar stationary phase, thus the shorter time it is retained. By virtue of polarity differences, components of the liquid sample are separated based on how long each is retained in the column. The time required for a component of a liquid sample to elute off the column is its retention time. For reverse-phase chromatography, the more polar the component, the shorter the retention time. The polarities of stationary and mobile phases are reversed in normal phase chromatography.

The column is the most important part of LC. For drug testing, columns can be 30–100 mm (3–10 cm) long with internal diameters of 2.5–5 mm. The functional group attached to the silica particles can be modified in various ways to create columns that can efficiently separate broad or specific classes of compounds. Many different types of columns are available from various vendors; Table 6.1 shows examples. Once in use, the life span of a column depends on the sample matrix being regularly injected into it. Biological samples have more components than drug products, so they tend to be "dirtier." Typically, a column can be used for 2000 to 4000 injections. The older the column gets, the higher its baseline pressure becomes. This can cause slight shifts in the retention time of drugs in LC-MS/MS.

Elution of a sample component from a column is indicated by the detector, which measures a physical or chemical property of a compound that will distinctly identify and differentiate it from the other sample components. For ordinary LC, this property can be absorbance (ultraviolet or visible spectrophotometer), refractive index (refractometer), or fluorescence (fluorometer). In LC-MS/MS, this property is the m/z of the compound's ion; thus, the mass spectrometer acts a sophisticated detector for the LC.

Mass spectrometry

The mass spectrometer operates by converting analyte molecules in a sample into ions and sorting them according to the m/z of their

TABLE 6.1 Types of HPLC columns commonly used in LC-MS/MS.

Type	Composition	Typical mobile phase used
Normal phase	Silica (polar stationary phase)	Nonpolar solvents such as dichloromethane, hexane, chloroform, or a mixture of the solvents with diethyl ether
Reverse phase	Bonded hydrocarbons such as C8 and C18 (nonpolar stationary phase) and other hydrophobic functional groups	Polar solvents such as water–methanol or water–acetonitrile mixture
Ion exchange	Positively or negatively charged functional groups covalently bound to a solid matrix (cellulose, agarose, polymethacrylate, polystyrene, polyacrylamide)	Volatile salt solutions, typically ammonium-based (nonvolatile salts in aqueous solution are incompatible with MS)
Size exclusion	Porous particles made up of a combination of polysaccharides and silica	Buffered aqueous solution with volatile salt

corresponding ions; drugs in a sample are first ionized into either positive or negative, depending on the polarity applied for ionization. Ions can also have unit or multiple charges. Mass sorting of the resulting ions then follows. Additionally, components of the mass spectrometer can induce fragmentation of the parent ion of an analyte by collisions with inert gas producing the mass spectrum of the analyte. These analyses are achieved using three important MS components: ion source, mass analyzer, and detector (Fig. 6.2).

Ion source

This is the interface between the liquid chromatograph and the mass spectrometer. Two processes occur in it: conversion of the liquid sample that elutes off the column to gas, and ionization of sample components. In contrast to what happens in GC-MS, these processes occur at atmospheric pressure, and only soft ionization is used where the parent ion and a wide variety of fragment ion sizes, including larger ions, are produced. Various types of ion sources and ionization techniques are available for LC-MS.

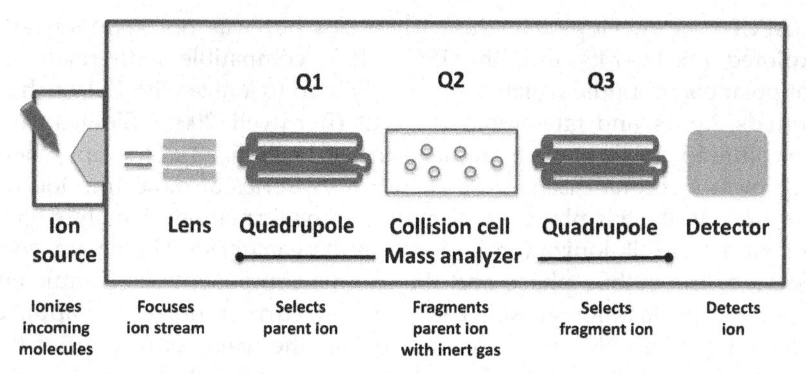

FIGURE 6.2 Components of a mass spectrometer and their functions.

The most common are electrospray ionization (ESI) and to a certain extent atmospheric pressure chemical ionization (APCI) and atmospheric pressure photoionization (APPI).

ESI was developed in 1984 and has since become the most commonly used ionization technique for drug analysis. This mode is compatible with slightly polar compounds, including endogenous metabolites, xenobionts (most prescription and recreational drugs), and peptides. Unlike other techniques, ESI produces multiply charged on top of singly charged ions (Rosenberg, 2003). Molecules from the liquid sample that elute off the LC are ionized by passing through a capillary that is charged between 3000 and 5000 V. Passage through the highly charged capillary nebulizes the liquid sample into small charged droplets that evaporate into a stream of gaseous charged particles because of the heat generated by the highly charged capillary and the application of heated inert gas such as nitrogen (Fig. 6.3) (Kebarle, 2000). Positive or negative voltage can be applied to the capillary, depending on whether positive or negative ions are desired for analysis.

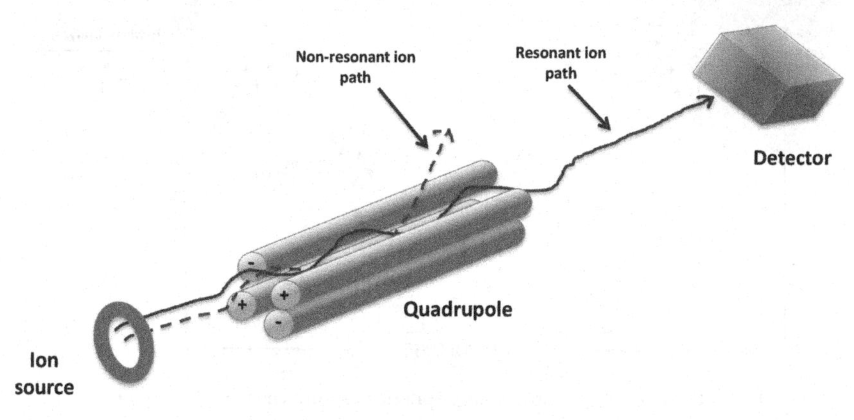

FIGURE 6.3 Schematic of electrospray ionization.

APCI was developed in the early 1970s but was not commercially explored for LC-MS until the 1990s. It is compatible with relatively nonpolar compounds, which are typically hard to ionize with ESI, such as steroids, lipids, and fat-soluble vitamins (Byrdwell, 2001). Eluates from the column pass through a capillary and are nebulized at the tip, where they meet a corona discharge that initiates a series of molecule–ion interactions in the gas phase, culminating in the formation of analyte ions. In contrast to ESI, ionization happens in the gas phase. The eluate, consisting of the mobile phase and the sample components, is completely vaporized prior to ionization. The discharge current ionizes the applied nitrogen gas into N_4^+, which in turn ionizes the main component of the evaporated mobile phase. The solvent ion formed then transfers the charge into or forms a charged adduct with the molecular components of the evaporated sample (Fig. 6.4). APCI usually produces singly charged ions (Rosenberg, 2003). Like ESI, ionization can be done in the positive or negative mode.

Mass analyzer

This is the primary component of the mass spectrometer that allows it to detect analytes. It takes ionized masses and separates them based on m/z. There are several types, defined by what the mass spectrometer is able to do. The most common are quadrupole, time of flight, ion trap, magnetic sector, and ion cyclotron. LC-MS/MS instruments use quadrupole and ion trap.

A quadrupole mass analyzer consists of four parallel metal rods with each opposing rod pair electrically connected (Fig. 6.5). Radiofrequency (RF) voltage is applied to one pair, direct current to the other. At a given

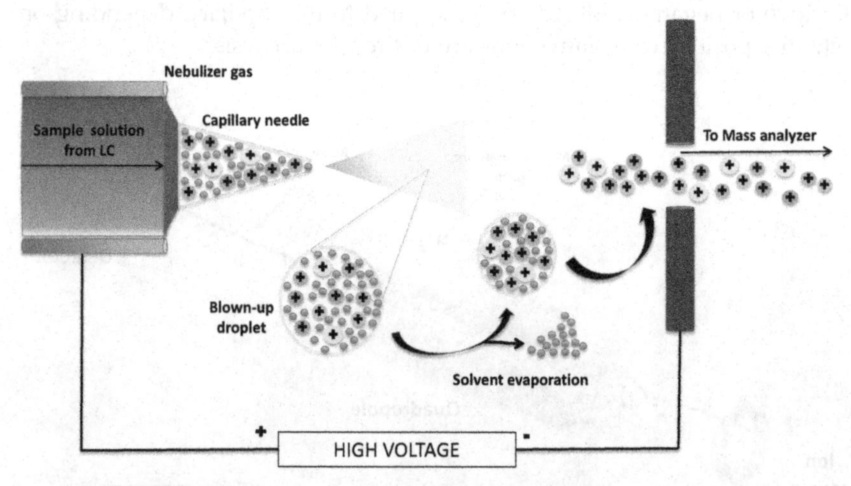

FIGURE 6.4 Schematic of atmospheric pressure chemical ionization.

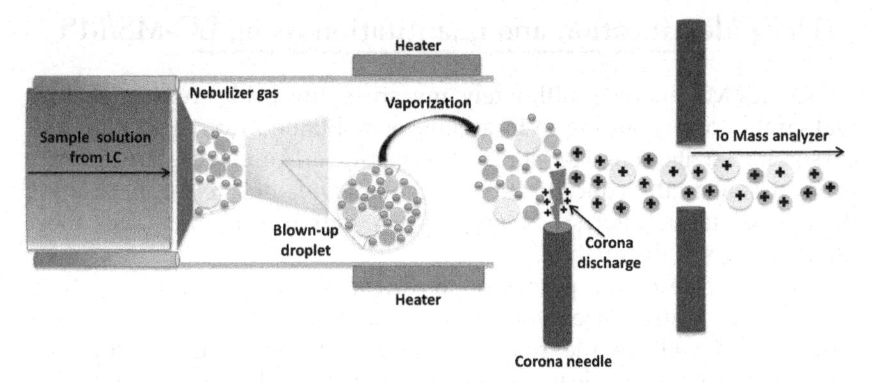

FIGURE 6.5 Schematic of a quadrupole mass analyzer.

RF and direct current combination, only ions of a particular m/z acquire a stable trajectory that allows its transmission through the quadrupole (resonant ion path). All other ions with unstable trajectories (nonresonant ion path) cannot pass the quadrupole and are sucked off in the vacuum applied in the mass spectrometer. By changing the applied direct current and RF combination, usually at a fixed ratio, the quadrupole can select a different m/z that can be transmitted to either the collision cell or the detector (Glish and Vachet, 2003).

An ion trap uses an oscillating electric field to store ions for release at a later time. Of the several types used in LC-MS/MS, the linear ion trap is most common. It creates a two-dimensional RF through a set of quadrupole rods to confine ions radially. Application of specific RF and voltages maintains the final position of the trapped ions within the center section of the trap. By adjusting the RF, voltage, and multifrequency resonance ejection waveforms, all ions except the desired ones are eliminated in preparation for mass analysis or fragmentation through the collision cell (Payne and Glish, 2005; March, 2009).

Mass detector

Ions carry the signal through the detector, which records the current generated. Because this current is weak, detectors typically consist of a multichannel plate detector or a secondary emission multiplier where the first dynode (converter plate) is flat to propagate or magnify the initial signal. The signal is then converted by a time-to-digital or analog-to-digital converter. Time-to-digital shows the time an event occurred (in this case, an ion hitting the detector); either the absolute time when the event occurred or the interval between two events is digitized. On the other hand, analog-to-digital converts an analog signal (e.g., sound, voltage, current) into a digital signal.

Drug identification and quantitation using LC-MS/MS

LC-MS/MS analysis ultimately measures the peak areas in a chromatogram corresponding to fragment ions obtained from a specific precursor ion at its characteristic retention time. These areas are directly proportional to the concentration of the analytes of interest, facilitating their quantitative analysis. Along with retention time, they also allow identification of the analyte.

The structure of the precursor ion dictates the specific way it fragments. The relative strengths of chemical bonds in a precursor ion's structure as well as the relative stability of the corresponding ions generated from their cleavage dictate their ease of fragmentation. The weaker the bond, the more easily it fragments, generating more of the corresponding fragment ion that forms from its cleavage; likewise, the more stable the generated ion, the more readily it forms. The relative abundance of fragment ions generated from a precursor ion is defined. This relative abundance can change according to the collision energy applied as some fragment ions further undergo fragmentation with higher collision energy. Thus, at a given collision energy, the ratio of the peak areas between two fragment ions is a constant number that can be used for the identification of their precursor ion. This ratio along with the retention time of the analyte's peak is used in confirming its identity. Generally, area ratios within 20% of the expected ratio and a retention time within 0.1–0.15 min of the expected are set as criteria for identifying or qualitatively confirming a drug in a sample.

As mentioned in the previous section, the fragment ion that reaches the detector is obtained from initially selecting a precursor ion in the first quadrupole, allowing it to fragment in the collision cell, and selecting the fragment ion in the second quadrupole. Thus, identification and quantification of the analyte is actually facilitated by the specific conversion of a precursor ion to a selected fragment ion. This conversion is referred to in LC-MS/MS as a transition. For LC-MS/MS methods, two transitions are used for each analyte of interest to allow its proper identification. For example, in the mass spectrum of methamphetamine (Fig. 6.6), two fragment ions (m/z 119.08498, m/z 91.05444) were obtained from its fragmentation at a collision energy of 12.5 eV. Thus, the transitions from the precursor ion (m/z 150.12752) to each of these fragment ions are incorporated in the LC-MS/MS method used to analyze methamphetamine.

The first transition forms the more abundant fragment ion and is used as the quantifier to optimize the method's sensitivity in quantitative analysis. The less abundant transition 2 is used as qualifier; along with the first transition, it allows the confirmation of methamphetamine through the characteristic peak area ratio of 4.22 between the two fragment ions or

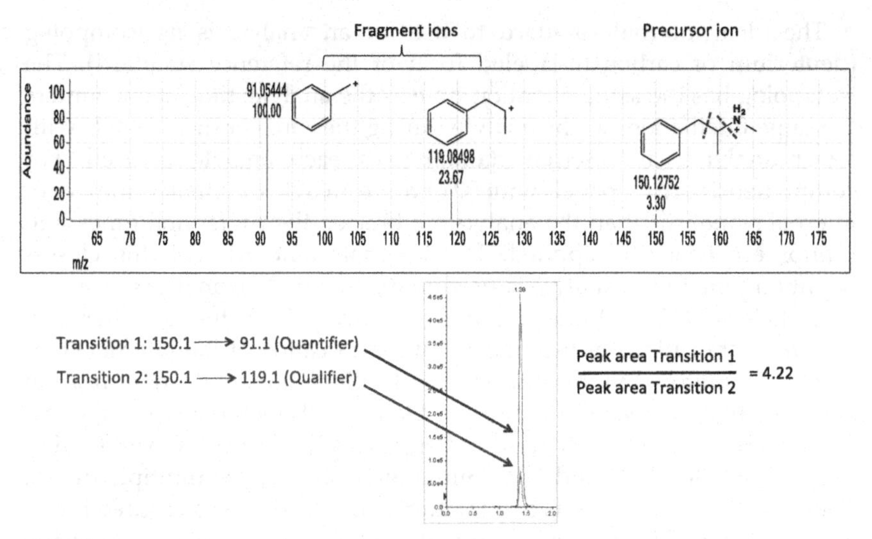

FIGURE 6.6 Mass spectrum of methamphetamine showing how fragment ions are used as transitions in facilitating its identification by LC-MS/MS.

transitions. At higher collision energies or for other drugs, there are usually more than two fragment ions obtained from the precursor ion. In such cases, the two most abundant fragment ions are usually selected as quantifier and qualifier.

For quantitative analysis, the use of external calibration is commonly applied in LC-MS/MS. The peak area measured for a confirmed analyte in a sample is compared with a calibration curve obtained from matrix blank samples spiked with various levels of known concentrations of the reference standard for the analyte (calibrant). Small fluctuations in electricity or the environment can affect the signal measured by the detector from the analyte in a calibrant. This can introduce random error in the peak area obtained for one or more calibrants. To exclude contributions from these random variations during the run and ensure that the change in peak area is directly proportional to the level of the reference standard in the calibrants, a known fixed concentration of an internal standard is also added in each calibration sample (and the test sample). The assumption is that any fluctuation incurred in measuring the signal of the reference standard proportionately affects the signal of the internal standard. The internal standard thus normalizes any random variation from run to run so that the only determinant in the changing peak area of the reference standard is its concentration. To accomplish this mathematically, the peak area ratios between the reference standard and the internal standard plotted against the concentrations of the calibrants are used to generate the calibration curve.

The ideal internal standard to use for an analyte is its isotopolog (deuterium or carbon-13-labeled form of the reference standard). The isotopolog has the same retention time as the analyte; thus, it encounters the same fluctuations as the analyte during the run. For drug panels with fewer analytes, an internal standard for each analyte is used. For comprehensive drug panels with 50 or more analytes, a smaller number of internal standards than the analytes to bracket their retention times into groups are generally applied. The internal standard with the closest retention time to the analyte is used in its quantitative analysis.

In LC-MS/MS methods for a single drug, both the qualifier and quantifier transitions of the analyte and a transition of the internal standard are monitored in the entire run to facilitate both drug identification and quantitative analysis. However, LC-MS/MS methods with a single analyte are less frequently used in drug analysis. Instead, most LC-MS/MS methods for TRD and NPS simultaneously analyze multiple drugs. Therefore, tens or hundreds of transitions may need to be measured for an analytical run. The process of monitoring multiple transitions during quantitative analysis using LC-MS/MS is referred to as multiple reaction monitoring (MRM), where each transition is treated as a reaction.

Because the mass spectrometer in an LC-MS/MS platform can only measure one transition at a time, it has to spend a specified amount of time—the dwell time—collecting data for each transition. To monitor all the transitions, the mass spectrometer has to cycle through all of them. The summation of dwell times to completely monitor all transitions in one pass is the method's cycle time. The data for fragment ions collected at each cycle comprise a data point in the corresponding chromatographic peak of the analyte. To allow creation of a Gaussian peak for an analyte and ensure good precision in its measurement, collection of a minimum of 10 data points is usually required. So for a typical peak in a chromatogram with a base width of 5 s (sec), a maximum cycle time of 500 ms (ms) is needed to get 10 data points.

For quantitative analysis, a dwell time of 5–10 ms is ideal to collect enough data for each transition in each cycle. This means only 50–100 transitions can be measured at a cycle time of 500 ms. Because two transitions need to be monitored for each analyte, only 25–50 analytes can reliably be measured in an LC-MS/MS method when the average base peak width is 5 s. This does not include internal standards, which in multidrug panels can be shared by multiple analytes. So when the number of analytes in a panel exceeds 50, scheduled MRM is typically applied, where the transitions of each analyte are only monitored close to its retention time, typically within 30 s. Instead of cycling through all the transitions at each cycle, only a fraction of overlapping transitions based on the analytes' retention times are monitored at any given time in the analysis. This significantly reduces the range of cycle times required in the analysis and makes possible the collection of 10 or more data points for each analyte's chromatographic peak. Dwell times and cycle times need to be optimized in an LC-MS/MS method once the transitions for each analyte are established.

Use in NPS analysis

Broad-spectrum NPS analysis

Because of the variety of NPS, analytical methods sensitive to multiple classes have been developed. The composition of these methods often reflects the number and variety of NPS available at the time in the region of the developer. Until 2017, most broad-spectrum tests were for synthetic cannabinoids, synthetic cathinones, and stimulants (amphetamines, phenethylamines, piperazines), the three most abundant NPS in the United States, Europe, Australia, and Japan. Most methods were developed in conventional biological matrices: serum or plasma, whole blood, and urine. A few were for hair.

Table 6.2 provides a sample of broad-spectrum NPS tests. The target biomarker is an important consideration. Like other recreational drugs, NPS are metabolized, so the parent may or may not be detectable in a biological matrix. Blood samples contain the parent of most NPS, although those that undergo rapid metabolism may be at very low levels requiring methods with sensitivity up to sub-ng/mL. A significant number of NPS can be detected in urine only through their metabolites. Urine contains the end products of metabolism of xenobionts, so for extensively metabolized NPS, unchanged parent may not be present. A few NPS are so rapidly metabolized that the parent is not detectable even in blood. This presents a challenge in developing analytical methods. Unlike traditional recreational drugs, the metabolism and pharmacokinetics of most NPS are unknown, so their primary metabolites may be unknown as well.

Using known metabolic pathways of similar recreational drugs, predictions can be made on potential NPS metabolites. They can be verified initially by HRMS through nontargeted data acquisition, notably using human liver microsomes. This method will be explained in more detail in the next chapter. Confirmation still requires that the reference standard for the putative metabolite be synthesized and run by the same LC-MS method used in the metabolic studies. The sheer number of NPS and the rapidity of their molecular evolution make a systematic study on the metabolism of every one of them impossible. Metabolic studies take time and effort. There are not enough laboratories worldwide to cover all circulating NPS and keep up with new structures. Nevertheless, targeted

TABLE 6.2 Selected published methods for the targeted analysis of multiple NPS classes.

Reference	Analytes	Matrix	Sample preparation	Highlights
Wohlfarth et al.(2010)	31 NPS	Plasma	Solid-phase extraction (SPE)	Rapid method; three transitions were used to distinguish isobaric compounds
Ambach et al. (2014)	64 NPS	Dried blood spot	Methanol extraction	Simple solvent extraction allows NPS detection in very small volume of blood
Odoardi et al. (2015)	78 NPS	Whole blood	Liquid–liquid microextraction (LLME)	Innovative sample preparation and rapid method; LLME requires minimal extracting solvent
Adamowicz and Tokarczyk (2016)	143 NPS	Whole blood	Protein precipitation	Simple sample preparation and rapid method; three transitions for each analyte monitored in scheduled MRM
Vaiano et al. (2016)	64 NPS	Serum	Protein precipitation	Simple sample preparation and rapid method
Lehmann et al. (2017)	69 NPS	Serum	Online solid-phase extraction	Fully validated quantitative method
Al-Saffar et al. (2013)	26 NPS	Urine	dilute-and-shoot	Simple preparation and rapid quantitative method

TABLE 6.2 Selected published methods for the targeted analysis of multiple NPS classes.—cont'd

Reference	Analytes	Matrix	Sample preparation	Highlights
Tang et al. (2014)	46 NPS and TRD	Urine	Enzymatic deconjugation and SPE	Mixed panel for NPS and TRD; three transitions for each analyte monitored in scheduled MRM
Strickland et al. (2019)	22 NPS	Urine	Enzymatic deconjugation and SPE	Very rapid LC-MS/MS method (4.5 min) with NPS targets updated based on surveillance testing
Boumba et al. (2017)	132 NPS	Hair	Acidic methanol extraction	Simple solvent extraction from pulverized hair
Nzekoue et al. (2021)	98 NPS	Hair	LLME in acidic and basic conditions	Solvent extraction from finely cut hair
Di Trana et al. (2020)	77 NPS and TRD	Whole blood, urine, oral fluid	Protein precipitation (whole blood), dilute-and-shoot (urine and oral fluid)	Simple sample preparation and rapid method for a mixed panel of NPS and TRD

methods for NPS and their known metabolites with available reference standards continue to be published.

Most broad-spectrum NPS analysis in blood uses protein precipitation or solid-phase extraction (SPE) for sample preparation. Gradient elution is typically applied in reverse-phase chromatography to optimize separation of a large number of drugs from multiple classes. Whole blood samples are more commonly used in forensic analysis. Adamowicz and Tokarczyk developed a method to analyze 143 NPS (Adamowics and Tokarczyk, 2016). The panel includes analytes that belong to several NPS classes: synthetic cannabinoids (34), synthetic cathinones (36), amphetamines and phenethylamines (26), tryptamines (18), piperazines (9), piperidines (2), aminoindanes (2), arylalkylamines (7), arylcyclohexyamines (3), and other drugs (6). Because a total 432 transitions were monitored, scheduled multiple reaction monitoring (MRM) was applied to optimize cycle time. Three transitions were used to identify each analyte. In standard practice, quantitative drug analysis requires monitoring of only two transitions for analyte. However, inclusion of a third allows differentiation of structural isomers, especially positional isomers that are more often observed in NPS than in traditional recreational drugs. The whole procedure was rapid (chromatographic run time 14 min) and simple using protein precipitation for sample preparation and gradient elution for chromatography. The limits of detection (LODs) established were 0.01–3.09 ng/mL. Application of the assay in routine forensic testing detected 112 positive samples for NPS in more than 1000 cases between 2011 and 2014.

Odoardi and colleagues used dispersive liquid–liquid microextraction for sample preparation to measure 78 NPS in whole blood (Odoardi et al., 2015). This is a variation of liquid–liquid extraction that uses a disperser solvent (e.g., acetone) in addition to the extracting solvent, allowing rapid and efficient extraction of a larger number of drugs from blood. The procedure is rapid and requires only microliter amounts of disperser and extracting solvents. The panel consists of synthetic cannabinoids, synthetic cathinones, amphetamines, phenethylamines, tryptamines, piperazines, benzofurans, and ketamine analogs, with LODs 0.2–2 ng/mL. In one of the 60 forensic samples initially tested, two synthetic cathinones and an aminoindane were identified. These two methods demonstrate that the sensitivity of targeted LC-MS/MS is adequate for the confirmation and quantification of NPS in blood.

Serum, plasma, and urine samples are commonly tested in clinical and forensic laboratories. As early as 2010, Wohlfarth and colleagues developed a method for the analysis of 31 NPS in plasma (Wohlfarth et al., 2010). Target analytes included amphetamines, phenethylamines, cathinones, tryptamines, and piperazines. SPE was used to prepare the sample, and LODs of 1–5 ng/mL were achieved for all analytes. Lehmann et al. developed a targeted LC-MS/MS for the quantitative analysis of 69 NPS and qualitative analysis of an additional 21 in serum (Lehmann et al., 2017). The 11-minute method used an automated online SPE connected to the LC-MS/MS employing a biphenyl column for elution gradient chromatography. The chromatographic run time was 9.3 min, with the rest of the analysis time spent on SPE and sample injection. LODs were 0.2–4 ng/mL. Application of the assay in 28 specimens yielded seven samples positive for NPS. A similar method by Vaiano et al. for the analysis of 64 NPS in serum achieved higher sensitivity across all analytes, with LODs 0.05–0.3 ng/mL. It required 200 μL of serum using protein precipitation for preparation (Vaiano et al., 2016). Reverse-phase chromatography was applied using gradient elution with a total run time of 15 min. Application of the method in three cases identified mephedrone along with methamphetamine in one case and AB-FUBINACA in another case.

Very few broad-spectrum NPS testing methods in urine have been published. An example is a targeted method that included 46 NPS (44 parent drugs and 2 metabolites) in a mixed panel consisting of traditional recreational drugs and NPS (Tang et al., 2014). Enzymatic deconjugation using H pomatia glucuronidase followed by SPE was used for sample preparation. Reverse-phase chromatography was applied, and the chromatographic eluates were ionized in the positive or negative polarity using an ESI source. Scheduled MRM was used for analyte detection, with three transitions for each drug. The LODs for NPS were 1–200 ng/mL. Application of the assay to 964 urine samples yielded only two cases with NPS, one with TFMPP and another with methcathinone. The low detection rate is likely because of the dynamic turnover of NPS in the recreational drug market but, more importantly, because the panel mostly included parent compounds of NPS.

Nontraditional biological matrices are also used in targeted NPS analysis of samples such as oral fluid and hair. Hair analysis is very useful in forensics, especially when temporal use of drug or chronic drug intake is sought, and many groups have developed and published broad-spectrum methods. Between 2007 and 2021, detection of 280 NPS was reported in hair. Boumba and colleagues validated a method for 132 NPS, which included analytes from synthetic cannabinoids, synthetic cathinones, amphetamines, phenethylamines, piperazines, piperidines, arylcyclohexylamines, benzofurans, and opioids (Boumba et al., 2017). Pulverization of hair in acidified methanol was used for sample preparation after it was washed three times in methanol and dichloromethane. Reverse-phase chromatography using gradient elution was used to separate the analytes, which were then ionized in the positive mode by an ESI source. Transitions from the analytes were monitored by scheduled MRM. LODs were 0.001–0.1 ng/mg hair. Using the assay in medicolegal cases, NPS was confirmed in 20 of 23 cases for which NPS have been detected in matched blood samples or suspected.

A method updated with NPS targets prevalent in the past couple of years was recently published by Nzekoue and colleagues (Nzekoue et al., 2021). Synthetic cannabinoids, synthetic cathinones, tryptamines, and new synthetic opioids (NSO) were among the target analytes. Hair samples (100–200 mg) were washed four times using water, methanol, and butanol. About 25 mg of washed hair was finely sliced into 0.5 mm lengths and extracted using the M3 reagent at 100°C for 1 h. Liquid–liquid microextraction using a mixture of dichloromethane and isopropyl alcohol was done in acidic and basic conditions to complete the sample preparation. Sample extracts were then analyzed by gradient elution in a reverse-phase column followed by ionization in the positive mode using ESI. Analytes were monitored by scheduled MRM. The LODs were 0.0006–0.0103 ng/mg. The method correctly identified synthetic cannabinoids, synthetic cathinones, and NSO in proficiency testing samples without incurring any false-positive or false-negative results.

Synthetic cannabinoids

This is the largest class of NPS. Along with synthetic cathinones, it was also the most abused in the first half of the past decade, which is why many analytical methods target these drugs and metabolites. The first-generation synthetic cannabinoids were less toxic than the current ones, and fatalities were rare. Urine is the standard sample in the clinic, so most initial methods were developed in urine.

Urine presents a challenge in the selection of the target analyte, especially for NPS. Because it often contains the metabolic products of xenobiotics, drug metabolites are easier to detect than their parents. With the scarcity of metabolic and pharmacokinetic data on NPS, including synthetic cannabinoids, many metabolites are unknown.

Initial methods were a combination of targeted and nontargeted and mostly supportive of metabolic and pharmacokinetic studies. Known parents with available reference standards were the analytes, and the methods were supported by metabolite prediction software and nontargeted data acquisition strategies using, for example, LC-HRMS to discover major metabolites of a synthetic cannabinoid that could then be used for clinical and forensic testing (Hutter et al., 2018).

The structures of synthetic cannabinoids are defined by four pharmacophores: central ring, tail, linker, and pendant ring or side chain (Fig. 2.8). Several functional groups are repeatedly used in the pharmacophores. For example, the tail is often one of five groups: alkyl chain, haloalkyl chain, cyanoalkyl chain, methylcyclohexyl ring, or the fluorobenzyl ring. The central ring is often indole, indazole, or azaindole, and the pendant ring can be naphthyl, tetramethylcyclopropyl, adamantyl, quinolinyl, or cumyl. The existence of these common functional groups helps to predict metabolic products of newly reported synthetic cannabinoids. Once a metabolic transformation of a functional group is discovered, that information can be applied to homologs or analogs, allowing reference standard manufacturers to prioritize specific predicted metabolites that would most likely be relevant targets for testing.

The aminoalkylindoles, exemplified by the JWH compounds, comprise the first- and second-generation synthetic cannabinoids. Initial metabolic studies revealed that the N-hydroxylation of the alkyl tails is their most common metabolic route; further oxidation of the initial hydroxylated metabolite at the terminal carbon to carboxylic acid follows (Fig. 2.14).

Other major metabolic products result from monohydroxylation of the indole ring and the pendant ring (e.g., naphthyl group). In certain instances, dihydroxylation either within or across functional groups is also observed. Glucuronidation of these phase I metabolites along with the parent extensively occur as well. Thus, in any synthetic cannabinoid analysis in urine, it is mandatory to have deconjugation as part of the sample preparation.

N-hydroxylation of the alkyl tail and its further oxidation is preserved in indazoles and azaindoles, which comprise later generations of synthetic cannabinoids. The linker is either carboxylate or carboxamide, while the pendant chain is expanded to add aminoacyl groups such as the 3-methyl-L-vallinate. The hydrolysis product of the esters and amides in these linkers and aminoacyl side chains also are the major metabolic routes in these third- and fourth-generation synthetic cannabinoids.

Table 6.3 summarizes targeted methods for synthetic cannabinoids and their metabolites. In 2013, Wohlfarth and colleagues published a qualitative assay to confirm nine first- and second-generation compounds and 20 of their metabolites in urine (Wohlfarth et al., 2013). Enzymatic hydrolysis followed by protein precipitation was used for sample preparation. Reverse-phase chromatography with gradient elution was used to separate the analytes. The method was developed in a tandem MS with linear ion trap that allowed MRM followed by information-dependent product ion scans to collect mass spectra of target analytes. The latter allows screening of the predicted metabolites. Analyte identification was done using library searches against a reference mass spectral library. The method had LODs of 1–10 ng/mL. Analysis of 10 urine samples identified 14 synthetic cannabinoids metabolites. Only AM-2201 was detected as parent.

A year later, Freijo et al. published a validated LC-MS/MS method that targets 14 synthetic cannabinoids and 15 of their metabolites (Freijo et al., 2014). A couple of third-generation synthetic cannabinoids were included. Similar to the previous method, enzyme deconjugation by *Helix pomatia* glucuronidase was done prior to LC-MS/MS. Reverse-phase chromatography with gradient elution and ionization in the negative mode using an ESI source was implemented. Two transitions for each analyte were assessed by MRM. LODs were 1–5 ng/mL. Application to samples collected from known users who had recently used the synthetic cannabinoid product Kush revealed the presence of AB-PINACA N-pentanoic acid metabolite and AB-PINACA N-(4-hydroxypentyl) metabolite but not AB-PINACA, the parent. AB-PINACA was, however, identified in a sample of Kush.

Because of the dynamic turnover of a synthetic cannabinoid product's molecular ingredient, targeted methods had to adjust their analytes. In 2019, Gaunitz and colleagues developed an LC-MS/MS method for the

TABLE 6.3 Selected published methods for the targeted analysis of synthetic cannabinoids (SCs).

Reference	Analytes	Matrix	Work-up	Highlights
Wohlfarth et al. (2013)	9 first- and second-generation SC and 20 of their metabolites	Urine	Enzymatic deconjugation followed by protein precipitation	Rapid, fully validated qualitative method; product ion scan on a QTRAP employed to determine predicted SC metabolites in urine by mass spectral library searches
Freijo et al. (2014)	14 SC and 15 of their metabolites including a couple of third generation	Urine	Enzymatic deconjugation followed by dilute-and-shoot	Simple sample preparation and rapid method; validated method
Scheidweiler and Huestis (2014)	20 SC and 33 of their metabolites	Urine	Enzymatic deconjugation followed by supported liquid extraction	Validated quantitative method except for nonchromatographically resolved alkyl hydroxyl isomeric metabolites
Gaunitz et al. (2020)	61 SC metabolites from 29 SC (first- to third generation)	Urine	Enzymatic deconjugation followed by solid phase extraction	Validated qualitative method
Ambroziak and Adamowicz (2018)	72 SC	Whole blood	Protein precipitation	Qualitative method validated only for 50 SC; three transitions per analyte were monitored by scheduled MRM

Ong et al. (2020)	Second- and third-generation SC	Whole blood	Supported liquid extraction	Validated quantitative method; higher levels of metabolites than parent compound observed in samples, similar to urine
Franz et al. (2018)	72 SC (first to third generation)	Hair	Ultrasonication with methanol of 1–2 mm cut hair	Validated semiquantitative method; method updates with newer SC done by partial revalidation
Cho et al. (2020)	18 SC and 41 of their metabolites (first to third generation)	Hair	Methanol extraction of finely cut hair	Validated quantitative method; the parent compound is present at much higher levels than metabolites for indazole carboxamides, while the reverse is true for some aminoalkyindoles

qualitative analysis of 61 metabolites from 29 parent compounds (Gaunitz et al., 2020). The panel added some metabolites of third-generation synthetic cannabinoids. Enzymatic hydrolysis using *H. pomatia* glucuronidase followed by SPE was used to prepare the urine samples. Reverse-phase chromatography using a biphenyl column and gradient elution followed by ionization in the positive mode with an ESI source was implemented. Scheduled MRM assessed 140 transitions for the analytes and internal standards. The 12.3 min method is quite sensitive, with LODs for all analytes of 0.025–0.5 ng/mL. Application to 61 urine samples from routine casework, of which 21 originated from autopsies, yielded only one positive result, ADB-PINACA N-pentanoic acid. Because the metabolite is shared by 5F-ADB-PINACA and ADB-PINACA and neither was detected in the sample, the drug taken by the subject was not established. This illustrates a potential problem in synthetic cannabinoid analysis caused by their very close structural similarities and shared metabolic pathways.

There are targeted methods in other biological matrices (Table 6.3). In 2019, Ong and colleagues published an LC-MS/MS method for the analysis of 29 compounds and their metabolites in whole blood (Ong et al., 2020). The panel consisted mainly of third-generation synthetic cannabinoids, with a few second-generation. The samples were prepared by supported liquid extraction. Reverse-phase chromatography using a biphenyl column and gradient elution followed by ionization in the positive mode with an ESI source was implemented. Two or three transitions for each analyte were monitored by scheduled MRM. LODs for the target analytes were 0.1–6 ng/mL. The validated method was used to analyze 564 antemortem and postmortem blood samples collected from coronial postmortem examinations, driving impairment cases, emergency department admissions, psychiatric care patients, and criminal cases. There were positive findings for 13 of the 29 synthetic cannabinoids and metabolites tested. AMB-FUBINACA (30, 5.3%) and its acid metabolite (81, 14.4%) and 5F-ADB (48, 8.5%) and its acid metabolite (53, 9.4%) had the highest detection frequencies. In contrast to urine, the parent compounds can be detected in whole blood. Although the metabolites are still present in higher concentration and detection frequency, some parent compounds can be found in whole blood as often as their major metabolites (e.g., 5F-ADB vs. 5F-ADB acid metabolite).

Targeted methods in hair and oral fluid have also been published. In a method from a Korean group, LODs of 0.0001–0.010 ng/mg in 20 mg hair samples were obtained for 18 synthetic cannabinoids and 41 of their metabolites, so both the parent and metabolites can be detected in hair as well as whole blood (Cho et al., 2020). In some compound classes, the parent was detected at much higher levels than its metabolite. This was observed for most indazoles. AB-CHMINACA (2.5–3637.4 pg/mg hair), for example, was detected at 10 to 200 times higher levels than its acid metabolite (0.5–75.9 pg/mg hair). In contrast, some aminoalkylindoles such as JWH-210 have the 5-hydroxypentyl metabolite present at a higher level than the parent. In a separate cohort from Germany, 163 of 294 hair samples tested positive for synthetic cannabinoids sampled between 2012 and 2016 (Franz et al., 2018). In 47% of those that tested positive, 4–14 were detected in each hair sample. As many as 17 were detected in one sample, indicative of the variety of synthetic cannabinoids circulating at a given time. Likewise, in cases where repeat analysis of the same person was done over a 15-month period, changes in the composition were observed, suggestive of the rapid molecular evolution of the composition of available synthetic cannabinoid products.

Stimulants, psychedelics, and dissociatives

These related pharmacological groups consist of several classes, including synthetic cathinones, amphetamines, phenethylamines, piperazines, piperidines, aminoindanes, benzofurans, tryptamines, arylcyclohexylamines, lysergamides, and cocaine analogs, and most targeted analytical methods group them accordingly. Even before the resurgence of NPS in the United States started in 2008, amphetamines, phenethylamines, tryptamines, and piperazines were available as designer drugs, a term that loosely referred previously to NPS. From about 2008, synthetic cathinones have been the dominant NPS stimulants in the Unites States and Europe. As with synthetic cannabinoids, the turnover of synthetic cathinone structures in the recreational market was dynamic, and the class was the initial focus of most targeted methods. Later, as new NPS classes with stimulant, psychedelic, and dissociative properties emerged, targeted panels covering these pharmacological classes all together were developed and published. Table 6.4 provides a sampling of targeted LC-MS/MS methods developed for synthetic cathinones and other classes of stimulants, psychedelics, and dissociatives.

With the emergence of bath salts and K2 (Spice) in the United States and Europe, emergency departments were inundated with intoxications related to synthetic cathinones. The need for testing these drugs provided the impetus for the development of targeted assays in urine, the standard biological matrix for emergency testing. In 2011, Bell and colleagues published a method that incorporated five synthetic cathinones (mephedrone, methedrone, methylone, butylone, MDPV) in a panel with three piperazines (BzP, mCPP, TFMPP). Prior to the emergence of synthetic cathinones, piperazines were the more popular designer stimulants (Bell et al., 2011). A simple dilute-and-shoot method was done to prepare the urine for LC-MS/MS. Reverse-phase chromatography with gradient elution followed by ionization in the positive mode using an ESI ion

TABLE 6.4 Selected published methods for the targeted analysis of NPS stimulants, psychedelics, and dissociatives.

Reference	Analytes	Matrix	Sample preparation	Highlights
Pichini et al. (2008)	1 amphetamine, 6 phenethylamines, 2 tryptamines, 1 piperazine	Urine	SPE at pH6 with or without prior enzymatic deconjugation	Quantitative method meant for hallucinogenic designer drugs; results obtained with deconjugation is higher for majority of analytes
Bell et al. (2011)	5 synthetic cathinones, 3 piperazines	Urine	Dilute-and-shoot	Qualitative method
Al-Saffar et al. (2013)	11 synthetic cathinones, 3, amphetamines, 2 phenethylamines, 3 tryptamines, 2 opioids, 2 piperidines, 1 piperazine, 1 aminoindane, 1 arylcyclohexylamine	Urine	Dilute-and-shoot	Simple and rapid (4 min run time) validated method for "legal high" drugs
Concheiro et al. (2013)	19 synthetic cathinones	Urine	Solid-phase extraction (SPE)	Quantitative method developed in LC-HRMS
Fan et al. (2020)	73 synthetic cathinones and related metabolites	Urine	Dilute-and-shoot	Validated quantitative method with sub-ng/mL LODs; revealed NPS penetrance in Asia
Wohlfarth et al. (2010)	14 amphetamines, 8 phenethylamines, 6 tryptamines, 5 piperazines, TRD	Serum	SPE	Qualitative method

Ammann et al. (2012)	19 synthetic cathinones, 6 amphetamines	Blood	Liquid–liquid extraction	Validated quantitative method but LOD not evaluated; LLOQ set at 10 ng/mL for all analytes
Swortwood et al. (2013)	9 synthetic cathinones, 10 amphetamines, 5 phenethylamines, 4 piperazines, 4 tryptamines	Serum	SPE	Validated quantitative method
Lau et al. (2020)	30 synthetic cathinones	Whole blood	SPE	Validated quantitative method for postmortem blood
Freni et al. (2019)	16 synthetic cathinones	Hair	Ultrasonication in 0.1 M HCl followed by SPE	Validated quantitative method; 20 mg hair required, segmental analysis done in some samples
Salomone et al. (2016)	10 synthetic cathinones, 3 amphetamines, 6 phenethylamines, 2 benzofurans, 1 piperazine, 3 arylcyclohexylamines, TRD	Hair	Methanol extraction of finely cut hair	Validated quantitative method; 25 mg hair required
Niebel et al. (2020)	32 synthetic cathinones, 3 piperazines	Hair	Methanol/acetonitrile extraction in acidic condition	Validated quantitative method
(2014)	5 synthetic cathinones, 1 amphetamine, 2 piperazines	Oral fluid	SPE	Validated quantitative method

source was implemented. LODs between 2 and 3.4 ng/mL were obtained for the synthetic cathinones and 6.5 and 29 ng/mL for the piperazines.

The dynamic turnover of synthetic cathinones in bath salts forced changes in the target analytes of panels. In 2013, Concheiro and colleagues published a method for 28 synthetic cathinones in urine. LODs were 0.25–0.5 ng/mL (Concheiro et al., 2013). The method was used to analyze 106 samples, of which four were confirmed to have synthetic cathinones. Two each were confirmed for methylone at 8.8–11.2 ng/mL, while the other cases had multiple synthetic cathinones (case 3: methylone, 1.7 ng/mL; pentedrone, 3.5 ng/mL; pentylone, 119.1 ng/mL; α-PVP, 806 ng/mL; case 4: α-PVP, 474.7 ng/mL; pyrovalerone, 2 ng/mL). The results from the last two cases were expected, as many bath salt products contain two or more synthetic cathinones (Schneir et al., 2014). The levels of synthetic cathinones also indicate that the assay had more than enough sensitivity to detect these drugs in urine. The low detection frequency is likely due to the low prevalence of synthetic cathinone use in the cohort.

The continuous expansion and changes in the synthetic cathinones available in the recreational drug market are reflected in a targeted method for 73 synthetic cathinones by Fan and colleagues published in 2020 (Fan et al., 2020). They used dilute-and-shoot as a preparation method. LC-MS/MS analysis was done through reverse-phase chromatography using a biphenyl column and gradient elution followed by ionization in the positive mode with an ESI ion source. Two transitions were monitored for each analyte by MRM. The range of LODs for all analytes was 0.1–0.5 ng/mL. The method was successfully applied to 67 urine samples, of which 32 tested positive. Twelve synthetic cathinones were identified: mephedrone, methylone, ethylone, butylone, dibutylone, ephylone, 4-chloromethcathinone, 4-chloroethcathinone, 4-chloro-N,N-dimethylcathinone, 4-ethylethcathinone, 4-methylpentedrone, and 4-methyl-α-ethylaminopentiophenone (4-MEAPP), along with one synthetic cathinone metabolite, 4-methylephedrine. Multiple synthetic cathinones were also observed in some samples, as many as seven in one case. These results demonstrate the penetrance of NPS to Asia, as the method and cohort analyzed were from Taiwan. The results obtained for synthetic cathinones demonstrate that parent compounds can be measured in urine, which is rarely the case for synthetic cannabinoids. The lack of requirement for deconjugation in the sample preparation also suggests that synthetic cathinones are not as extensively conjugated as synthetic cannabinoids and their metabolites.

Targeted methods for synthetic cathinones were also developed in other biological matrices. As early as 2012, Ammann and colleagues published an LC-MS/MS method for 25 synthetic cathinones in blood (Ammann et al., 2012). Liquid–liquid extraction with 1-chlorobutane and isopropyl alcohol was used for sample preparation. LC-MS/MS was implemented by reverse-phase chromatography using gradient elution followed by ionization in the positive mode using an ESI ion source. Three transitions for each analyte were monitored by MRM. The LOD for the analytes was not systematically evaluated, but the limit of quantitation (LOQ) for all analytes was set at 10 ng/mL. A year later, Swortwood and colleagues published a method incorporating eight synthetic cathinones in a serum test panel for amphetamines, phenethylamines, tryptamines, and piperazines (Swortwood et al., 2013). Serum samples were prepared by SPE. Reverse-phase chromatography using gradient elution followed by ionization in the positive mode with an ESI ion source was implemented in the LC-MS/MS analysis. Two transitions for each analyte were monitored by MRM. LODs were 0.01–0.1 ng/mL. Application to two forensic cases confirmed methylone in one and MDPV, BZP, TFMPP, 5-methoxy-DiPT, MDMA, and MDA in the other.

Freni and colleagues published a targeted method for 16 synthetic cathinones in hair in 2019 (Freni et al., 2019). The method required 20 mg of hair, which was washed twice with dichloromethane and methanol before pulverization and SPE. LC-MS/MS was done similarly to other published methods. LODs were 0.0001–0.0025 ng/mg. Application to 17 postmortem cases confirmed synthetic cathinones in two samples. One case had ethcathinone in its first 2.5 cm segment, while the other had eight synthetic cathinones (3,4-DMMC, 4-FMC, 4-MEC, α-PVP, α-PHP, methcathinone, methedrone, pentedrone) in both its first and second 2.5-cm segments.

Test panels combining different NPS classes with stimulant, psychedelic, and dissociative properties have also been developed. In 2008, Pichini and colleagues published an LC-MS/MS method for the confirmation of 10 "hallucinogenic designer drugs" in urine that includes an amphetamine (MDMA), phenethylamines (2C–B, D, E, I, T2, B-Fly), tryptamines (4-OH-DiPT, 4-Acetoxy-DiPT), and a piperazine (mCPP). The LODs were 9–16 ng/mL (Pichini et al., 2008). The method was applied to 32 urine samples donated by hallucinogenic designer drug users in a program setup by a Spanish nongovernmental organization that promotes harm reduction. All target analytes were confirmed in the samples of their respective users. Interestingly, the analysis was done before and after enzymatic hydrolysis of samples. Except for 4-acetoxy-DiPT, all analytes were considerably present as conjugated compared with their free form, indicating that a deconjugation protocol prior to LC-MS/MS can greatly improve detection.

Five years after the resurgence of synthetic cathinones, Al-Saffar et al. published an LC-MS/MS method to confirm drugs commonly found in "legal highs" in urine. The panel consists of synthetic cathinones, amphetamines, phenethylamines, tryptamines, aminoindane, piperazine, piperidine, and arylcyclohexylamine (Al-Saffar et al., 2013). LODs were

0.5–50 ng/mL. The method was applied to 1335 samples collected in a year that were submitted for clinical testing of "Internet drugs." 13 of the target analytes were detected in 87 samples at varying frequencies; the majority were synthetic cathinones.

Targeted methods were developed in other biological matrices as well. The method published by Swortwood et al. mentioned previously is an example. Additionally, Salomone et al. published a targeted LC-MS/MS method to confirm stimulant, psychedelic, and dissociative NPS in hair (Salomone et al., 2016). Samples (25 mg) were washed twice with dichloromethane and methanol prior to cutting into 1–2 mm segments and extraction with methanol at 55°C for 15 h. An LC-MS/MS assay similar to published methods, comprised of reverse-phase chromatography with gradient elution and ionization in the positive mode using an ESI source, was implemented. Two transitions for each analyte were monitored by MRM. LODs were 0.0009–0.017 ng/mg hair. The method was applied to 23 samples from confirmed MDMA and ketamine abusers (Group A) and 54 samples that previously tested negative for traditional recreational drugs from individuals in a driver's license recovery program (Group B). Six samples (5 Group A, 1 Group B) tested positive for at least one NPS. Mephedrone, methylone, MDPV, α-PVP, 4-methylethcathinone, 4-fluoroamphetamine, methoxetamine, and diphenidine were identified. One sample from Group A had six NPS. The levels ranged from 7.7 to 440 pg/mg hair.

New synthetic opioids and designer benzodiazepines

Although fentanyl analogs circulated as designer drugs in the late 1970s and early 1980s, their furious resurgence in the middle of the past decade is incomparable. Fentalogs have become the predominant NPS in the recreational drug market. Because of their potency, they now cause more fatalities than any other NPS class. Stringent regulatory actions came down fast. This has given rise to other synthetic opioids that do not share the same scaffold. Parallel with this resurgence is the introduction of designer benzodiazepines.

Because fentanyl analogs have been around for so long, the first targeted assays using LC-MS/MS were for traditionally abused opiates. They involve reverse-phase chromatography using gradient elution for analyte separation followed by ionization in the positive mode with an ESI ion source. Detection of two transitions for each analyte was done by MRM. Table 6.5 gives published targeted methods.

In 2009, Gergov et al. published a method that incorporated seven fentalogs in an LC-MS/MS assay for 25 opioids in urine and whole blood (Gergov et al., 2009). LODs for the fentalogs were 0.01–0.2 ng/mL in both matrices. Because of the relative potencies of the fentalogs, sub-ng/mL

TABLE 6.5 Selected published methods for the targeted analysis of new synthetic opioids.

Reference	Analytes	Matrix	Sample preparation	Highlights
Gergov et al. (2009)	7 fentalogs, common opioids	Urine	Liquid–liquid extraction (LLE)	Validated quantitative method; concordant with multiple reference methods
Cooreman et al. (2010)	4 fentalogs, fentanyl, norfentanyl	Plasma, urine	LLE	Validated quantitative methods
Moody et al. (2018)	19 new synthetic opioids, 17 new synthetic opioids	Whole blood, serum/ plasma, urine	Protein precipitation followed by solid phase extraction	Validated quantitative method; isomeric fentalogs could not be distinguished
Bergh et al. (2018)	27 fentanyl analogs	Whole blood	LLE	Validated quantitative method; isomeric fentalogs were chromatographically separated
Fogarty et al. (2018)	18 fentanyl analogs	Whole blood	Solid-phase extraction	Validated quantitative method
Seymour et al. (2019)	13 fentanyl analogs	Dried blood spot	Methanol/ acetonitrile extraction	Validated quantitative methods; analytes distributed into two methods
Adamowicz et al. (2020)	43 new synthetic opioids	Whole blood	LLE under basic condition	Validated as quantitative method for 21 drugs
Freni et al. (2020)	19 fentanyl analogs, fentanyl	Hair	Acid extraction of ground hair followed by SPE	Validated quantitative method; 25 mg required
Ramírez Fernández et al. (2020)	16 new synthetic opioid	Hair	Methanol extraction of pulverized hair followed by SPE	Validated quantitative method of segmented hair; 20 mg required
Qin et al. (2020)	37 new synthetic opioids, fentanyl, norfentanyl	Hair	Cryogenic grinding of pulverized hair with methanol/ acetonitrile	Validated quantitative method; 20 mg required, segmental analysis done

sensitivity is required, especially for blood. Application of the method in autopsy samples did not detect any of the fentalogs in the cases. However, the method was able to quantify other opioids, which agreed within a 3% −24% bias with the results obtained by the laboratory in two other reference methods. The lack of detection of the fentalogs therefore was not due to the method's lack of sensitivity.

A year later, Cooreman and colleagues published a validated method for fentanyl, norfentanyl, alfentanil, sufentanil, remifentanil, and 3-methylfentanyl in urine and plasma (Cooreman et al., 2010). Liquid–liquid extraction was used to prepare both samples. LOQs for all analytes except sufentanil were 0.1 ng/mL in both matrices; the LOQ for sufentanil was 0.2 ng/mL. Although the method was not applied to actual samples at the time of publication, the authors argued that validated LOQs were below their reference ranges.

The resurgence of new synthetic opioids brought back the fentalogs from the 1970s and 1980s and expanded the variety of fentalogs and other types of opioids available in the recreational drug market, including those with extreme potency. Carfentanil, for example, is 10,000 times more potent than morphine. Because of their relative potencies, NSO intoxications are often fatal, so most of the methods for whole blood are used in forensic death investigations.

Bergh and colleagues published a method in 2018 for the analysis of 27 fentanyl analogs in whole blood (Bergh et al., 2018). Liquid–liquid extraction with ethyl acetate and heptane was used for sample preparation, and a biphenyl column was employed in reverse-phase chromatography. The method was extremely sensitive, with LODs of 0.0015–0.004 ng/mL. It also separated structural isomers, which is important for the fentalogs, as numerous isomers exist that share common fragment ions. The method confirmed and quantified target analytes in two cases. In one, 2.5 ng/mL fentanyl and 5.3 ng/mL alfentanil were confirmed. In the other, 0.01 ng/mL acetylfentanyl and 28 ng/mL cyclopropylfentanyl were detected. Two years later, Adamowicz and colleagues published a validated method for 38 fentalogs and 5 other opioids in whole blood (Adamowicz et al., 2020). Other NSOs were AH-7921, U-47700, U-49900, W-15, and W-18. Blood samples at small volume (0.2 mL) were extracted with ethyl acetate under basic conditions. Three transitions were monitored for each analyte by scheduled MRM. The LODs were 0.01–0.2 ng/mL. The method was applied in forensic cases investigated in Krakow, Poland in 2018. Benzylfentanyl (67–110 ng/mL), 4-fluoroisobutyrylfenatnyl (74 ng/mL), and despropionyl-p-fluorofentanyl (6.5 ng/mL) were confirmed in three cases. Drugs from other NPS classes were found in all the cases.

Moody and colleagues published methods in 2018 for the analysis of 19 NSOs in whole blood and serum and 17 in urine for use in forensics

(Moody et al., 2018). The target analytes were those reported in the United States at the time. Four analytes, MT-45, AH7921, U-47700, and U-50488, were not fentalogs. Protein precipitation and SPE were used for sample preparation, and two transitions were monitored for each analyte by MRM. LODs were 0.0125–0.25 ng/mL. Notably, the method could not distinguish between two pairs of structural isomers (butyrylfentanyl vs. isobutyrylfentanyl and p-fluorobytyrlfentanyl vs. FIBF). The method was applied to 2758 samples between October 2016 and September 2017. Eleven target analytes were confirmed. The five with the highest detection frequency were 4-ANPP (1549), 2-furanylfentanyl (1228), carfentanil (697), FBF/FIBF (563), and U-47700 (543).

Various groups have also developed methods for NSO in hair. Qin et al., for example, published a method for 37 compounds (Qin et al., 2020). The samples (20 mg) were washed three times with water and acetone before pulverization and extraction with methanol. Two transitions were monitored for each analyte by MRM. The LODs were 0.0005–0.0025 ng/mg hair. The method was successfully applied in two cases: fentanyl was detected at 8.02 pg/mg at the 0–3 cm segment, and sufentanil was detected in each of three 3-cm segments at 183.91, 131.68, and 31.48 pg/mg hair.

Methods for designer benzodiazepines began to be published in the latter half of the past decade. Drugs in this group bear close resemblance to their prescription counterparts, which means that a lot of designer benzodiazepines cross-react with the immunoassay. In such a case, confirmatory testing becomes more important. Table 6.6 presents a sampling of targeted methods. Some are highlighted in the following.

In 2016, Pettersson Bergstrand and colleagues published a validated method for 11 designer benzodiazepines in urine (Pettersson Bergstrand et al., 2016). Dilute-and-shoot was used for preparation. A phenyl column was used in reverse-phase chromatography, and two transitions of each analyte were monitored by MRM. The LODs were 1–10 ng/mL. The method was applied in a cohort of 390 samples collected in Sweden from February 2014 to November 2015. They tested positive to the CEDIA immunoassay but confirmed negative in an LC-MS/MS assay for prescription benzodiazepines. Of these samples, 40% had at least one of the target designer benzodiazepines. Eight of the 11 designer benzodiazepines were detected. Flubromazolam (n = 96, range 5.4–1500 ng/mL), meclonazepam (n = 45, range 1.6–190 ng/mL), flubromazepam (n = 14, range 2.7–30 ng/mL), etizolam (n = 11, range 5.8–270 ng/mL), and pyrazolam (n = 9, range 3.2–920 ng/mL) were the top five in frequency.

TABLE 6.6 Selected published methods for the targeted analysis of designer benzodiazepines.

Reference	Analytes	Matrix	Sample preparation	Highlights
Pettersson Bergstrand et al. (2016)	11 designer benzodiazepines	Urine	Enzymatic deconjugation followed by dilute-and-shoot	Validated quantitative method; confirmed cross-reactive of most analytes with CEDIA immunoassay for prescription benzodiazepines
Mei et al. (2019)	11 designer benzodiazepines	Whole blood	Solid-phase extraction	Validated quantitative method for postmortem blood
Garcia et al. (2021)	17 designer benzodiazepines	Whole blood	Protein precipitation	Validated quantitative method for 15 of 17 target analytes
Mastrovito et al. (2021)	11 designer benzodiazepines, one metabolite	Whole blood	Liquid–liquid extraction	Validated quantitative method

Recently, Mastrovito and colleagues published a validated method for 11 designer benzodiazepines in whole blood (Mastrovito et al., 2021). Liquid–liquid extraction was used to prepare samples, and two transitions were monitored for each analyte by MRM. LODs were 0.06–0.5 ng/mL. The method was applied to 1262 forensic casework samples from the United States that tested positive in immunoassay but negative in the confirmatory LC-MS/MS assay for prescription benzodiazepines. The presence of at least one designer benzodiazepine was confirmed in 705 of the samples. Nine of the 11 analytes were detected. Etizolam (n = 607, range 2–900 ng/mL), flualprazolam (n = 337, range 2–990 ng/mL), flubromazolam (n = 136, range 2.8–760 ng/mL), delorazepam (n = 20, range 5.9–520 ng/mL), and clonazolam (n = 18, range 5–26 ng/mL) had the highest detection frequencies.

Closeup: the lesser-known THCs

As a research associate in Dr. Roy Gerona's lab at the University of California San Francisco, I analyzed drug cases from hospitals across the country to identify NPS. Identifying individual drugs from complex biological samples can be tricky, as individual molecular signals need to be teased apart from the noise that comes from all the other compounds in the human body. To accomplish this, we use liquid chromatography to separate the molecular components of the sample by polarity, followed by mass spectrometry to ionize and detect the separated molecules by their mass. We also use LC-MS/MS to fragment individual parent ions to generate their fragment ions, which facilitates their unequivocal confirmation and identification.

This powerful analytical tool allowed me to witness changes in the American NPS landscape firsthand, such as the decline of cases involving the opioid analog U-47700 following a Chinese ban and the continued surge of fentalogs despite worldwide restrictions on fentanyl and its analogs. While my prior laboratory job required rote adherence to analysis workflows, the nature of work in the Gerona lab meant that we had to track these global events to know what to look for. I was fortunate to have this experience early in my scientific career, and to learn that sometimes an unexplained result can lead to the most important findings.

One such case was referred to the lab in early 2021 by Dr. Peter Akpunonu at the University of Kentucky. A 2-year-old girl was hospitalized and intubated after ingesting "CBD gummies" that her father had left within reach. We were sent serum and urine samples from the patient, as well as the gummies. Having closely followed trends in cannabis analog usage, we had acquired a library of possible drug compounds that could be used as reference standards when analyzing unknown drug products and biological samples in a targeted manner. We were particularly interested in investigating a match with delta-8-THC, given its recent popularity in states that, like Kentucky, ban recreational cannabis.

Delta-8 is a psychoactive compound found in cannabis, an isomer of the primary psychoactive compound delta-9 (Fig. 6.7). Although delta-8 is half as potent as delta-9, the former has become popular following the passage of the 2018 Farm Bill. That bill relaxed rules regarding hemp production in the United States, allowing delta-8 retailers to market the compound as a product of hemp rather than cannabis, which remains scheduled (Erickson, 2021). But isolating large amounts of delta-8 from hemp are impractical. Instead, manufacturers isolate the much more

continued

Closeup: the lesser-known THCs (*cont'd*)

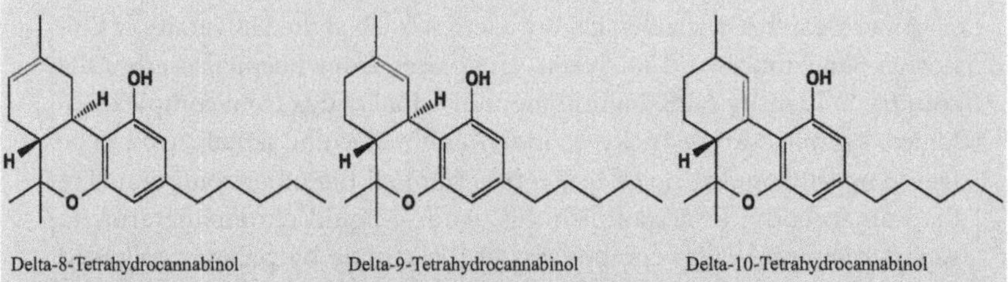

Delta-8-Tetrahydrocannabinol Delta-9-Tetrahydrocannabinol Delta-10-Tetrahydrocannabinol

FIGURE 6.7 Structural differences between delta-8, delta-9, and delta-10-THC.

prevalent cannabidiol and use an organic solvent, an acid, and sometimes a heavy metal to catalyze a ring closure reaction, forming delta-8.

As a drug being produced in the gray area between legal and illegal, there is no regulation or standardization with regard to how delta-8 is manufactured. This worries public health officials, as there is no way to know if contaminant solvents, by-products, or other psychoactive compounds also exist in products sold as "delta-8-THC." Even if it is pure delta-8, little is known about its effects, particularly in high doses or in children (Babalonis et al., 2021).

Hospital urine drug screens cannot differentiate between delta-8 and delta-9, hindering clinicians' ability to learn which substance their patient actually consumed. Hence, our lab plays an important role by using LC-MS to accurately identify and quantify drug levels in the patient. Clinical data collected at the hospital can then be interpreted alongside the drug levels, contributing to understanding of the new drug in different contexts.

We quickly identified delta-8 as the primary drug in the gummies, urine, and serum samples, and worked with the medical providers to publish the case report in *The American Journal of Case Reports* (Akpunonu et al., 2021). More than a year later, a case of a 20-year-old man who presented with status epilepticus (prolonged seizure for more than 5 min or multiple seizures within 5 min) from smoking a vape pen was referred to our lab from the same group at Kentucky. Analysis of the contents of the vape pen revealed delta-8 and yet another lesser-known THC, delta-10 (Fig. 6.6). This is a clear indication that even in this space, the products purported to have THC are evolving in molecular composition. The legal future of delta-8 and delta-10 is uncertain, but at least we were able to expand what we know about this newly popular drug.

Andrew Reckers
Ph.D. Student, Integrated Program in Cellular, Molecular, and Biomedical Studies, Columbia University, New York, NY, United States

References

Adamowicz P, Tokarczyk B. Simple and rapid screening procedure for 143 new psychoactive substances by liquid chromatography-tandem mass spectrometry. Drug Test Anal 2016;8: 652–67.

Adamowicz P, Bakhmut Z, Mikolajczyk A. Screening procedure for 38 fentanyl analogues and five other new opioids in whole blood by liquid chromatography-tandem mass spectrometry. J Appl Toxicol 2020 Aug;40(8):1033–46. https://doi.org/10.1002/jat.3962. Epub 2020 Feb 26. PMID: 32103530.

Akpunonu P, Baum RA, Reckers A, Davidson B, Ellison R, Riley M, et al. Sedation and Acute Encephalopathy in a Pediatric patient following ingestion of delta-8-Tetrahydro cannabinol gummies. PMID: 34762615 Am J Case Rep 2021 Nov 11;22:e933488. https://doi.org/10.12659/AJCR.933488. PMCID: PMC8594112.

Al-Saffar Y, Stephanson NN, Beck O. Multicomponent LC-MS/MS screening method for detection of new psychoactive drugs, legal highs, in urine-experience from the Swedish population. J Chromatogr B Anal Technol Biomed Life Sci 2013 Jul 1;930:112–20. https://doi.org/10.1016/j.jchromb.2013.04.043. Epub 2013 May 7. PMID: 23727875.

Ambach L, Hernández Redondo A, König S, Weinmann W. Rapid and simple LC-MS/MS screening of 64 novel psychoactive substances using dried blood spots. Drug Test Anal 2014 Apr;6(4):367–75. https://doi.org/10.1002/dta.1505. Epub 2013 Jul 19. PMID: 23868723.

Ambroziak K, Adamowicz P. Simple screening procedure for 72 synthetic cannabinoids in whole blood by liquid chromatography-tandem mass spectrometry. Epub 2018 Jan 31. PMID: 29963203 Forensic Toxicol 2018;36(2):280–90. https://doi.org/10.1007/s11419-017-0401-x. PMCID: PMC6002442.

Ammann D, McLaren JM, Gerostamoulos D, Beyer J. Detection and quantification of new designer drugs in human blood: Part 2 - designer cathinones. J Anal Toxicol 2012 Jul; 36(6):381–9. https://doi.org/10.1093/jat/bks049. Epub 2012 May 16. PMID: 22593565.

Babalonis S, Raup-Konsavage WM, Akpunonu PD, Balla A, Vrana KE. Δ^8-THC: legal status, widespread availability, and safety concerns. PMID: 34662224 Cannabis Cannabinoid Res 2021 Oct;6(5):362–5. https://doi.org/10.1089/can.2021.0097. PMCID: PMC8664123.

Bell C, George C, Kicman AT, Traynor A. Development of a rapid LC-MS/MS method for direct urinalysis of designer drugs. Drug Test Anal 2011 ;3(7–8):496–504. https://doi.org/10.1002/dta.306. Epub 2011 Jul 11. PMID: 21744513.

Bergh MS, Bogen IL, Wilson SR, Øiestad ÅML. Addressing the fentanyl analogue epidemic by multiplex UHPLC-MS/MS analysis of whole blood. Ther Drug Monit 2018 Dec;40(6): 738–48. https://doi.org/10.1097/FTD.0000000000000564. PMID: 30157097.

Boumba VA, Di Rago M, Peka M, Drummer OH, Gerostamoulos D. The analysis of 132 novel psychoactive substances in human hair using a single step extraction by tandem LC/MS. Forensic Sci Int 2017;279:192–202.

Byrdwell WC. Atmospheric pressure chemical ionization mass spectrometry for analysis of lipids. Lipids 2001 Apr;36(4):327–46. https://doi.org/10.1007/s11745-001-0725-5. PMID: 11383683.

Cho B, Cho HS, Kim J, Sim J, Seol I, Baeck SK, et al. Simultaneous determination of synthetic cannabinoids and their metabolites in human hair using LC-MS/MS and application to human hair. Forensic Sci Int 2020 Jan;306:110058. https://doi.org/10.1016/j.forsciint.2019.110058. Epub 2019 Nov 16. PMID: 31786516.

Concheiro M, Anizan S, Ellefsen K, Huestis MA. Simultaneous quantification of 28 synthetic cathinones and metabolites in urine by liquid chromatography-high resolution mass spectrometry. Anal Bioanal Chem 2013 Nov;405(29):9437–48. https://doi.org/10.1007/s00216-013-7386-z. Epub 2013 Oct 3. PMID: 24196122.

Cooreman S, Deprez C, Martens F, Van Bocxlaer J, Croes K. A comprehensive LC-MS-based quantitative analysis of fentanyl-like drugs in plasma and urine. J Separ Sci 2010 Sep; 33(17–18):2654–62. https://doi.org/10.1002/jssc.201000330. PMID: 20658494.

de Castro A, Lendoiro E, Fernández-Vega H, Steinmeyer S, López-Rivadulla M, Cruz A. Liquid chromatography tandem mass spectrometry determination of selected synthetic cathinones and two piperazines in oral fluid. Cross reactivity study with an on-site immunoassay device. J Chromatogr A 2014 Dec 29;1374:93–101. https://doi.org/10.1016/j.chroma.2014.11.024. Epub 2014 Nov 26. PMID: 25482853.

Di Trana A, Mannocchi G, Pirani F, Maida NL, Gottardi M, Pichini S, et al. A comprehensive HPLC-MS-MS screening method for 77 new psychoactive substances, 24 classic drugs and 18 related metabolites in blood, urine and oral fluid. J Anal Toxicol 2020;44:769–83.

Erickson B. Delta-8-THC craze concerns chemists. C&EN 30 Aug 2021;99(31).

Fan SY, Zang CZ, Shih PH, Ko YC, Hsu YH, Lin MC, et al. A LC-MS/MS method for determination of 73 synthetic cathinones and related metabolites in urine. Epub 2020 Jul 31 Forensic Sci Int 2020 Oct;315:110429. https://doi.org/10.1016/j.forsciint.2020.110429. PMID: 32784041.

Fogarty MF, Papsun DM, Logan BK. Analysis of fentanyl and 18 novel fentanyl analogs and metabolites by LC-MS-MS, and report of fatalities associated with methoxyacetylfentanyl and cyclopropylfentanyl. J Anal Toxicol 2018 Nov 1;42(9):592–604. https://doi.org/10.1093/jat/bky035. PMID: 29750250.

Franz F, Jechle H, Angerer V, Pegoro M, Auwärter V, Neukamm MA. Synthetic cannabinoids in hair - Pragmatic approach for method updates, compound prevalences and concentration ranges in authentic hair samples. Epub 2018 Jan 2 Anal Chim Acta 2018 May 2;1006:61–73. https://doi.org/10.1016/j.aca.2017.12.029. PMID: 30016265.

Freijo Jr TD, Harris SE, Kala SV. A rapid quantitative method for the analysis of synthetic cannabinoids by liquid chromatography-tandem mass spectrometry. J Anal Toxicol 2014 Oct;38(8):466–78. https://doi.org/10.1093/jat/bku092. PMID: 25217534.

Freni F, Bianco S, Vignali C, Groppi A, Moretti M, Osculati AMM, et al. A multi-analyte LC-MS/MS method for screening and quantification of 16 synthetic cathinones in hair: application to postmortem cases. Forensic Sci Int 2019 May;298:115–20. https://doi.org/10.1016/j.forsciint.2019.02.036. Epub 2019 Feb 28. PMID: 30897447.

Freni F, Moretti M, Radaelli D, Carelli C, Osculati AMM, Tronconi L, et al. Determination of fentanyl and 19 derivatives in hair: application to an Italian population. J Pharm Biomed Anal 2020 Sep 10;189:113476. https://doi.org/10.1016/j.jpba.2020.113476. Epub 2020 Jul 15. PMID: 32693203.

Garcia L, Tiscione NB, Yeatman DT, Richards-Waugh L. Novel and nonroutine benzodiazepines and suvorexant by LC-MS-MS. J Anal Toxicol 2021 May 14;45(5):462–74. https://doi.org/10.1093/jat/bkaa109. PMID: 33988239.

Gaunitz F, Kieliba T, Thevis M, Mercer-Chalmers-Bender K. Solid-phase extraction-liquid chromatography-tandem mass spectrometry method for the qualitative analysis of 61 synthetic cannabinoid metabolites in urine. Drug Test Anal 2020 Jan;12(1):27–40. https://doi.org/10.1002/dta.2680. Epub 2019 Oct 18. PMID: 31412168.

Gergov M, Nokua P, Vuori E, Ojanperä I. Simultaneous screening and quantification of 25 opioid drugs in post-mortem blood and urine by liquid chromatography-tandem mass spectrometry. Forensic Sci Int 2009 Apr 15;186(1–3):36–43. https://doi.org/10.1016/j.forsciint.2009.01.013. Epub 2009 Feb 20. PMID: 19232849.

Glish GL, Vachet RW. The basics of mass spectrometry in the twenty-first century. Nat Rev Drug Discov 2003 Feb;2(2):140–50. https://doi.org/10.1038/nrd1011. PMID: 12563305.

Hutter M, Broecker S, Kneisel S, Franz F, Brandt SD, Auwarter V. Metabolism of nine synthetic cannabinoid receptor Agonists encountered in clinical casework: major in vivo phase I metabolites of AM-694, AM-2201, JWH-007, JWH-019, JWH-203, JWH-307, MAM-2201, UR-144 and XLR-11 in human urine using LC-MS/MS. Curr Pharmaceut Biotechnol 2018;19(2):144–62. https://doi.org/10.2174/1389201019666180509163114. PMID: 29745330.

Kebarle P. A brief overview of the present status of the mechanisms involved in electrospray mass spectrometry. J Mass Spectrom 2000 Jul;35(7):804–17. https://doi.org/10.1002/1096-9888(200007)35:7<804::AID-JMS22>3.0.CO;2-Q. PMID: 10934434.

Lau T, Concheiro M, Cooper G. Determination of 30 synthetic cathinones in postmortem blood using LC-MS-MS. J Anal Toxicol 2020 Oct 12;44(7):679–87. https://doi.org/10.1093/jat/bkaa071. PMID: 32591789.

Lehmann S, Kieliba T, Beike J, Thevis M, Mercer-Chalmers-Bender K. Determination of 74 new psychoactive substances in serum using automated in-line solid-phase extraction-liquid chromatography-tandem mass spectrometry. J Chromatogr B Analyt Technol Biomed Life Sci 2017 Oct 1;1064:124–38. https://doi.org/10.1016/j.jchromb.2017.09.003. Epub 2017 Sep 7. PMID: 28922649.

March RE. Quadrupole ion traps. Mass Spectrom Rev 2009 ;28(6):961–89. https://doi.org/10.1002/mas.20250. PMID: 19492348.

Mastrovito RA, Papsun DM, Logan BK. The development and validation of a novel designer benzodiazepines panel by LC-MS-MS. J Anal Toxicol 2021 May 14;45(5):423–8. https://doi.org/10.1093/jat/bkab013. PMID: 33476376.

Mei V, Concheiro M, Pardi J, Cooper G. Validation of an LC-MS/MS method for the quantification of 13 designer benzodiazepines in blood. J Anal Toxicol 2019 Oct 17;43(9):688–95. https://doi.org/10.1093/jat/bkz063. PMID: 31436813.

Moody MT, Diaz S, Shah P, Papsun D, Logan BK. Analysis of fentanyl analogs and novel synthetic opioids in blood, serum/plasma, and urine in forensic casework. Drug Test Anal 2018 Sep;10(9):1358–67. https://doi.org/10.1002/dta.2393. Epub 2018 May 3. PMID: 29633785.

Niebel A, Krumbiegel F, Hartwig S, Parr MK, Tsokos M. Detection and quantification of synthetic cathinones and selected piperazines in hair by LC-MS/MS. Forensic Sci Med Pathol 2020 Mar;16(1):32–42. https://doi.org/10.1007/s12024-019-00209-z. Epub 2019 Dec 18. PMID: 31853826.

Nzekoue FK, Agostini M, Verboni M, Renzoni C, Alfieri L, Barocci S, et al. A comprehensive UHPLC-MS/MS screening method for the analysis of 98 New Psychoactive Substances and related compounds in human hair. J Pharm Biomed Anal 2021 Oct 25;205:114310. https://doi.org/10.1016/j.jpba.2021.114310. Epub 2021 Aug 8. PMID: 34391138.

Odoardi S, Fisichella M, Romolo FS, Strano-Rossi S. High-throughput screening for new psychoactive substances (NPS) in whole blood by DLLME extraction and UHPLC-MS/MS analysis. J Chromatogr B Analyt Technol Biomed Life Sci 2015 Sep 1;1000:57–68. https://doi.org/10.1016/j.jchromb.2015.07.007. Epub 2015 Jul 17. PMID: 26209771.

Ong RS, Kappatos DC, Russell SGG, Poulsen HA, Banister SD, Gerona RR, et al. Simultaneous analysis of 29 synthetic cannabinoids and metabolites, amphetamines, and cannabinoids in human whole blood by liquid chromatography-tandem mass spectrometry - a New Zealand perspective of use in 2018. Drug Test Anal 2020 Feb;12(2):195–214. https://doi.org/10.1002/dta.2697. Epub 2019 Nov 28. PMID: 31595682.

Payne AH, Glish GL. Tandem mass spectrometry in quadrupole ion trap and ion cyclotron resonance mass spectrometers. Methods Enzymol 2005;402:109–48. https://doi.org/10.1016/S0076-6879(05)02004-5. PMID: 16401508.

Peters FT. Recent advances of liquid chromatography-(tandem) mass spectrometry in clinical and forensic toxicology. Clin Biochem 2011;44:54–65.

Pichini S, Pujadas M, Marchei E, Pellegrini M, Fiz J, Pacifici R, et al. Liquid chromatography-atmospheric pressure ionization electrospray mass spectrometry determination of "hallucinogenic designer drugs" in urine of consumers. J Pharm Biomed Anal 2008 Jun 9;47(2):335–42. https://doi.org/10.1016/j.jpba.2007.12.039. Epub 2008 Jan 4. PMID: 18262381.

Pettersson Bergstrand M, Helander A, Beck O. Development and application of a multi-component LC-MS/MS method for determination of designer benzodiazepines in urine. J Chromatogr, B: Anal Technol Biomed Life Sci 2016 Nov 1;1035:104–10. https://doi.org/10.1016/j.jchromb.2016.08.047. Epub 2016 Sep 19. PMID: 27697727.

Qin N, Shen M, Xiang P, Wen D, Shen B, Deng H, et al. Determination of 37 fentanyl analogues and novel synthetic opioids in hair by UHPLC-MS/MS and its application to authentic cases. PMID: 32665579 Sci Rep 2020 Jul 14;10(1):11569. https://doi.org/10.1038/s41598-020-68348-w. PMCID: PMC7360565.

Ramírez Fernández MDM, Wille SMR, Jankowski D, Hill V, Samyn N. Development of an UPLC-MS/MS method for the analysis of 16 synthetic opioids in segmented hair, and evaluation of the polydrug history in fentanyl analogue users. Forensic Sci Int 2020 Feb;307:110137. https://doi.org/10.1016/j.forsciint.2019.110137. Epub 2019 Dec 31. PMID: 31927248.

Rosenberg E. The potential of organic (electrospray- and atmospheric pressure chemical ionisation) mass spectrometric techniques coupled to liquid-phase separation for speciation analysis. J Chromatogr A 2003 Jun 6;1000(1–2):8 41–89. https://doi.org/10.1016/s0021-9673(03)00603-4. PMID: 12877203.

Salomone A, Gazzilli G, Di Corcia D, Gerace E, Vincenti M. Determination of cathinones and other stimulant, psychedelic, and dissociative designer drugs in real hair samples. Anal Bioanal Chem 2016 Mar;408(8):2035–42. https://doi.org/10.1007/s00216-015-9247-4. Epub 2015 Dec 17. PMID: 26680593.

Scheidweiler KB, Huestis MA. Simultaneous quantification of 20 synthetic cannabinoids and 21 metabolites, and semi-quantification of 12 alkyl hydroxy metabolites in human urine by liquid chromatography-tandem mass spectrometry. J Chromatogr A 2014 Jan 31;1327: 105–17. https://doi.org/10.1016/j.chroma.2013.12.067. PMCID: PMC3963402.

Schneir A, Ly BT, Casagrande K, Darracq M, Offerman SR, Thornton S, et al. Comprehensive analysis of "bath salts" purchased from California stores and the internet. Clin Toxicol 2014 Aug;52(7):651–8. https://doi.org/10.3109/15563650.2014.933231. PMID: 25089721.

Seymour C, Shaner RL, Feyereisen MC, Wharton RE, Kaplan P, Hamelin EI, et al. Determination of fentanyl analog exposure using dried blood spots with LC-MS-MS. J Anal Toxicol 2019 May 1;43(4):266–76. https://doi.org/10.1093/jat/bky096. PMID: 30462229.

Strickland EC, Cummings OT, Mellinger AL, McIntire GL. Development and validation of a novel all-inclusive LC-MS-MS designer drug method. J Anal Toxicol 2019 Apr 1;43(3): 161–9. https://doi.org/10.1093/jat/bky087. PMID: 30462231.

Swortwood MJ, Boland DM, DeCaprio AP. Determination of 32 cathinone derivatives and other designer drugs in serum by comprehensive LC-QQQ-MS/MS analysis. Anal Bioanal Chem 2013 Feb;405(4):1383–97. https://doi.org/10.1007/s00216-012-6548-8. Epub 2012 Nov 23. PMID: 23180084.

Tang MH, Ching CK, Lee CY, Lam YH, Mak TW. Simultaneous detection of 93 conventional and emerging drugs of abuse and their metabolites in urine by UHPLC-MS/MS. J Chromatogr B Analyt Technol Biomed Life Sci 2014 Oct 15;969:272–84. https://doi.org/10.1016/j.jchromb.2014.08.033. Epub 2014 Aug 30. PMID: 25203724.

Vaiano F, Busardo FP, Palumbo D, Kyriakou C, Fioravanti A, Catalani V, et al. A novel screening method for 64 new psychoactive substances and 5 amphetamines in blood by LC-MS/MS and application to real cases. J Pharm Biomed Anal 2016;129:441–9.

Viette V, Hochstrasser D, Fathi M. LC-MS (/MS) in clinical toxicology screening methods. CHIMIA Int J Chem 2012;66:339–42.

Wohlfarth A, Weinmann W, Dresen S. LC-MS/MS screening method for designer amphetamines, tryptamines, and piperazines in serum. Anal Bioanal Chem 2010 Apr;396(7): 2403–14. https://doi.org/10.1007/s00216-009-3394-4. Epub 2010 Jan 13. PMID: 20069283.

Wohlfarth A, Scheidweiler KB, Chen X, Liu HF, Huestis MA. Qualitative confirmation of 9 synthetic cannabinoids and 20 metabolites in human urine using LC-MS/MS and library search. Epub 2013 Mar 18. PMID: 23458260 Anal Chem 2013 Apr 2;85(7):3730–8. https://doi.org/10.1021/ac3037365. PMCID: PMC3874406.

CHAPTER

7

NPS analysis using high-resolution mass spectrometry

HRMS platforms and role in NPS analysis

With 60–100 new drugs added each year, the sheer number of new psychoactive substances (NPS) and their rapid molecular evolution have imposed unique analytical requirements for drug testing:

- Comprehensive coverage of target analytes for the wide variety of NPS classes
- Capability to perform nontargeted data acquisition to enable discovery of previously unreported NPS
- Capacity for rapid method updates to respond to the quick phase of their molecular evolution

Immunoassays and targeted testing using gas chromatography–mass spectrometry (GC-MS) and liquid chromatography–tandem MS (LC-MS/MS) fall short in addressing these requirements. Immunoassays are limited in the breadth of their analyte coverage, while targeted analysis using GC-MS and LC-MS/MS can analyze only for prior known drugs. Furthermore, the strict guidelines for method development and validation prevent rapid updates. In the past decade and a half, the emergence of high-resolution MS (HRMS) as the method of choice for NPS has shown promise in responding to these challenges.

Like other MS assays, HRMS has analytical specificity and sensitivity equivalent to LC-MS/MS; it is far superior to immunoassays in these properties. It measures the exact mass of molecular ions instead of their nominal mass measured by GC-MS and LC-MS/MS. *Exact mass* is the monoisotopic mass of a molecular ion that is calculated from the exact masses of the most abundant isotopes of its constituent

Designer Drugs
https://doi.org/10.1016/B978-0-12-811764-4.00016-1

Fentanyl
($C_{22}H_{28}N_2O$)

Element	Nominal mass	Exact mass
carbon (x22)	12.01	12.0000
hydrogen (x28)	1.008	1.00783
nitrogen (x2)	14.007	14.0031
oxygen	15.999	15.9949
Total	336.58	336.2203

FIGURE 7.1 Difference between nominal and exact mass for fentanyl.

elements. *Nominal mass*, in contrast, is the sum of the mass number of the primary isotope of each atom in the molecular ion. For example, fentanyl ($C_{22}H_{28}N_2O$) has an exact mass of 336.2203 atomic mass units (amu) and a nominal mass of 336.58 amu (Fig. 7.1). The ability to measure exact mass facilitates assignment of the matched molecular formula from which the degree of unsaturation or double-bond equivalents of a putative formula match can be calculated and a suspect compound (drug) can be assigned. With detection of all ionizable compounds in a sample and the assignment of their molecular formulas, HRMS can perform nontargeted data acquisition, a requirement for nontargeted analysis.

The specificity of HRMS platforms, which is key to their utility in nontargeted testing, depends primarily on their resolution. Mass resolution is a measure of the ability to separate two narrow mass spectral peaks (Pasin et al., 2017). One common way to measure it is the "full width at half maximum" (FWHM) method, where the mass (m) associated with the peak is divided by the peak width at 50% of the peak height ($\Delta m@50\%$). A mass spectrometer is considered high resolution when its $m/\Delta m@50\% > 10{,}000$ (Fig. 7.2) (Xian et al., 2012). Mass resolutions between 20,000 and 1,000,000 can be achieved depending on the HRMS platform used. In contrast, LC-MS/MS has resolutions between 1000 and 10,000 (Table 7.1).

There are three analytical platforms capable of HRMS: time-of-flight (TOF), orbitrap (OT), and Fourier transform ion cyclotron resonance (FTICR). In clinical and forensic laboratories, the most commonly used is hybrid quadrupole time-of-flight mass spectrometry (QTOF/MS). This is typically appended to LC, which adds another dimension of specificity through an analyte's retention time, which is a function of the analyte's relative polarity. In some cases, it can also be appended to GC, especially when the target analytes are relatively hydrophobic, such as anabolic steroids and long-chain fatty acids.

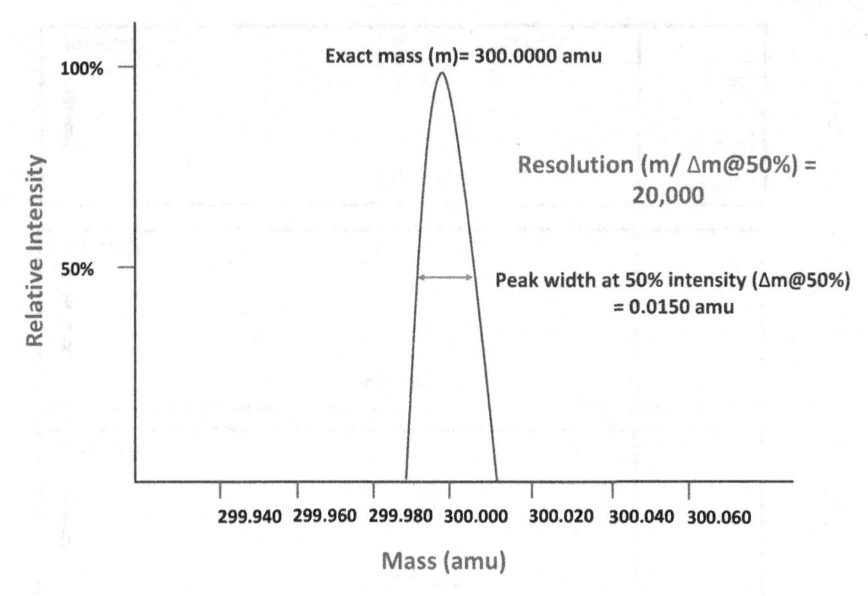

FIGURE 7.2 Full width at half maximum method for the calculation of a mass spectrometer's resolution.

TABLE 7.1 Comparison of mass resolving power of various LC-MS platforms.

Platform	Mass resolving power (FWHM)
Quadrupole ion trap	1000
Time-of-flight/quadrupole time-of-flight	10,000–40,000
High-resolution time-of-flight	40,000–60,000
Fourier transform orbitrap	100,000
Fourier transform ion cyclotron resonance	1,000,000

Instrumental analysis techniques

The key components of HRMS are the same as other types of MS: ion source, mass analyzer, and detector. Like LC-MS/MS, HRMS relies on soft ionization techniques, which include electrospray ionization (ESI), atmospheric pressure chemical ionization (APCI), and atmospheric pressure photoionization (APPI). The principles behind these ionization sources are discussed in Chapter 6.

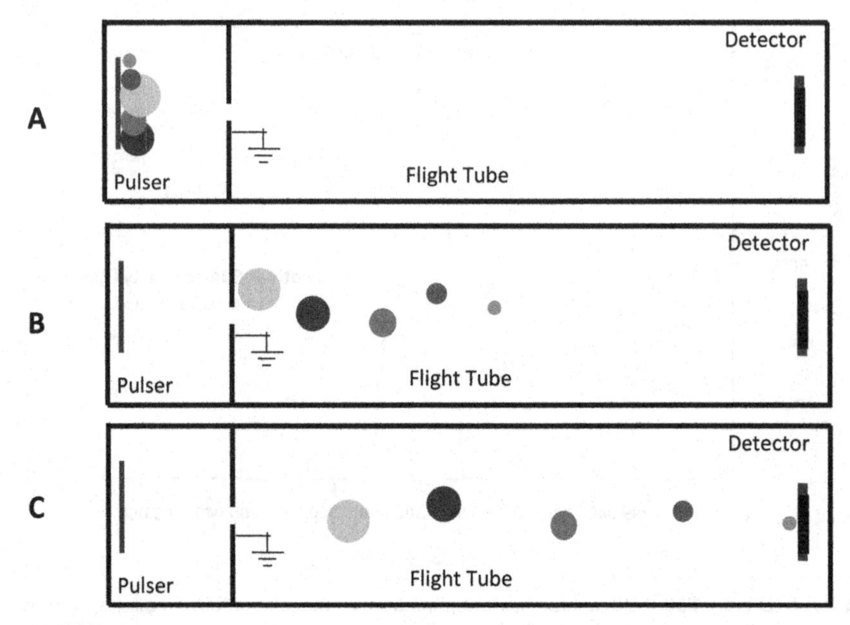

FIGURE 7.3 Schematic of the principle behind TOF/MS. (A) Ions are aligned at the pulser before being repelled by an electric jolt. (B and C) Ions travel in the flight tube according to their size.

For mass analyzers, HRMS can be facilitated in several ways: TOF, OT, and FTICR. TOF is the most common and the only one discussed in detail here. In TOF/MS, ions are accelerated by an electric field of known strength in a flight tube. This acceleration results in ions of similar charges acquiring the same kinetic energy as they traverse the tube (Fig. 7.3). Because the kinetic energy of any moving particle is equivalent to half its mass times the square of its velocity, the velocity of the ion as it moves in the electric field is inversely related to its mass-to-charge ratio (m/z). Heavier ions (higher m/z) travel slower than lighter ions (lower m/z). TOF/MS measures the time it takes for an ion to travel between the pulser (point of origin for ion acceleration) and the detector; this is dependent on the ion's velocity and is proportional to its m/z (heavier ion, longer time of flight). TOF/MS allows measurement of the exact mass of the parent or precursor ion. The measured exact mass by a TOF/MS is technically referred to as accurate mass.

Although the accurate mass of a precursor ion allows formula assignment, its corresponding identity is not always easy to decipher. Several formulae may fit a given accurate mass and each formula can have multiple isomers. To allow confirmation of the identity of a compound from a given formula, information on fragment ions that

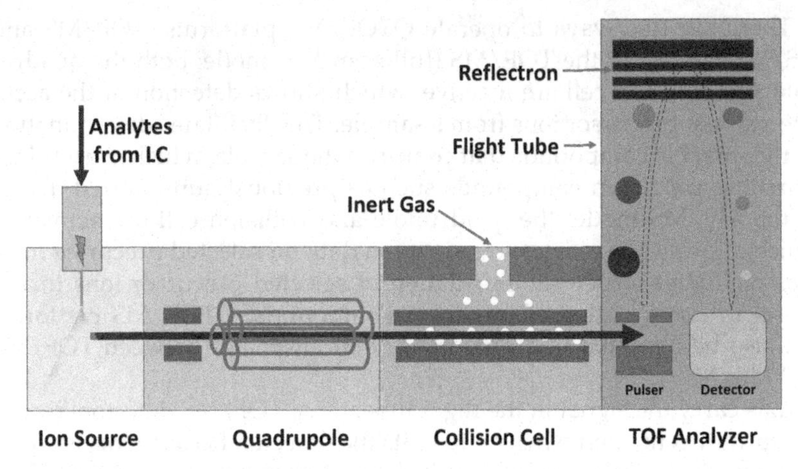

FIGURE 7.4 Schematic of a quadrupole time-of-flight mass spectrometer.

can be generated from the precursor is required. This can be done by placing a quadrupole and a collision cell before the TOF to form the hybrid QTOF/MS platform (Fig. 7.4). As previously discussed, the quadrupole allows selection of a specific ion from a mixture of precursor (parent) ions. It serves as a selective gate that allows only a specific accurate mass designated by the user to pass through and continue on its path. The selected precursor ion is bombarded with inert gas (usually nitrogen) in the collision cell. Fragmentation of the selected precursor ion is induced by collision with the inert gas at a defined energy. This happens in specific ways based on the relative labilities (strength) of the bonds in the precursor ion's molecular structure. The structure of the precursor ion dictates the specificity of its fragmentation allowing QTOF/MS to collect structural data. The type and number of fragment or product ions obtained from the collision cell also depend on the energy of the inert gas, which can be controlled based on the selected operational parameters applied. The fragment ions obtained from the collision cell eventually enter the flight tube, where their accurate masses are measured by TOF/MS.

The collection of fragment ion accurate masses comprises the mass spectrum of the molecule associated with the precursor ion. Unlike GC-MS, the mass spectrum generated from a compound is not universally similar across QTOF/MS platforms. With soft ionization technique, the number and relative intensity of fragment ions generated vary according to the collision energy applied and the QTOF/MS platform used. Thus, each QTOF/MS platform creates and maintains spectral libraries that are not directly usable in another platform.

There are two ways to operate QTOF/MS platforms: TOF/MS and MS/MS modes. In the TOF/MS (full-scan MS) mode, both the quadrupole and collision cell are inactive, which allows detection of the accurate mass of precursor ions from a sample. This facilitates determination of all ionizable compounds that comprise the sample, which is useful for searching unknown compounds such as previously unreported drugs. In the MS/MS mode, the quadrupole and collision cell are activated, which allows collection of fragment ion data on selected precursor ions. This facilitates structural elucidation of selected precursor ions that is useful in confirmation of unknown compounds. QTOF/MS platforms can also be operated in a dual mode that alternates between TOF/MS and MS/MS.

Ions carry the signal in the flight tube; thus, TOF/MS detectors record the current generated when an ion hits the detector. Because this is weak, detectors typically consist of a multichannel plate detector or a secondary emission multiplier, where the first dynode (converter plate) is flat to propagate or magnify the initial electrical signal. The detected electrical signal is then converted using a time-to-digital converter (TDC) or an analog-to-digital converter (ADC). TDC provides a digital representation of the time an event occurred (in this case, an ion hitting the detector); either the absolute time when the event occurred or the time interval between two events is digitized. On the other hand, ADC converts an analog signal (e.g., sound, voltage, current) into a digital signal. In TOF/MS, an ADC converts an input current to a number that represents the magnitude of the current (signal intensity). Thus, the TOF of the ion is converted to its accurate mass. The larger the ion, the slower its TOF (~ larger TOF value) and the higher the accurate mass measured.

Data acquisition strategies

The primary advantage of HRMS is its versatility in data acquisition. As mentioned earlier, the two common modes at which it can be operated are full-scan MS (TOF/MS) and MS/MS.

Full scan is the simplest acquisition mode carried by all HRMS platforms. It measures all intact ion (precursor) masses that reach the detector from the sample. This measurement provides the accurate mass expressed as m/z of all ions detected (Maurer, 2010). Most ions have unit charge, so their m/z is equivalent to their mass. Advances in optics and detectors in the past two decades have allowed these measurements with accuracies of less than 2 parts per million (ppm) in mass error, although the typical tolerance threshold used in practice for identifying a compound is 5–10 ppm.

Mass accuracy is expressed in mass error as ppm. This is equivalent to the ratio of the difference between the measured accurate mass and theoretical exact mass of an ion (Δm), and its exact mass (m) multiplied by 1,000,000 (Δm/m × 1,000,000). At 2–10 ppm, mass measurements are accurate to about 0.001 amu or lower for the compounds with exact masses of 100–500 amu, which are most commonly found in biological samples. At such levels of accuracy, it becomes possible to generate molecular formulae that match the accurate masses measured. With generation of formula matches, the degree of unsaturation associated with an accurate mass can also be calculated. This allows what is called nontargeted data acquisition, where formulae for all ions generated from a sample can be assigned (Wu et al., 2012). This contrasts with how LC-MS/MS typically acquires data through targeted analysis, where only data for known target analytes included in the method are acquired.

The unbiased data acquisition approach facilitated through non-targeted data acquisition has advantages in data query that are not possible for targeted data acquisition. For one, nontargeted collects comprehensive data on all ionizable components of a sample. This allows discovery of previously unreported compounds that may be present. The discovery procedure may be complex; nonetheless, it provides an avenue to discover newly released NPS that have not been previously analyzed and reported. Retrospective data analysis can also be done with full-scan MS acquisition. This is very powerful, as it allows interrogation of previously acquired data on analytes that may not be known during the time of analysis without rerunning the sample. Retrospective data analysis is a great tool for drug surveillance in that it allows querying the temporal origins of a newly discovered NPS (Gundersen et al., 2020).

Although TOF/MS has been implemented in the analysis of NPS, one big limitation is its inability to provide structural data. This can be critical, because NPS can have structural isomers. In cases where the isomers differ substantially in structure and polarity, the chromatographic separation that precedes QTOF/MS may be able to differentiate them with their retention time. However, in cases where these isomers are positional, relying on retention time differences from chromatography alone may not be enough to distinguish them from one another. MS/MS acquisition can provide fragment ions data for the structural elucidation of the precursor ion. This can be done in either tandem-in-space, which is common with QTOF/MS, or tandem-in-time, common with linear ion orbitrap. For either strategy, MS/MS can be operated in data-dependent acquisition (DDA, information-dependent) or data-independent acquisition (DIA, information-independent) mode.

Data-dependent acquisition

A full-scan MS survey is initially done, after which MS/MS scans are triggered on a select number of precursor ions from the preceding MS scan that satisfy predetermined criteria such as signal intensity threshold (Fig. 7.5). The MS/MS scans can be nontargeted or targeted.

Nontargeted scans are triggered based on the abundance of the precursor ions. The number of MS/MS events triggered is set by the user. A higher number of scans can be collected to increase the coverage of precursor ions with structural data. However, the number of MS + MS/MS scans that comprise the cycle time of each complete set of scans should be considered in conjunction with the chromatographic peak width of the target analyte. To get the ideal Gaussian curve for a peak that can be accurately quantified, 10–12 sampling points are necessary. Hence, for a typical peak width of 0.1 min (6 s) (minutes and seconds), for example, the cycle time cannot exceed 500 ms. If each scan time is set at 50 ms, then one full scan and nine MS/MS scans can be obtained to get 12 sampling points.

Limitations of nontargeted MS/MS in DDA include (1) oversampling of abundant irrelevant compounds such as background endogenous metabolites or contaminants, (2) oversampling of highly abundant

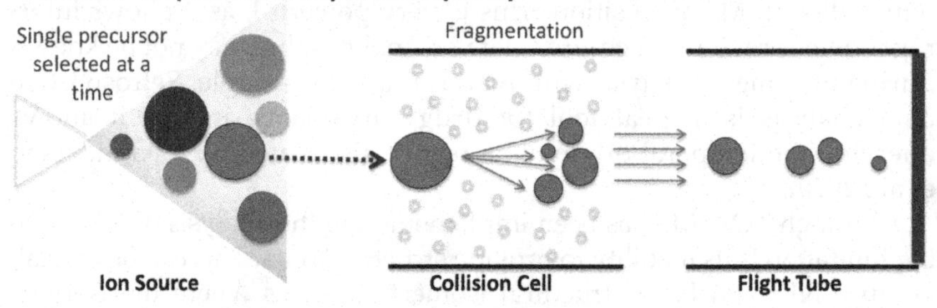

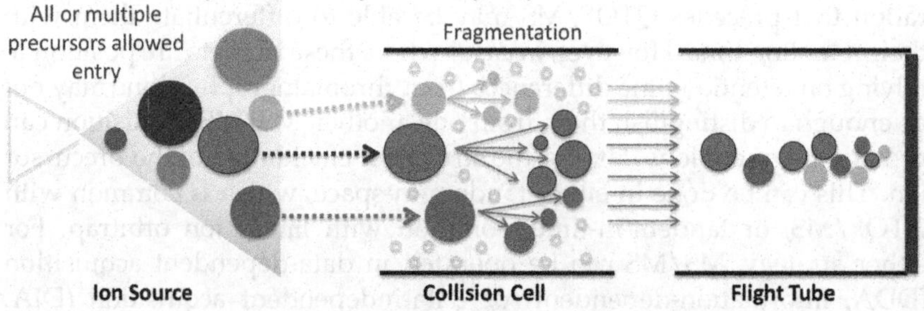

FIGURE 7.5 Schematic illustrating the difference between two data acquisition strategies.

relevant compounds that have already been previously selected, and (3) failure to select potentially relevant analytes of low abundance for fragmentation (Allen and McWhinney, 2019). These limitations can be prevented by the active exclusion of previously selected precursor ions held at a specified time period and/or using an exclusion list of background and contaminant precursor ions. The exclusion list can be appended in the method so that the precursor selection algorithm used in data acquisition can implement it. With active exclusion, on the other hand, selection of an abundant precursor ion is suspended for however long the user chooses to implement it. For example, with active exclusion held at 5 s, the maximum number by which a highly abundant fragment ion can be selected for fragmentation is only two with the average peak analyte peak width of 6 s.

Targeted precursor selection for MS/MS scans can also be done in DDA. An inclusion list of precursor ions' m/z with or without their characteristic retention time is incorporated in the method. Each time a precursor ion on the list is detected in the MS scan above a set abundance threshold, an MS/MS scan of the precursor ion is induced. This method requires a list of known targets, and if retention time is included, previous chromatographic characterization of the reference standard for the target analyte should have been done. Incorporation of the inclusion list can increase the chances of obtaining structural data for each target analyte if it is present in the sample. Moreover, because the number of MS/MS scans triggered is much less, the analysis is a lot simpler. However, this method does not allow retrospective structural analysis of previously unreported NPS. So if a promising suspect is obtained from the TOF/MS scan, the analysis of the sample will have to be rerun to get structural data for it. If a certain period has passed since the analysis was done, the sample might no longer be available for reanalysis.

Targeted and nontargeted precursor selection for MS/MS scans can also be run together. In this method, a full-scan MS survey is followed by a targeted MS/MS scan using an inclusion list, and then by a nontargeted MS/MS scan on n-selected precursors (Gonzalez- Marino et al., 2016; Kleis et al., 2021a,b). This combined approach can be very effective in ensuring that known target analytes are not missed, but at the same time provide the opportunity to discover previously unreported NPS. Careful monitoring of the cycle time, however, should be done to ensure that the method can provide both qualitative and quantitative data. Furthermore, the large amount of data extracted requires huge data storage capacity. It is common to get more than a gigabyte of data from a single sample run, in contrast to only several megabytes from an LC-MS/MS run.

To maximize both unbiased survey of precursor ions and fragment data selection of promising precursor leads in an analysis, a combination of approaches can be done using DDA. For example, a laboratory can perform DDA with nontargeted scans as its primary method to allow a more unbiased selection of precursor ions to fragment. If the initial analysis reveals a promising precursor ion that was not fragmented, a follow-up run with targeted precursor selection using an inclusion list of promising precursor leads can then be pursued. This approach is especially useful when dealing with very potent drugs that have very low levels in biological matrices. Often, these drugs have low abundances; hence, they are not typically selected in DDA with nontargeted scans.

Data-independent acquisition

In DIA, no predetermined criteria are imposed in triggering MS/MS scans. Instead, all detected precursor ions are subjected to collision-induced dissociation (CID) (Fig. 7.5). This can be done in a number of ways, depending on the available platform. Most common is to do rapid switching between low-energy and high-energy channels to collect data for both MS and MS/MS, respectively. Examples for different HRMS platforms are All Ions MS/MS (Agilent Technologies), MS^E (Waters), and broadband CID (Bruker). An alternative method was developed by AB Sciex using sequential window acquisition of all theoretical spectra (SWATH), which involves sequential CID of mass range increments. An initial MS scan of precursor ions is done, followed by MS/MS scans to generate precursor ions in small mass range isolation windows (~20 Da) of the precursor ions. Using many consecutive isolation windows, the entire defined mass range of the precursor ions is stepped through continuously during each scan cycle.

DIA provides the most comprehensive MS/MS data of all non-targeted acquisition strategies and yields the best opportunity for discovering previously unreported NPS. However, because MS/MS scans of multiple precursor ions are happening simultaneously, the fragment or product ion data generated are chimeric spectra. Analysis and interpretation of these spectra can be very challenging if the precursor ions cannot be linked to the correct product ions (Oberacher and Arnhard, 2016).

Data processing techniques

Data processing varies according to the aim of an analysis and the platform employed. Vendors of different HRMS platforms have developed their own software to process data obtained from LC-HRMS. Thus, familiarization with the specific software that accompanies a given platform is necessary when transitioning from one platform to another. The MS/MS data obtained for a compound from LC-HRMS analysis are also not universal. This is in wide contrast to mass spectral data from GC-MS, where standardization allows the same MS/MS data to be collected regardless of the vendor platform. An NIST library of GC-MS spectra for drugs of abuse is available that can be universally used in different GC-MS platforms. The equivalent does not exist for LC-HRMS. The qualitative composition of the MS/MS spectra and the relevant abundances of product ions obtained from an LC-HRMS run are highly dependent on the collision energy used for fragmentation. The design of the HRMS platform may also influence the mass spectral data; HRMS vendors sell their own spectral library for use in their own platforms. A crowd-sourced NPS mass spectral library recently became available online. Curators of the library make it available in multiple formats to allow compatibility with various HRMS platforms' data processing software.

Depending on the aim, three different workflows used in LC-HRMS are available: targeted analysis, suspect screening, and nontargeted analysis. The workflows differ in whether the target analytes are known a priori or not, and whether reference standards for the target analytes are available or not. Table 7.2 summarizes the requirements and differences of the three workflows.

Targeted analysis

This involves the interrogation of LC-HRMS data based on compound data collected from its available reference standard. Typically, a library is created for all target compounds, where their exact mass, retention time, and mass spectral data are compiled. Targeted analysis can be used for

TABLE 7.2 Types of analysis facilitated by QTOF/MS. Differences on requirements, criteria for matching, and the type of results are depicted.

Analysis type	Targeted	Suspect screening	Nontargeted
Reference availability (in-house)	Reference standards Drug database Mass spectral library	Drug database Mass spectral library	None
Match criteria	Accurate mass Isotope cluster Retention time (RT) Mass spectral library	Accurate mass Isotope cluster RT plausibility; relative RT (if RT available in literature)	Accurate mass Isotope cluster RT plausibility
Results generated	Confirmed drug Quantitative level of confirmed drug	Suspect drug Relative levels of suspect drug across samples	Suspect drug

screening (qualitative) and confirmation (quantitative). Identification and confirmation of a drug depends on the criteria set by the user. Although there are no universal guidelines for what is considered a compound match, the usual criteria include mass error <5−10 ppm (mass match), retention time match <0.1−0.2 min, target score (overall indicator of mass, retention time, and precursor ion's isotope pattern match) ≥ 70, and a product ion match to the precursor's spectral data. The weighting of the criteria for the target score varies from one vendor to the other and can be reconfigured by the user. When a chromatographic column, for example, has been used for a while, the weight of the retention time match's contribution to the target score can be temporarily relaxed. As the column ages and junk from all sample runs accumulates, retention time drift in the analytes usually occurs.

There is even more diversity of implemented rules on spectral library matching. Vendor software uses different algorithms for spectral matching, which can significantly affect the compound's scoring profile. Generally, a requirement for matches to two or more product ions is imposed. There are two types of spectral matching: forward and reverse. Forward matching is done by searching the product ions obtained for a putative precursor ion from a sample in the product ions of the compound in the library (reference spectra). The opposite search where the product ions of the compound in the library are searched in the product ions generated for the putative match in the sample (reference spectra) is reverse spectral matching. Because penalties in scoring are imposed on ions that do not have matches on the reference spectra and fragment ions data from a sample are seldom very clean unlike data from a reference standard, reverse spectral matching is usually given more weight.

As a screening assay, LC-HRMS offers significantly better specificity and coverage than immunoassays, the most common method used for screening in clinical and forensic laboratories. It also allows nontargeted analysis, which is not the case for either immunoassays or LC-MS/MS. More importantly, for qualitative analysis, a targeted panel's method can easily be updated and expanded with data from new reference standards. A disadvantage of targeted analysis that LC-HRMS shares with other platforms is the unavailability and cost of some reference standards. Reference standards for newly discovered NPS may not be available right away. The ease by which some can be procured may also be limited with scheduled drugs and NPS. And procuring hundreds of reference standards, some of which will have limited temporal relevance, is costly. For quantitative analysis, the linear dynamic range for some analytes may not be as wide as that established for LC-MS/MS, either because the LC-HRMS assay has lower sensitivity for the analyte or the signal for the analyte saturates right away.

Suspect screening

An alternative way of performing targeted screening is by interrogation of the acquired LC-HRMS data with a database or library of suspect compound-specific data in the absence of a reference standard (Krauss et al., 2010). The database can be formulas and exact masses compiled by a vendor or another third party; inclusion of mass spectral data converts the database to a library. The obvious difference from targeted analysis is the lack of data for the retention time of the suspect. So the criteria for evaluating the suspect rely on mass error, isotopic pattern match through the target score, and spectral library match if spectra of the suspect are available. At the end of suspect screening, the compound match remains a suspect and not a confirmed compound. A reference standard of the suspect needs to be procured and run using the same method for the suspect compound. Procurement and analysis of the reference standard converts the confirmation of the suspect to a targeted analysis.

Appending suspect screening to a targeted analysis significantly broadens the coverage of the analytical method without additional investment in reference standards. This is a cost-effective way to triage the NPS that may be relevant in a specific patient population. Having a drug or NPS as a suspect first allows a laboratory to decide whether an investment in buying its reference standard is worthwhile based on the frequency with which the suspect is screened in the patient population. The plausibility of the retention time of the suspect can also be assessed first to determine the likelihood of its confirmation. The molecular structure of the suspect should allow an analyst to assess whether its observed retention time falls within the range expected based on the chromatographic method applied. If a chromatographic method for a panel that includes the NPS has been reported in the literature and another compound in the panel (anchor compound) is also in the panel of the laboratory, the relative retention time of the NPS to the anchor compound can also be used to assess the likelihood that it will be confirmed by a reference standard.

The use of suspect screening is limited by the coverage of the database or mass spectral library available to the analyst. For NPS analysis, the challenge in mass spectral libraries sourced from vendors is the lag time between when an NPS appears in the recreational drug market and when it is added to the vendor's library. Especially for rapidly evolving NPS classes, even a yearly update of the mass spectral library will not suffice to make it very useful. Therefore, most laboratories develop their own in-house library to stay up to date with NPS. It is also noteworthy to reiterate that a mass spectral library developed from one platform cannot be directly used in another platform.

Nontargeted analysis

Nontargeted data acquisition allows the collection of comprehensive data on a sample that may provide an avenue to discover previously unreported NPS. In such a case, the potential NPS or drug is not part of a database or library used in targeted analysis and suspect screening. The workflow necessary for this discovery process is nontargeted analysis (Oberacher and Arnhard, 2016).

Nontargeted analysis is often mistakenly equated to nontargeted data acquisition. The latter is just part of the workflow for nontargeted analysis. Data analyzed through suspect screening or even targeted analysis can also be acquired through nontargeted means. Nontargeted analysis is applied when no prior information about the compound in a sample is known. Strictly speaking, nontargeted analysis follows the same workflow used in metabolomics, where all the precursor ions are identified in the treatment samples and compared with all the precursor ions identified in the control samples. The comparison allows discovery of changes in the general metabolic profile induced by a treatment and allows the identification of potential biomarkers that can be associated with the treatment. A statistical treatment group can be a drug taken, a regimen prescribed, or a disease. General metabolic profiling, however, can pull out many compounds or metabolites that may not have direct relevance to clinical and forensic toxicology. Thus, metabolomic workflows may complicate nontargeted analysis without producing "high-value compounds" directly related to toxicology.

A biased approach to nontargeted analysis that allows discovery of new compounds and metabolites related to NPS has found more success in NPS analysis. Two diametrically opposed methods are usually employed: top-down and bottom-up (Pasin et al., 2017; Maurer, 2020). In top-down, nontargeted interrogation of LC-HRMS data is done by the selection of most abundant peaks from the total ion chromatogram (TIC) by visual inspection. Once a peak is selected, a formula match can be assigned to the accurate mass of the most abundant precursor ion associated with the peak, after which other structural information such as the degree of unsaturation of the putative compound can be calculated. The mass spectral data of the putative compound is usually compared for similarity with the mass spectral data of known NPS, and a careful analysis of structural assignments to product ion data allows ruling out or assignment of the putative identity of the unknown compound.

The top-down approach has an obvious limitation: the degree of analytical complexity can rise exponentially when the TIC presents too many peaks that require analysis. Thus, the approach works well with simple TICs, when obvious peaks of interest are limited to one or a few such as those obtained from drug product analysis (Schneir et al., 2014).

For biological samples, the TICs obtained are far too complex to be reasonably amenable to this approach. Furthermore, in a significant number of cases, the peak that may have toxicological relevance is small and buried in highly abundant endogenous metabolites and contaminants, especially for some potent drugs in the synthetic cannabinoid and new synthetic opioid classes of NPS.

In the bottom-up approach, class-specific mass spectral information is used to interrogate LC-HRMS data. Such information may be common product ions or neutral losses characteristic of an NPS class. For fentalogs, for example, the phenethylpiperidinium ion ($m/z = 188.1434$) is a common product ion shared. Hence, a precursor ion search for it can lead to the discovery of a novel analog belonging to the fentanyl analog class. Neutral loss filtering works the same way. The bottom-up approach relies heavily on the generation of MS/MS data, so the DIA mode, which produces a significantly larger number of MS/MS data, would be more compatible with it. A limitation is the requirement for data processing software capable of generating a large number of extracted ion chromatograms for a large number of common product ions if a comprehensive analysis of all NPS classes is needed. Moreover, with the current practice of scaffold hopping in generating new subclasses of NPS, the newer generation of NPS may not share a common product ion with an NPS class. This is demonstrated in the emergence of new scaffolds used for new synthetic opioids such as the nitazenes that do not share the common product ions found in fentalogs.

Laboratory application of LC-HRMS analysis

Our laboratory is among the first to explore the utility of LC-TOF/MS in drug analysis in biological samples. Our method has evolved from being focused on unknown drugs in emergency intoxication cases referred to a poison center from 2010 to NPS surveillance across the United States since case referrals started to come in 2012 (Wu et al., 2012; Trecki et al., 2015). Now we use LC-QTOF/MS for the screening, confirmation, and quantitative analysis of NPS, TRD, prescription drugs, dietary supplement stimulants, and drug precursors, adulterants, and impurities in surveillance samples from clinical and forensic cases. Although our main work is NPS analysis, NPS is hardly present by itself in most intoxication and fatal cases; a comprehensive analysis of each case requires analysis of other drugs.

We analyze NPS in three stages: screening by TOF/MS, confirmation by MS/MS (QTOF/MS), and quantitative analysis. TOF/MS is used to generate matched compounds using precursor ion data, while MS/MS provides confirmation via mass spectral matching; quantitative analysis of the confirmed drug is by the isotope dilution method. To generate MS/MS spectra, data-dependent acquisition with active exclusion set at 30 s is used. The three most abundant peaks are automatically chosen from each TOF/MS scan for precursor ions to undergo an MS/MS scan. A chosen accurate mass for a precursor ion is excluded for MS/MS spectral generation for 30 s to acquire as many MS/MS spectra of individual precursor ions as possible. Using this nontargeted data acquisition strategy, we are able to perform both targeted analysis and suspect screening. To facilitate data processing by targeted analysis, we employ a laboratory-developed mass spectral library of more than 1200 drugs and other substances for which the laboratory has reference standards and mass spectral data. For suspect screening, we also developed databases for various pharmacological classes of NPS (stimulants, opioids and depressants, psychedelics and dissociatives, and synthetic cannabinoids) and dietary supplement bioactive compounds. Both our targeted mass spectral library and suspect databases are updated monthly to keep current with the molecular evolution of NPS classes and other evolving drug trends.

A screening match by TOF/MS is generated with the following criteria: mass error $\leq$10 ppm, retention time match within 0.15 min, and a target score $\geq$70. Confirmation by MS/MS is achieved with a mass spectral library match score of at least 70 with at least two fragment ion matches to the reference standard's MS/MS spectrum. Clearly, even though the basic premise for HRMS is matching the measured accurate mass of a compound to its exact mass in a database, a combination of tools is required to safeguard against false-positive matches. This is necessary because a given accurate mass can have multiple formula match assignments within a 10 ppm mass error, and each formula match can further have multiple isomers. Chromatographic data from retention time matching helps in differentiating isomers to a great extent. Additionally, the target score, which is a weighted composite of the matches between accurate mass, isotopic abundance and spacing pattern, and retention time, further helps in differentiating isomers and a few isobaric compounds with very close exact masses. Ultimately, the mass spectral library match provides confirmation of a compound. This is similar to the specificity provided by matching the ion ratios of two transitions in LC-MS/MS except that several fragment ions and their relative abundances are being matched instead of just two. Mass spectral matching also takes into consideration the accurate masses of the fragment ions aside from their relative abundances.

Often our laboratory also performs suspect screening using our in-house suspect databases as well as those from vendors such as Agilent's Forensic Tox Database and Library and open-source libraries such as HighResNPS (Gundersen et al., 2020). Our criteria are the same for matching as for targeted analysis except the use of retention time matching and mass spectral library matching if we lack the mass spectrum of a suspect NPS. Instead of retention time matching, we use retention time plausibility as an additional criterion for screening a suspect. The structure of a suspect primarily dictates its retention time. The laboratory has compiled the expected retention time ranges of specific classes and subclasses of NPS based on their molecular structure and the characteristics of the chromatography column. These expected retention times are compared with the suspect's to assess its retention time plausibility. In a few instances, where the retention times of a suspect and another drug or internal standard similar to ours are published, we have also used relative retention time as a criterion for suspect matching. Tentative matches or suspects are generated at the end of suspect screening. To confirm a suspect, its reference standard needs to be purchased and the mass spectral library match between the reference standard and the suspect also assessed.

Some NPS are rapidly metabolized; therefore, only their metabolites can be detected and measured in biological samples. In such cases, our laboratory also performs suspect screening of the predicted metabolites of a suspect NPS (Schwartz et al., 2015; Tyndall et al., 2015; Adams et al., 2017; Kasper et al., 2019). This is more frequently true for synthetic cannabinoids, especially when only urine samples are available. In this type of screening, we predict the possible metabolites of a suspect NPS based on known metabolic pathways published for similar classes of drugs. The formulae and accurate masses for the predicted metabolites are compiled in a database that is then used to query the total ion chromatograms generated from sample runs. The matching criteria used for analysis include mass error, target score, and retention time plausibility. Retention time plausibility in this type of analysis is based on the predicted retention time from the metabolic transformation of the parent drug for which the retention time is known. Because the structures of the predicted metabolites are known, it is also easy to predict a common fragment ion among the metabolites and the parent drug. This fragment ion can be searched in the MS/MS spectrum obtained for a suspect metabolite as a further confirmation. Although our platform does not allow it, this experiment can be done in other LC-QTOF/MS as a precursor ion scan.

In both targeted analysis and suspect screening, there are instances when a match or suspect obtained from screening does not generate an MS/MS spectrum in the MS/MS run. Two common instances when this happens are (1) the match or suspect has such a low concentration that its precursor ion is not selected for fragmentation after a TOF/MS scan, and (2) the structure of the suspect or match is too stable to fragment. In both instances, we perform additional runs of the sample using a modified data-dependent acquisition strategy. For matches or suspects that have low concentration, we apply DDA with targeted scan by adding the accurate mass of the match or suspect in an inclusion list that is appended to the acquisition method. By doing this, when the accurate mass within the mass error window set for the match or suspect is detected at a given time range, the suspect is prioritized for fragmentation regardless of its relative signal intensity. Thus, the method is forced to obtain an MS/MS spectrum of the suspect or match, provided it meets a threshold signal intensity that we set. For drugs with very stable structure, we are successful in generating an MS/MS spectrum by running the acquisition at much higher collision energies.

Once a match or suspect is confirmed, we run a quantitative analysis of the drug by the isotope dilution method. This is similar to what is done for LC-MS/MS except that the precursor ion signal is quantified instead of a chosen fragment ion, and the mass error and retention time of the precursor ion are used as qualifier instead of fragment ions ratio and retention time. With comprehensive targeted panels comprised of more than 1200 drugs, it is impractical and very costly to buy the internal standard for each drug analyte. Instead, a set of internal standards that brackets the retention times of the analytes in the panel is used. For our method, 14 internal standards bracket the analytes with retention times between 0.5 and 9.6 min in a 14-min method. The internal standard with the closest retention time to the drug being quantified is used for quantitative analysis.

Use in NPS analysis

LC-HRMS has been implemented in drug analysis in the past decade and a half. Because of its ability to facilitate nontargeted analysis, its utility has been advantageously exploited for NPS. However, because NPS intoxication commonly involves TRD, most LC-HRMS methods involve comprehensive drug panels that include them. Most published methods rarely focus on a specific NPS class. Thus, in contrast with LC-MS/MS methods, a survey of published LC-HRMS methods according to approach and strategy makes more sense than according to NPS class.

LC-TOF/MS

Most LC-HRMS methods initially published involve only TOF/MS. A few have been applied to biological samples. LC-TOF/MS has been used in qualitative and quantitative analysis of drugs. One such method was developed in hair for the simultaneous targeted screening and quantification of drugs of abuse and prescription drugs (Nielsen et al., 2010).

The panel consisted of 52 drugs, including amphetamines, opioids, benzodiazepines, cocaine, ketamine, antipsychotics, antidepressants, and analgesics. The analytes in the 17-minute method were ionized by ESI in the positive mode, and accurate mass data were collected for ions with m/z 50 to 1000 using nontargeted data acquisition. The criteria used for screening were mass tolerance within 15 mDa (mDa equivalent to 15–150 ppm for m/z 100–1000) and retention time match within 0.2 min. The same raw data obtained for screening were used for quantification, but the criteria for peak selection were adjusted to a mass tolerance within 30 mDa and retention time match within 0.05 min. A set of five internal standards was used in quantification. Although the mass tolerance for the assay was wide, the method demonstrated good selectivity and achieved limits of detection (LODs) of 0.01–0.10 ng/mg for 10 mg hair samples. Application in 15 autopsy hair samples identified 79 positive hits consisting of 29 different analytes. The mean mass accuracy obtained was 1.82 mDa (~7.28 ppm for m/z 250). An evaluation of the software used for data processing showed a higher tendency for false-negative findings on some of the target analytes. The authors also noted that the number of false positives can significantly increase with a larger database of drugs used unless narrower tolerance criteria are implemented.

Guale et al. demonstrated that the data obtained from LC-TOF/MS are verifiable by previously established and clinically validated methods (Guale et al., 2013). They used LC-TOF/MS in screening for common drugs of abuse, prescription drugs, synthetic cannabinoids, and synthetic cathinones in whole blood, serum, and urine. With broad matching criteria for compound identification (mass error <15 ppm, retention time [RT] < 0.15 min, target score ≥ 50), the validated assay identified all drugs in 21 cases that were previously confirmed by immunoassay, GC-MS, and LC-MS/MS, except for barbital, which does not ionize well in the positive ionization mode that was used. The targeted screening method consisted of 95 drugs. Accurate mass measurement distinguished isobaric compounds that cannot be chromatographically separated using a 15 ppm mass error match criterion, while differences in RT differentiated isomeric compounds with RT match within 0.15 min. The authors did not report any NPS in the 21 cases even though the method included 22 synthetic cannabinoids and 8 synthetic cathinones,. However, this is because the targeted methods being used for comparison cannot detect synthetic cannabinoids and cathinones, rather than a failure of LC-TOF/MS.

Lung et al. reported a more comprehensive LC-TOF/MS targeted analysis supported by suspect screening and nontargeted analysis (Lung et al., 2016). The targeted panel consisted of 325 drugs of abuse, prescription drugs, and NPS, while the laboratory-developed suspect database consisted of ~2000 NPS and their metabolites. The targeted panel included 10 NPS classes: synthetic cannabinoids, cathinones, phenethylamines, amphetamines, piperazines, tryptamines, arylcyclohexylamines, aminoindanes, benzofurans, and new synthetic opioids. The method implemented narrower matching criteria for drug identification: mass error <10 ppm, RT < 0.10 min, target score ≥ 70. Applied to 23 serum samples obtained from agitated emergency department patients who received haloperidol and benzodiazepine, the authors identified seven NPS on top of common drugs of abuse, including three synthetic cannabinoids, one synthetic cathinone, one piperazine, one ketamine analog, and a previously unreported opioid analog, herkinorin. The study demonstrated the extensive breadth of analyte coverage possible with HRMS methods, allowing discovery of emerging drugs in biological samples. It also underscored that compound identification beyond targeted analysis can be simultaneously accomplished by suspect screening and nontargeted analysis in the same sample run.

Another study used broad-spectrum LC-TOF/MS as a targeted screening panel for TRD, prescription drugs, and NPS in whole blood from forensic cases (Papsun et al., 2017). Two new synthetic opioids, U-47700 and furanyl fentanyl, were identified in 11 cases, which were all then confirmed and quantified by LC-MS/MS, demonstrating that LC-TOF/MS can be a very effective screening method paired with LC-MS/MS quantitation. Unfortunately, the total number of drugs in the panel and the criteria for matching of the LC-TOF/MS method were not reported.

LC-QTOF/MS

As previously noted, full-scan MS is limited by its inability to provide structural data. With the proliferation of isomeric NPS, methods that include MS/MS became a necessity. Most HRMS methods that are currently employed use LC-QTOF/MS and take advantage of both its full-scan MS and MS/MS scan capabilities.

Data-dependent acquisition strategies

The analytical utility of QTOF/MS in drug analysis was first demonstrated in simpler matrices. Andres-Costa et al. used DDA to monitor and quantify 42 common illicit drugs and their metabolites in wastewater (Andres-Costa et al., 2016). Precursor ions that exceeded 1000 counts within a mass error tolerance of 10 mDa of the targeted precursor ions were selected for MS/MS scan at 40 eV. An alternative method with multiple collision energies for low and high intensity peaks was also used on a case-by-case basis. Suspect screening was performed concurrently using a database comprised of ~2000 pharmaceuticals, drugs of abuse, NPS, and pesticides. The authors demonstrated the utility of LC-QTOF/MS in quantitative analysis and its versatility to acquire qualitative data simultaneously. The validated quantitative method established lower limits of quantification between 1 and 100 ng/mL for the 42 target analytes; great concordance was observed between the quantitative data obtained from LC-QTOF/MS and a previously validated LC-MS/MS method. Furthermore, more than 200 additional analytes were identified on top of the illicit drugs that were quantified.

NPS analysis by LC-QTOF/MS in more complex biological samples was demonstrated in quite a few studies; both DDA and DIA had been employed as acquisition methods. One method was developed for 500 NPS in serum (Kleis et al., 2021a,b). Automated MS/MS acquisition (DDA mode) was applied consisting of a full-scan MS followed by the selection of the three most abundant precursor ions in the prior MS scan for an MS/MS scan. A precursor ion threshold of 1000 counts was imposed for MS/MS as well as active precursor ion exclusion held for 0.09 min after one spectrum of the precursor ion was previously collected. A preferred list was included to force the precursor ion selection algorithm to prioritize ions in the targeted list prior to selecting ions of high abundance. MS/MS scans were triggered based on the preferred list using the following parameters: maximum mass error of 20 ppm within 0.5 min of the expected RT and at predetermined fixed collision energy for each target analyte.

To demonstrate sensitivity, matrix blanks were fortified with 1 ng/mL and 10 ng/mL of each target analyte. Ninety-six percent (470/492) of the target analytes were detected at 10 ng/mL, 88% (432/492) at 1 ng/mL, a clear demonstration of sensitivity for the simultaneous analysis of ~500 NPS. The combination of targeted and nontargeted precursor ion selection is typical of current DDA methods using LC-HRMS. This allows the best sensitivity for target compounds but at the same time provides avenues for discovery of new drugs. If the cost can be made affordable and the required expertise more accessible, LC-QTOF/MS would provide a superior alternative to immunoassays. Nonetheless, it is still limited in distinguishing certain types of isomers that are difficult to separate chromatographically, such as those that differ only in the position of a functional group around the aromatic ring. DDA methods for comprehensive NPS analysis and specific NPS classes have been reported in whole blood, serum, urine, oral fluid, hair, and even meconium (Broecker et al., 2012; Kronstrand et al., 2014; Montesano et al., 2017; Gundersen et al., 2019; Kim et al., 2021; Kleis et al., 2021a,b; Lopez-Rabuñal et al., 2021).

Data-independent acquisition strategies

Other groups developed LC-QTOF/MS methods that incorporated NPS as target or suspect analytes in wastewater using DIA. Baz-Lomba et al. published a method for the targeted quantitative analysis of 51 psychoactive substances consisting of 45 TRD and 6 NPS (Baz-Lomba et al., 2016). Data acquisition was done using MS^E. In this format, both the precursor and fragment ions are generated simultaneously by alternating the collision energy applied between low and high. The low collision energy scan allows transmission of intact precursor ions to the TOF mass analyzer, while the high collision energy scan induces fragmentation and allows transmission of fragment or product ions to the TOF mass analyzer. No prior selection of precursor ions is done at the high collision energy scan, so all precursor ions undergo fragmentation. The method was able to quantify 24 psychoactive substances in 15 samples, including two NPS. With All Ions MS/MS, another DIA method, Causanilles used LC-QTOF/MS as a qualitative screening method for NPS (Causanilles et al., 2017). An in-house database of more than 2000 entries consisting of TRD, prescription drugs, NPS and their metabolites, and transformation products was used as reference for data analysis by the following match criteria: mass error $\leq$10 ppm for precursor ions and $\leq$20 ppm for fragment ions; retention time deviation between precursor and fragment ions $\leq$0.1 min; isotope abundance and spacing match $\geq$50; and $S/N \geq 3$. Eight were identified as suspects—fentanyl and seven NPS—of which five were later confirmed with reference standards.

DIA methods that give the most comprehensive LC-HRMS data have also been implemented in NPS analysis for biological samples. The method by which DIA is applied is platform-dependent. Pope et al., for example, used MS^E to perform nontargeted MS/MS data acquisition with DIA at a mass range of 50–1200 m/z (Pope et al., 2021). An MS scan was initially acquired at 6 eV in the first low energy function, followed by a collision energy ramp from 10 to 40 eV in the second high energy function to generate product ions. Using a vendor-sourced drug mass spectral library that contains 1400 analytes (prescription drugs, drugs of abuse, NPS), a compound match was defined as mass error <5 ppm, RT < 0.5 min, and a minimum of one high energy fragment ion eluting at the same retention time as its putative precursor ion. The LC-QTOF/MS method had 1267 positive drug findings in 217 urine samples that had previously undergone screening. This number is more than double the positive findings obtained in the same samples by a previously validated GC-MS method.

The same acquisition method was also applied in serum samples (Grapp et al., 2018). A vendor library consisting of TRD and NPS with 1279 positively ionizing substances and 79 negatively ionizing substances along with drugs added by the laboratory was used in the targeted analysis. Thus, each sample was run in both positive and negative modes. The method was validated for a subset of 31 representative analytes. The LODs achieved for 25 of the 31 analytes ranged from 2 to 25 ng/mL. Higher LODs were observed for negatively ionizing substances (e.g., 140 ng/mL for pentobarbital). The criteria for compound identification were mass error <5 ppm for positively ionizing substance, mass error <10 ppm for negatively ionizing substances, retention time error <0.35 min, isotope pattern match error <10 ppm, isotope intensity match <20% root mean square error, and match to at least two fragment ions. When the method was applied to 247 antemortem serum and 12 postmortem femoral blood samples, 950 compounds comprising 185 different drugs and metabolites were identified and their concentrations were quantified. Three synthetic cathinones, one tryptamine, and three synthetic cannabinoids were among those identified. When compared with results obtained from a previously validated GC-MS serum screening, the LC-QTOF-MS method identified 240% (335 vs. 141) more compounds in 100 serum samples.

A DIA method in a different format was used by Kinuya et al., who implemented an All Ions MS/MS method for the analysis of NPS and drugs of abuse in serum and urine (Kinyua et al., 2015). MS data were acquired using three scan segments with different collision energies (0, 15, 35 eV). The first segment (0 eV) provided the precursor ion, and the second and third provided the product ions at low (15 eV) and high (35 eV) collision energies. Similar studies using DIA have been published for NPS analysis in blood, urine, and hair (Kinyua et al., 2016; Sundstrom et al., 2016; Mollerup et al., 2017).

An alternative method to implement DIA was used by Scheidweiler et al., who implemented SWATH in the screening of 47 synthetic cannabinoid metabolites in urine (Scheidweiler et al., 2015). SWATH facilitates the sequential acquisition of MS/MS scans in small mass range windows. In their validated method, a TOF/MS scan (mass range 100 to 1000 Da) was acquired at 10eV collision energy. SWATH MS/MS scans (mass range 30 to 1000) acquired spectra using 30 eV collision energy $\pm$ 15 eV spread. 36 Da SWATH isolation windows ranging from 228 to 408, which covers the most common molecular weights of synthetic cannabinoids, were set up. The criteria for compound identification were mass error <15 ppm, RT error <2%, isotopic pattern <80% difference from the theoretical, and library fit >60%. At least two fragments in the sample and the reference MS/MS spectra were also required for verification. Despite a long cycle time of 900 msec, the validated method showed very good sensitivity at 0.25–5 ng/mL for the 47 metabolite targets. Because the method acquired comprehensive MS/MS data between 228 and 408 Da, any future emerging synthetic cannabinoid that falls within the mass range can be detected without the need for method modification.

Advantages and limitations of LC-HRMS

The DDA and DIA modes implemented in the targeted analysis and suspect screening of traditional recreational drugs and NPS clearly demonstrate the advantage of LC-HRMS in providing a wide scope of analyte coverage. Its ability to facilitate nontargeted data acquisition has allowed acquisition of all ionizable ions from biological samples. This comprehensive data, in turn, can be interrogated for a wide range of drugs and NPS, including those that may not be initially known to the analyst.

Another significant advantage is facilitation of retrospective data analysis without the need for sample reanalysis. Gundersen et al. showed the extent of this capability. Fragment ion data for synthetic cannabinoids (n = 251), new synthetic opioids (n = 88), and designer benzodiazepines (n = 26) were obtained from a crowdsourced database, HighResNPS. The data were converted into a personalized library compatible with their instrumentation. In applying the new library to 1314 samples previously analyzed between 2014 and 2018, the group had five new findings, which were supported by MS/MS library match (Gundersen et al., 2020). The ability to mine previously acquired LC-HRMS raw files with emerging NPS is a powerful tool not only for clinical and forensic laboratories but more importantly for drug surveillance. Such capability allows accurate monitoring of the temporal origins of specific NPS.

Finally, an advantage of HRMS that is compatible with the rapid molecular evolution of NPS is the ease with which it can be updated. One can rapidly build upon an LC-HRMS screening method with procurement of new reference standards. For example, the LC-TOF/MS that was used by Lung et al. in 2013 consisting of 325 target analytes has been converted to an LC-QTOF/MS method that currently screens for more than 1200 target analytes: NPS, drugs of abuse, prescription drugs, dietary supplement stimulants, and NPS precursors, adulterants, and impurities (Lung et al., 2016; DEA, 2021). In the past 9 years, the targeted panel has added more than 750 NPS, including anticipated synthetic cannabinoids and new synthetic opioids that have not been released yet in the recreational drug market but have been synthesized by collaborators of the laboratory. Moreover, as more information on the metabolism of newly released NPS are published, suspect databases can be constantly updated with relevant and predicted NPS metabolites that can be extremely useful in urine NPS analysis. Such additions, modifications, and rapid updates cannot be easily accommodated by platforms that are primarily used in targeted analysis. It is therefore not surprising that the discovery of almost all NPS in the past decade was facilitated through LC-HRMS, and leading NPS surveillance programs across the world all employ LC-HRMS.

Despite its advantages, LC-HRMS is not without limitations. Its implementation in clinical and forensic laboratories remains challenging primarily because the expertise required to interpret the results of LC-HRMS runs is still highly specialized. It certainly requires constant and thorough guidance from qualified analytical, clinical or forensic chemists who have extensive mass spectrometry background. This is further complicated by the fact that the standard guidelines for validating and implementing LC-HRMS assays in the clinical laboratory are still limited due to its relative novelty.

Although the sensitivity and linear dynamic range of LC-QTOF/MS have not been comprehensively compared with LC-MS/MS, initial published data indicate that it has lower sensitivity and narrower linear dynamic range for some analytes. Thus, for some analytes, LC-QTOF/MS may still not be as sensitive as LC-MS/MS in quantitative targeted analysis. This limitation can be a challenge for very potent NPS like some synthetic cannabinoids and new synthetic opioids where serum toxic levels may be in the sub-ng/mL range. However, the huge current interest on HRMS platforms is providing impetus for this platform's vendors to improve its detection and analytical range capabilities.

As previously pointed out, mass spectral libraries generated by soft ionization in LC-HRMS are platform-specific. This is in contrast with GC-MS where mass spectral libraries are universally applicable across platforms. Hence, users are beholden to their instrument's platforms for commercially available mass spectral libraries or they have to devote time to generate their own mass spectral library for use in data analysis. More recently, there are online open sources for mass spectral libraries that are available in multiple formats to allow wider access. One such open source repository is HighResNPS.

Finally, LC-HRMS is not free from other limitations shared with other mass spectrometry platforms. Its cost can be prohibitive to small clinical or forensic laboratories, similar to LC-MS/MS. Furthermore, in assembling comprehensive targeted libraries for screening and quantitative analysis, the cost of reference standards can add up. This is exaggerated by the short life span of NPS in the recreational drug market, which implies a fast turnover of relevant analytes that would require constant procurement of the more relevant reference standards. The ability to acquire comprehensive data on samples also means large data file size for each sample run that is typically 1 GB or more in size. This would require added cost on the provision for huge data capacity for the laboratory and a system of extensive data archival.

Notwithstanding these limitations, LC-HRMS is by far the most compatible platform to use if the ability to screen and analyze NPS in the clinical or forensic laboratory is needed. NPS has become an integral part of the recreational drug market. Hence, the challenge that it imposes in the laboratory testing of drugs is here to stay. The utility of LC-HRMS for NPS analysis has already been demonstrated by several surveillance studies across the world. In the past decade, the discovery of almost all NPS was facilitated through LC-HRMS. With continuous innovations in this platform to improve its sensitivity and bring down its cost, a more aggressive push for educating laboratory scientists in its use and HRMS data interpretation, and the development of more detailed guidelines in the validation and implementation of LC-HRMS assays, it has the potential to be part of routine drug testing in clinical and forensic laboratories.[1]

Closeup: How it all started: A brief history of high-resolution mass spectrometry

Science historians have credited John J. Thompson, who won the Nobel Prize in physics in 1906, as the first to construct a mass spectrometer, in 1913 (Griffiths, 2008). The key to this development was his discovery of stable nonradioactive isotopes. The early instruments separated ions by using dual magnetic and electrostatic sectors. As such, these instruments were very large and occupied much floor space.

In the early 1950s, quadrupole mass spectrometers were developed. This detector was soon coupled with gas chromatographs as a means to separate complex biological samples into individual components. In later years, an interface was developed to link liquid chromatographs to the quadrupole analyzer. Today, tandem mass spectrometers, which contain three quadruple sectors (two quadrupoles and a collision cell), are widely used for targeted drug confirmation, quantitative drug analysis, and qualitative comprehensive unknown drug analysis. In the latter case, a comparison of mass fragmentation patterns against pretested standards or library spectra is used for identification (Wu and Colby, 2016).

The first high-resolution mass spectrometer (HRMS) was actually developed a few years before quadrupole MS. The early instruments used time-of-flight (TOF) as the means for detection. Instead of separating ions by their mass-to-charge ratio, ions are sent through a "flight tube," and the time it takes for an ion to traverse the tube is measured (Fig. 7.6A). Naturally, larger fragments take longer to reach the detector than smaller ones. The mass resolution is a function of the distance traveled. Current HRMS instruments can resolve ions to 0.001 atomic mass units.

[1] A significant portion of this chapter is adopted from a review of Gerona and French (2022).

Closeup: How it all started: A brief history of high-resolution mass spectrometry (*cont'd*)

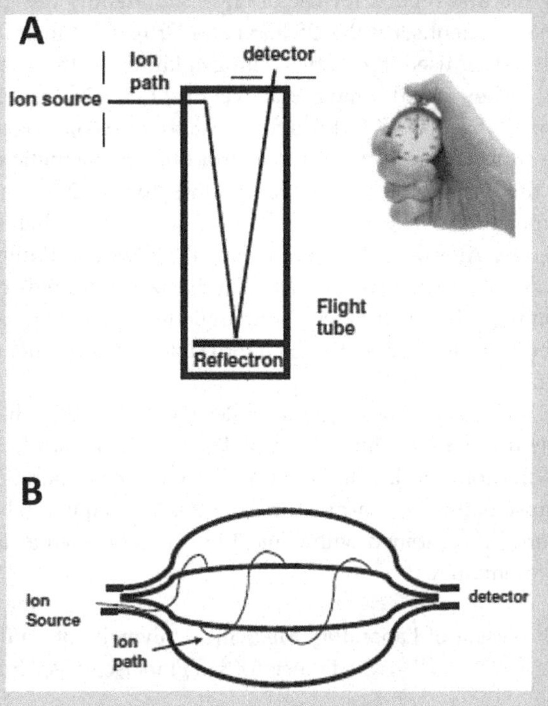

FIGURE 7.6 Schematic of high-resolution mass spectrometer using (A) time-of-flight or (B) orbitrap.

Following soft ionization that produces the molecular ion and minimizes fragmentation, the exact molecular mass of the molecular ion of any compound can be measured, enabling the determination of the exact molecular formula. Analysts can tentatively identify unknown compounds by assigning a formula match without comparing the mass spectrum of previously tested standards or identification through MS libraries, as the analysis currently performed using tandem MS. Now there are alternative high-resolution MS detectors, such as the orbitrap (Fig. 7.6B).

HRMS did not become popular until many years after the implementation of quadrupole MS. The key development was MALDI—matrix-assisted laser desorption ionization—in the 1980s. Today, MALDI-TOF analyzers are used for the identification of higher-molecular-weight compounds such as peptides and proteins, and bacteria.

continued

Closeup: How it all started: A brief history of high-resolution mass spectrometry (*cont'd*)

Among the first reports for use of HRMS to identify designer drugs was in 1968 by scientists at the US Food and Drug Administration. Martin and Alexander used it for a hallucinogen known to them only by the label "STP." They found a molecular weight of 209 Da and an empiric formula of $C_{12}H_{19}NO_2$ (Martin and Alexander, 1968). Together with other data from infrared spectroscopy and nuclear magnetic resonance analysis, they determined the compound to be 2,5-dimethoxy-4-methylamphetamine (DOM). This is a substituted amphetamine first synthesized by Alexander Shulgin in 1963 (Shulgin and Shulgin, 1991). Shulgin was a medicinal chemist who synthesized hundreds of psychoactive compounds, including 3,4-methylenedioxymethamphetamine (MDMA or Ecstasy) and tested their physiological and mental effects on himself.

HR-MS for detection of small molecules (i.e., $m/z < 1000$) has become particularly useful in the modern era of designer drugs, adulterants, and herbal medications. Today, there are many more compounds available that can cause pathologic harm. Data from this technique can be particularly powerful if combined with clinical information such as a patient's medical presentation.

Alan H.B. Wu
Professor of Laboratory Medicine, University of California San Francisco, San Francisco, CA, United States

References

Adams AJ, Banister SD, Irizarry L, Trecki J, Schwartz M, Gerona R. "Zombie" outbreak caused by the synthetic cannabinoid AMB-FUBINACA in New York. N Engl J Med 2017 Jan 19;376(3):235–42. https://doi.org/10.1056/NEJMoa1610300.

Allen DR, McWhinney BC. Quadrupole time-of-flight mass spectrometry: a paradigm shift in toxicology screening applications. Clin Biochem Rev 2019;40(3):135–46.

Andrés-Costa MJ, Andreu V, Picó Y. Analysis of psychoactive substances in water by information dependent acquisition on a hybrid quadrupole time-of-flight mass spectrometer. J Chromatogr, A 2016;1461:98–106.

Baz-Lomba JA, Reid MJ, Thomas KV. Target and suspect screening of psychoactive substances in sewage-based samples by UHPLC-QTOF. Anal Chim Acta 2016 Mar 31;914: 81–90. https://doi.org/10.1016/j.aca.2016.01.056.

Broecker S, Herre S, Pragst F. General unknown screening in hair by liquid chromatography-hybrid quadrupole time-of-flight mass spectrometry (LC-QTOF-MS). Forensic Sci Int 2012;218(1–3):68–81.

Causanilles A, Kinyua J, Ruttkies C, van Nuijs ALN, Emke E, Covaci A, et al. Qualitative screening for new psychoactive substances in wastewater collected during a city festival using liquid chromatography coupled to high-resolution mass spectrometry. Chemosphere 2017 Oct;184:1186–93. https://doi.org/10.1016/j.chemosphere.2017.06.101.

Gerona RR, French D. Drug testing in the era of new psychoactive substances. Adv Clin Chem 2022;111:217–63. https://doi.org/10.1016/bs.acc.2022.08.001.

Gonzalez-Marino I, Gracia-Lor E, Bagnati R, Martins CP, Zuccato E, Castiglioni S. Screening new psychoactive substances in urban wastewater using high resolution mass spectrometry. Anal Bioanal Chem 2016;408:4297–309.

Grapp M, Kaufmann C, Streit F, Binder L. Systematic forensic toxicological analysis by liquid-chromatography-quadrupole-time-of-flight mass spectrometry in serum and comparison to gas chromatography-mass spectrometry. Forensic Sci Int 2018 Jun;287:63–73. https://doi.org/10.1016/j.forsciint.2018.03.039.

Griffiths J. A brief history of mass spectrometry. Anal Chem 2008 Aug 1;80(15):5678–83.

Guale F, Shahreza A, Walterscheid JP, Chen HH, Arndt C, Kelly AT, et al. Validation of LC-TOF-MS screening for drugs, metabolites, and collateral compounds in forensic toxicology specimens. J Anal Toxicol 2013;37(1):17–24.

Gundersen POM, Spigset O, Josefsson M. Screening, quantification, and confirmation of synthetic cannabinoid metabolites in urine by UHPLC-QTOF-MS. Drug Test Anal 2019;11(1): 51–67.

Gundersen POM, Broecker S, Slørdal L, Spigset O, Josefsson M. Retrospective screening of synthetic cannabinoids, synthetic opioids and designer benzodiazepines in data files from forensic post mortem samples analysed by UHPLC-QTOF-MS from 2014 to 2018. Forensic Sci Int 2020;311:110274.

Kasper AM, Ridpath AD, Gerona RR, Cox R, Galli R, Kyle PB, Parker C, Arnold JK, Chatham-Stephens K, Morrison MA, Olayinka O, Preacely N, Kieszak SM, Martin C, Schier JG, Wolkin A, Byers P, Dobbs T. Severe illness associated with reported use of synthetic cannabinoids: a public health investigation (Mississippi, 2015). Clin Toxicol (Phila) 2019 Jan;57(1):10–18. https://doi.org/10.1080/15563650.2018.1485927. Epub 2018 Jul 10. PMID: 29989463.

Kim NS, Lim NY, Choi HS, Lee JH, Kim H, Baek SY. Application of a simultaneous screening method for the detection of new psychoactive substances in various matrix samples using liquid chromatography/electrospray ionization tandem mass spectrometry and liquid chromatography/quadrupole time-of-flight mass spectrometry. Rapid Commun Mass Spectrom 2021;35(10):e9067.

Kinyua J, Negreira N, Ibáñez M, Bijlsma L, Hernández F, Covaci A, et al. A data-independent acquisition workflow for qualitative screening of new psychoactive substances in biological samples. Anal Bioanal Chem 2015;407:8773–85.

Kinyua J, Negreira N, Miserez B, Causanilles A, Emke E, Gremeaux L, et al. Qualitative screening of new psychoactive substances in pooled urine samples from Belgium and United Kingdom. Sci Total Environ 2016;573:1527–35.

Kleis JN, Hess C, Germerott T, Roehrich J. Sensitive screening of synthetic cannabinoids using substances in serum using liquid-chromatography quadrupole time-of-flight mass liquid chromatography quadrupole time-of-flight mass spectrometry after solid phase extraction. Drug Test Anal 2021a;13(8):1535–51.

Kleis J, Hess C, Germerott T, Roehrich J. Sensitive screening of new psychoactive spectrometry. J. Anal. 2021b;46(6):592–9.

Kronstrand R, Brinkhagen L, Birath-Karlsson C, Roman M, Josefsson M. LC-QTOF-MS as a superior strategy to immunoassay for the comprehensive analysis of synthetic cannabinoids in urine. Anal Bioanal Chem 2014;406(15):3599–609.

Krauss M, Singer H, Hollender J. LC-high resolution MS in environmental analysis: from target screening to the identification of unknowns. Anal Bioanal Chem 2010;397:943–51.

López-Rabuñal A, Di Corcia D, Amante E, Massano M, Cruz-Landeira A, de-Castro-Ríos A, et al. Simultaneous determination of 137 drugs of abuse, new psychoactive substances, and novel synthetic opioids in meconium by UHPLC-QTOF. Anal Bioanal Chem 2021; 413(21):5493–507.

Lung D, Wilson N, Chatenet FT, LaCroix C, Gerona R. Non-targeted screening for novel psychoactive substances among agitated emergency department patients. Clin Toxicol 2016; 54(4):319–23.

Martin RJ, Alexander TG. Analytical procedures used in FDA laboratories for the analysis of hallucinogenic drugs. JAOAC (J Assoc Off Anal Chem) 1968;51:159–63.

Maurer HH. Perspectives of liquid chromatography coupled to low- and high-resolution mass spectrometry for screening, identification, and quantification of drugs in clinical and forensic toxicology. Ther Drug Monit 2010;32(3):324–7.

Maurer HH. Pitfalls in drug testing by hyphenated low- and high-resolution mass spectrometry. Drug Test Anal 2020;12(2):172–9.

Mollerup CB, Dalsgaard PW, Mardal M, Linnet K. Targeted and non-targeted drug screening in whole blood by UHPLC-TOF-MS with data-independent acquisition. Drug Test Anal 2017;9(7):1052–61.

Montesano S, Vannutelli G, Massa M, Simeoni MC, Gregori A, Ripani L, et al. Multi-class analysis of new psychoactive substances and metabolites in hair by pressurized liquid extraction coupled to HPLC-HRMS. Drug Test Anal 2017;9(5):798–807.

Nielsen MK, Johansen SS, Dalsgaard PW, Linnet K. Simultaneous screening and quantification of 52 common pharmaceuticals and drugs of abuse in hair using UPLC-TOF-MS. Forensic Sci Int 2010 Mar 20;196(1–3):85–92. https://doi.org/10.1016/j.forsciint.2009.12.027.

Oberacher H, Arnhard K. Current status of non-targeted liquid chromatography-tandem mass spectrometry in forensic toxicology. TrAC, Trends Anal Chem 2016;84:94–105.

Papsun D, Hawes A, Mohr ALA, Friscia M, Logan BK. Case series of novel illicit opioid-related deaths. Acad Forensic Pathol 2017 Sep;7(3):477–86. https://doi.org/10.23907/2017.040.

Pasin D, Cawley A, Bidny S, Fu S. Current applications of high-resolution mass spectrometry for the analysis of new psychoactive substances: a critical review. Anal Bioanal Chem 2017;409(25):5821–36.

Pope JD, Black MJ, Drummer OH, Schneider HG. Urine toxicology screening by liquid chromatography time-of-flight mass spectrometry in a quaternary hospital setting. Clin Biochem 2021;95:66–72.

Scheidweiler KB, Jarvis MJ, Huestis MA. Nontargeted SWATH acquisition for identifying 47 synthetic cannabinoid metabolites in human urine by liquid chromatography-high-resolution tandem mass spectrometry. Anal Bioanal Chem 2015;407(3):883–97.

Schneir A, Ly BT, Casagrande K, Darracq M, Offerman SR, Thornton S, et al. Comprehensive analysis of "bath salts" purchased from California stores and the internet. Clin Toxicol 2014 Aug;52(7):651–8. https://doi.org/10.3109/15563650.2014.933231.

Schwartz MD, Trecki J, Edison LA, Steck AR, Arnold JK, Gerona RR. A common source outbreak of severe delirium associated with exposure to the novel synthetic cannabinoid ADB-PINACA. J Emerg Med 2015 May;48(5):573–80. https://doi.org/10.1016/j.jemermed.2014.12.038.

Shulgin A, Shulgin A. PiHKAL [phenethylamines i have known and loved]: a chemical love story. Transform Press; 1991.

Sundström M, Pelander A, Simojoki K, Ojanperä I. Patterns of drug abuse among drug users with regular and irregular attendance for treatment as detected by comprehensive UHPLC-HR-TOF-MS. Drug Test Anal 2016;8(1):39–45.

Trecki J, Gerona RR, Schwartz MD. Synthetic cannabinoid-related illnesses and deaths. N Engl J Med 2015 Jul;373(2):103–7. https://doi.org/10.1056/NEJMp1505328. PMID: 26154784.

Tyndall JA, Gerona R, De Portu G, Trecki J, Elie MC, Lucas J, et al. An outbreak of acute delirium from exposure to the synthetic cannabinoid AB-CHMINACA. Clin Toxicol 2015;53(10):950–6. https://doi.org/10.3109/15563650.2015.1100306.

U.S. Drug Enforcement Administration, Diversion Control Division. DEA TOX: quarterly report –4th quarter 2021. Springfield, VA: U.S. Drug Enforcement Administration; 2021.

Wu AH, Gerona R, Armenian P, French D, Petrie M, Lynch KL. Role of liquid chromatography-high-resolution mass spectrometry (LC-HR/MS) in clinical toxicology. Clin Toxicol 2012;50(8):733–42.

Wu AHB, Colby J. High-resolution mass spectrometry for untargeted drug screening. Methods Mol Biol 2016;1383:153–66.

Xian F, Hendrickson CL, Marshall AG. High resolution mass spectrometry. Anal Chem 2012; 84:708–19.

CHAPTER

8

NPS surveillance and epidemiology

The rapid diversification of new psychoactive substances (NPS) has continued to impose significant challenge to public health and law enforcement. The increasing potency in some NPS classes has caused an alarming number of fatal intoxications. In response, surveillance systems focused on NPS were introduced in the past 2 decades using different combinations of strategies. This chapter introduces key strategies and describes how surveillance systems have employed them to identify and track NPS in different contexts and at different times.

Sampling strategies

Multiple methods are used to determine trends in drug use and harm. The most common for NPS are surveys, law enforcement drug seizures, poison center calls, coroners' data, drug screens (from clinic, workplace, and driving under the influence of drugs [DUID] evaluation), and drug checking services (Peacock et al., 2019). Surveillance data inform NPS detection and help establish the appropriate public health and regulatory response to trends, especially those with high potential for harm. Each strategy has its advantages, biases, and limitations; each relies on self-report or laboratory confirmation or both.

Surveys vary in scale from national to regional to specialized subgroups; they can be cross-sectional or occasionally longitudinal. They collect self-reports on recent or historical use and other information such as demographics, motivation for use, route of administration, frequency, pattern of use, perception on safety and legality, sources, price, and adverse event experiences. Depending on scale, surveys can inform NPS use and trends for a whole state or representative subgroups. National surveys can reveal large-scale data not only on the scope of the population

Designer Drugs
https://doi.org/10.1016/B978-0-12-811764-4.00015-X

surveyed but also on the depth of detailed data that can be derived. National surveys take time to complete, and data compilation and analysis can be a long process (Khaled et al., 2016). This time requirement is not compatible with the rapid evolution of NPS in the recreational drug market. Moreover, because of the massive resources required, surveys are infrequent, further weakening their temporal relevance. National surveys also tend to be biased against marginalized populations such as the homeless, prisoners, and those in very rural areas; missing them leads to underestimation of trends. Nonetheless, national surveys give the most comprehensive large-scale data on trends. Well-coordinated cross-sectional national surveys have improved on bias, but they are few and far between. Those on specialized subpopulations can also limit bias and provide more detailed data. However, methodological differences in sampling and survey strategies limit comparability of data across types of survey (Pirona et al., 2017). Because of the wide-scale uncertainty about the composition of NPS products and the intentional misrepresentation or adulteration in fake prescriptions and other drug products with NPS, surveys based solely on self-reports can have limited accuracy about drugs available to and used by the sampled population.

Analysis of drugs seized by law enforcement provides valuable data on the actual NPS circulating. The quality of the data is highly dependent on the capacity of the laboratory performing the analysis. Most laboratories can do targeted GC-MS, LC-MS/MS, and to some extent LC-QTOF/MS. In laboratories without high-resolution mass spectrometry, the analysis is limited by the depth and scope of the available drug panel, and the latest NPS are missed.

Data from poison center calls can reflect population use, especially for higher potency NPS. Data so derived can be a powerful tool in forecasting harm from specific NPS. However, the method also relies heavily on self-report, so accuracy suffers. Because only drugs that cause untoward symptoms are reported, the method is biased toward the more potent ones.

Coroners' reports are invaluable in establishing emerging trends for potent NPS. They give more weight, especially if the information is used to create NPS alerts and execute emergency scheduling, primarily because the drugs are associated with fatalities and are most likely laboratory confirmed. Similar to analysis of seized drugs, coroner data are only as good as the breadth and scope of the drug panel used by the laboratory.

Drug screens in healthcare settings and impaired driving and workplace testing are vital to providing toxicological data for traditional recreational drugs. But their utility in NPS is limited if only immunoassays are used, which is typically the case. Without screening and/or confirmatory assays using mass spectrometry that specifically targets NPS, very limited data on NPS use can be derived. Furthermore, data on what drug was intended to be taken can be lacking, as the people being tested are typically concerned about punitive consequences.

Drug checking has recently emerged as a rapid and inexpensive method. But it has little utility in NPS surveillance, primarily because most of the tests in the kits are based on chemical reactions that target a broad range of functional groups that can be found in traditional recreational drugs and NPS alike. The tests lack specificity and sensitivity—are prone to high rates of false-positive and false-negative results—with respect to NPS.

From multiple NPS surveillance systems established and published, it has become evident that a single surveillance method is not enough to cover trends and use. Triangulation from multiple methods is the most effective approach to accurately determine trends and forecast those that likely threaten public health (Madras, 2017).

Surveillance systems

The earliest system was established in Europe even before the NPS resurgence started in 2008. The early warning system (EWS) was created by the European Union (EU) in 1997 for synthetic drugs, and its focus on NPS was strengthened in 2005. The system is operated by the European Monitoring Center for Drugs and Drug Addiction (EMCDDA) in close cooperation with Europol. Along with these two agencies, it is comprised of 29 national early warning systems (27 EU states, Norway, and Turkey), Europol's law enforcement networks, the European Medicines Agency, the European Commission, and other partners (Evans-Brown and Sedefov, 2018). Apart from EWS, Sweden ran the STRIDA project between 2010 and 2016. A collaboration between Karolinska Institutet, the Karolinska University Laboratory, and the Swedish Poison Information Center (PIC) in Stockholm, STRIDA offered comprehensive toxicological analysis of cases referred by emergency departments and intensive care unit as well as PIC consultations on suspected NPS- or unknown drug-related overdoses (Helander et al., 2020). STRIDA (which is both an acronym of the Swedish project name and Swedish for fight or combat) stands out as one of its kind in Europe in proactively using laboratory confirmation of NPS-related intoxication.

Global NPS monitoring was launched by the United Nations Office on Drugs and Crime (UNODC) in June 2013 as a response to the increasing global threat. Dubbed as the Early Warning Advisory (EWA), the project aims to monitor, analyze, and report global trends as a framework for evidence-based policy responses (CND, 2013). With the data it collects, the project also aims to contribute to a better understanding of global NPS use and distribution patterns and serve as an information repository and a platform for providing technical assistance to its 189 member states. EWA is administered by the UNODC Global Synthetics Monitoring: Analyses, Reporting and Trends (SMART) program, which operates in Vienna and has teams based in Thailand and Panama (UNODC, 2022). As its general aim, Global SMART seeks to improve the capacity of targeted member states to generate, manage, analyze, report, and use information on synthetic illicit drugs.

In the United States, no formal systematic and focused NPS monitoring program was established right away. Data on use were initially obtained from poison control center calls. A developing trend in calls was first reported in the 2010 Annual Report of the American Association of Poison Control Centers (AAPCC) National Poison Data System, after which the monthly statistics of increasing calls related to K2 (synthetic cannabinoids) and bath salts (synthetic cathinones) became the data source for most agencies following NPS in the United States. Another source is the National Forensic Laboratory Information System (NFLIS), a US Drug Enforcement Administration (DEA) program that collects results of forensic analyses from local, regional, and national forensic laboratories across the United States. In August 2011, NFLIS issued its first special report on the synthetic cannabinoids and cathinones reported to it in 2009 and 2010. It continues to issue special reports on emerging NPS threats. The informal collaboration among researchers at the DEA, the Centers for Disease Control and Prevention (CDC), and the University of California San Francisco (UCSF), which reported on multiple mass intoxications involving synthetic cannabinoids between 2012 and 2017, also contributed significantly to providing initial information to the public health threat posed by synthetic cannabinoids. This collaboration worked with local experts where mass intoxications were reported, and published laboratory-confirmed data on synthetic cannabinoid intoxications across the United States (Trecki et al., 2015). Currently, there are several drug monitoring systems that report on NPS use and intoxication, including the National Drug Early Warning System (NDEWS), the Center for Forensic Science Research and Education (CFSRE), and the DEA Toxicology Testing Program (DEA TOX).

Surveillance programs were also established in other regions of the world to respond to the serious local public health threat posed by NPS. In most countries, this is dovetailed with existing illicit drug surveillance programs. In some East Asian countries, the use of national surveys initially served as a major data source. An example is the National Household Survey on Health and Substance Abuse in Taiwan (Feng and Li, 2020). In Japan, data from their Poison Information Center were supplemented by data collected from the analysis of NPS products. The Data Search System for New Psychoactive Substances was launched by Japan's

National Institute of Health Sciences in 2014 (Tanaka et al., 2016). The database provides information on NPS sold in the country, including their mass spectral and pharmacological data.

The first national prevalence rates for NPS use in Australia were collected in 2013 through the National Drug Strategy Household Survey (AIHW, 2014). This program is conducted every 3 years. Likewise, NPS data are available on the Drug Trends program coordinated by Australia's National Drug and Alcohol Research Center. Drug Trends triangulates data from various sources to forecast emerging substance use and harm. In New Zealand, a similar approach was taken by the Institute of Environmental Science Research through its Border to Grave project, which aims to link data from NPS seized at its borders with use data from the community (motor vehicle accidents, criminal case work, emergency department cases) and data obtained by environmental scientists on drugs in wastewater (IESR, 2022).

Apart from the United States, Canada is among the largest NPS markets in the Americas. It first included NPS in its biennial Youth Smoking Survey in 2010 (Health Canada, 2012). In lieu of an EWS, the Canadian Center on Substance Use and Addiction established the Canadian Community Epidemiology Network on Drug Use with the primary task of informing the Canadian public on emerging drug use trends. The Network regularly issues drug alerts on trends that often focus on NPS. In Latin America, the Inter-American Drug Abuse Control Commission (known by its Spanish acronym CICAD) issues bulletins on NPS and emerging drugs in the Americas. Only Argentina, Chile, Colombia, and Uruguay have national systems, which comprise the EWS for the Americas (Spanish acronym SATA).

Global surveillance and UNODC

UNODOC was established in 1997 as a merger between the United Nations Center for International Crime Prevention and the United Nations International Drug Control Program. It has mandates in three areas: drugs, crime, and international terrorism. As the central drug control agency, it coordinates and leads all UN drug control activities. It is the repository of technical expertise and advises member states on international and national control.

Following the increasing global spread and public health threat posed by NPS between 2008 and 2012, UNODC established the EWA in 2013 to facilitate the timely sharing of comprehensive information including trend analysis, analytical methods for identification, NPS mass spectra, and reference documents (Tettey et al., 2018; UNODC, 2020, 2021b). EWA is administered by the UNODC Global SMART program created in 2008 to respond to the rise of synthetic drugs, which then consisted primarily of amphetamine-type stimulants (ATS) (UNODC Global SMART Programme, 2013).

The EWA initially collected data on NPS found in seized materials. It served as a voluntary online data system that gathers and collates submissions from forensic drug testing laboratories across the world (UNODC, 2011). In 2016, it added toxicology data collection to identify the most persistent and harmful NPS that pose the greatest public health threat. The data are also used in the prioritization of substances for placement under international control and in legislative responses at the national level. Currently, the EWA gathers NPS data from the following sources:

- the global survey by the Global SMART program (2012, 2014, 2016, 2019);
- the Annual Report Questionnaire submitted by all member states;
- law enforcement data collected through the UNODC Individual Drug Seizures Database;
- the UNODC International Collaborative Exercises (ICE) program, with over 280 national forensic laboratories in 84 countries;
- reports from regional networks of forensic science institutions;
- Global SMART teams using the Drug Abuse Information Network for Asia and the Pacific;
- the Drug Abuse Information Network in Latin America and the Caribbean in partnership with the Organization of American States; and
- interagency meetings with EMCDDA, DEA, the Organization of American States, the International Narcotics Control Board, the World Anti-Doping Agency, the World Customs Organization, and the World Health Organization.

NPS data from this immense network are collated and communicated in various platforms. In the same year as the creation of EWA, NPS was first covered in the World Drug Report, an annual publication of UNODC about global drug market trends. From 2009 to 2021, the World Drug Report documented a sevenfold increase in the cumulative number of NPS, from 166 to 1127 (Fig. 8.1) (UNODC, 2022).

The Global SMART program issues the Global SMART Updates, which are special reports of 12–15 pages providing targeted information on emerging market trends, patterns of supply, demand and use, and health implications related to ATS and NPS. Started in 2009, these briefs are in their 26th volume. One on synthetic drugs not under international control was published in 2011, and there was a closer look at NPS in 2012. Results of the first global survey on NPS undertaken in 2012 were published in 2013. There were reports on specific NPS classes in 2015 (synthetic cannabinoids) and 2017 (new synthetic opioids [NSO] and designer benzodiazepines).

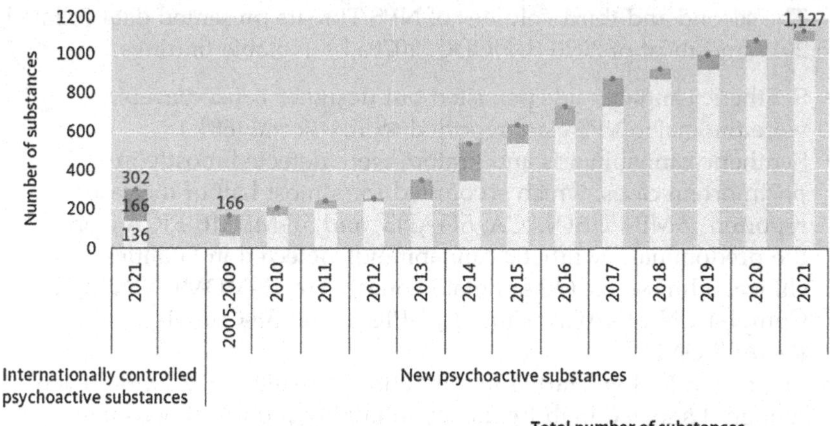

FIGURE 8.1 Cumulative number of NPS reported to UNODC from 2005 to 2021 compared with the number of internationally controlled psychoactive substances. *Reproduced from World Drug Report 2022 (UNODC, 2022).*

In 2019, the Global SMART program also started issuing Current NPS Threats, which summarizes data on the toxicology cases resulting from NPS use reported to the EWA Toxicology Portal (Tox-Portal). Created in 2016, the Tox-Portal is an online tool developed in collaboration with The International Association of Forensic Toxicologists (TIAFT) to enhance global NPS data collection from seized products. Forensic and toxicology laboratories submit data from postmortems, clinical cases, and other casework, including drug-impaired driving.

The first volume on NPS Threats summarized data collated from 2016 to 2018 on 367 cases from 29 countries (UNODC, 2019). Key findings:

- Synthetic cannabinoids (21%, 34%, 38%) and NSO (36%, 29%, 4%, mostly U-47700 and fentalogs) comprise more than half of the cases reported in those 3 years.
- Fatalities involving cannabinoids trended upward, while new NPS and synthetic cannabinoid emergence trended downward overall.
- Two synthetic cannabinoids, 5F-ADB and AMB-FUBINACA, persisted across Europe, the Americas, Asia, and Oceania; they were the most reported of their class in drug seizures and in biological fluid detection reported by Tox-Portal.
- Polydrug use, consisting of four or more agents, comprises more than half of the fatalities.
- Designer benzodiazepines, particularly etizolam and flubromazolam, were notable in DUID. Most of these cases were in the United States. Designer benzodiazepines continued to increase in drug seizures from 2014 to 2018.

The second and third volumes of NPS Threats presented data collated in 2019 and most of 2020 (UNODC, 2020a,b). Notable findings:

- Synthetic cannabinoids persisted but designer benzodiazepines were the major NPS class reported to Tox-Portal (68%).
- Synthetic cannabinoids and kratom were detected mostly in postmortem cases, which accounted for almost half of the cases reported. AMB-FUBINACA, 5F-ADB, and 5F-MDMB-PICA were the predominant synthetic cannabinoids detected in fatalities and clinical admissions. Other notable ones were 4F-MDMB-BINACA, Cumyl-4-CN-BINACA, Cumyl-5F-PICA, and 5F-Cumyl-PeGACLONE.
- The rise in kratom started in 2018 and intensified in 2019. But even though it showed high frequency in fatality reports, it is usually determined to be noncontributory or of moderate contribution to death. It is usually found in combination with other drugs such as opioids, ATS, benzodiazepines, and alcohol. The antihistamine diphenhydramine was frequently present in kratom cases, purportedly because it provides relief to the itching induced by kratom and is thought to potentiate kratom's effects. Kratom was also detected with high frequency in DUID cases.
- Most designer benzodiazepine reports were associated with DUID, doubling in number from 2018 to 2019, and most were in the United States and Canada, with etizolam, flualprazolam, and flubromazolam predominating.

The latest volume of NPS Threats summarized data collected in the later part of 2020 and 2021 (UNODC, 2021a). Notable findings:

- Designer benzodiazepines continued to be the most frequent NPS class reported to Tox-Portal, with 69% of 1900 instances. Synthetic cannabinoids, NSO, kratom, and ketamine were also frequently reported.
- Between October 2020 and April 2021, there were more than 1500 cases involving 58 NPS. They were 55% of DUID, 25% of postmortem, 7% of clinical, and 15% of other case types (e.g., drug-facilitated sexual assault).
- Designer benzodiazepines were 49% of postmortem, 32% of clinical, and 85% of DUID. As of January 2021, 28 designer benzodiazepines have been reported to UNODC. Flualprazolam ($n = 494$), etizolam ($n = 371$), flubromazolam ($n = 271$), and clonazolam ($n = 152$) were the most frequent. Clonazolam was detected mostly in DUID, the other three across all case types.
- Postmortem frequencies were designer benzodiazepines 49%, synthetic cannabinoids 16%, NSO 14%, stimulants and

hallucinogens 4%, and others 17%. Kratom predominated in the other classes.

- Stimulants and hallucinogens were detected only in postmortem and clinical. Of the stimulants detected, synthetic cathinones consisting of N-ethylheptedrone, N-ethylhexedrone, mephedrone, benzylone, alpha-PVP, and alpha-PHP were reported to have medium or high contribution to death.
- Of the hallucinogens, 25E-NBOH, 25I-NBOMe, and 2C-E were reported solely from South America, while 3-MeO-PCP, 4-MeO-PCP, 3-MeO-PCE, and 5-MeO-DMT were reported from Europe.
- Of NSO in postmortem, 63% involved acetylfentanyl and 11% involved carfentanil. Both were already internationally controlled, in 2016 and 2018, respectively. Emerging NSO brorphine and nitazenes (2-benzylbenzimidazoles) were also reported for the first time.
- Synthetic cannabinoids were reported in 172 cases. In 45 postmortem cases, 87% were accounted for by 4F-MDMB-BINACA, 5F-MDMB-PICA, and MDMB-4en-PINACA. The same three and AMB-FUBINACA were the most frequent in clinical.
- Polydrug use remains a predominant feature in NPS casework; 79% in postmortem involved more than one drug.

Aside from the Global SMART Update and Current NPS Threats, Global SMART issues other short publications as part of the EWA. Early on, they annually published ATS market trends in Asia and the Pacific. With the advent of NPS, this morphed into an annual Synthetic Drugs Assessment in East and Southeast Asia. From time to time, they also publish global synthetic drugs assessment and regional synthetic drugs assessment in Latin America and Africa. Other informative briefs aimed to improve public understanding of NPS have also been published, such as "The Challenge of NPS" (UNODC Global SMART Programme, 2013), "Categories of NPS Sold in the Market" (UNODC, 2014), and "The Role of Drug Laboratories in EWS" (UNODC Global SMART Programme, 2020).

Europe

The Early Warning System

NPS surveillance in Europe is the earliest and most developed in the world. EMCDDA was created in 1993 to combat the growing drug problem with policies based on scientific evidence rather than ideology. It opened in Lisbon in 1995 with 15 member countries, and since then has expanded to 29, 27 of which belong to the European Union. It established the EWS in 1997 (Joint Action 97/396/JHA, 1997). In administering the EWS, it collaborates with other European agencies, including the Europol and its law enforcement networks, the European Medicines Agency, the European Commission, and other partners. The EWS is the first in a three-step legal framework designed to facilitate the EU's ability to rapidly detect, assess, and respond to the public health and social threats caused by NPS. NPS risk assessment and control measures comprise the other components of the legal framework (EMCDDA, 2022a,c). In close cooperation with Europol, the EMCDDA also handles NPS risk assessment, while the European Commission proposes control measures.

The focus of the EWS on NPS was strengthened in 2005 (Council Decision, 2005/387/JHA, 2005). This multiagency, multidisciplinary network is tasked with collating, analyzing, assessing, and communicating the information reported by its member states and agencies to enable the EMCDDA to prepare and issue an initial report on an NPS or class of NPS that could pose significant health and social risks at the EU level. It gathers data from forensic and toxicology laboratories that perform chemical identification of NPS on caseworks related to drug seizures by law enforcement, poisonings (clinical cases in emergency departments and medicolegal death investigations), samples collected from people who use NPS, and test purchases such as those from online marketplaces. Timely communication of reports from the laboratories allows rapid reporting of event-based information on NPS appearance and the harms they cause at the national level. Aggregated data on seizures by law enforcement and poisonings in annual reports complement these data. Each member state bears responsibility for organizing and operating its own national early warning system (NEWS). Although each functions independently to address their country's specific needs, all are linked by a common format, guidelines, and reporting tools to the EMCDDA (2022a,c).

An effective workflow for reporting, review, and formal notification of newly identified NPS has been established through the EWS. A substance typically identified from a law enforcement drug seizure is reported to the EMCDDA by a member state's NEWS. The report usually includes chemical and analytical information as well as the circumstances of the seizure event. Submission of analytical data is required to facilitate the verification of the newly reported NPS, as most of these substances do not initially have available reference standards. The EMCDDA reviews data in the report and gathers more information that might be relevant in the literature. Once the substance is confirmed as an NPS, EMCDDA issues a formal notification to its network on behalf of the reporting member state. The notification includes the following information: (1) name and other relevant identifiers of the NPS, (2) its physical and chemical properties, (3) analytical methods used for its identification, (4) its known pharmacology and toxicology, (5) circumstances of its detection, and (6) any other relevant information. The formal notification process is the pillar of the EWS's success. The process serves three important purposes: it ensures that members of the network are alerted as soon as possible when an NPS is identified in Europe, it allows the network to identify and assess any potential threats and establish and implement any response measures that might be necessary, and it allows forensic and toxicology laboratories in the network to add the new substance to their respective analytical panels, facilitating its monitoring.

Once it has been formally notified, the EMCDDA starts monitoring the substance for any reports of harm on the EWS network of interlinked systems, including event-based data, toxicovigilance, signal management, and open source information. Depending on the level of harm observed, the EWS can mount a response ranging from placing the NPS under intensive monitoring to public health alerts to preparation of an initial report leading to a risk assessment.

The EWS has issued initial reports on 30 NPS since 2005. An initial report contains detailed information about a specific NPS that the EMCDDA considers to pose a significant threat. A report gives information on the following:

- physical and chemical properties of the NPS and the precursors and methods used for its synthesis, manufacture, and extraction;
- pharmacological and toxicological properties;
- patterns of use;
- nature, number, and scale of incidents where it may have caused health and social problems;
- involvement of criminal groups in its manufacture and distribution;
- known human or veterinary medical use;
- commercial and industrial use, the extent of such use, and its use for scientific research and development;
- any restrictive measures associated with it in member states; and
- any previous or current assessment of it within the system established by the 1961 and 1971 United Nations Drug Conventions.

Of the 30 initial reports, 25 were issued with Europol from 2005 to 2018 when they were called the EMCDDA-Europol Joint Reports (EMCDDA, 2022b). Initial reports were issued for 11 NSO, 7 synthetic cannabinoids, 5 synthetic cathinones, 3 amphetamines, 2 piperazines, 1 phenethylamine, and 1 arylcyclohexylamine (Table 8.1). This number is a small fraction of the 884 NPS that are being monitored by the EMCDDA as of June 2022.

Through the EWS, the EMCDDA was able to track the trends and developments in NPS markets, use, and threats in Europe. Between 2005 and 2008, fewer than 20 NPS each year were notified for the first time to EMCDDA (Fig. 8.2). The number rose above 20 in 2009 and grew

TABLE 8.1 NPS categorized by class that EMCDDA issued initial reports on.

Class	Drug
New synthetic opioid	Isotonitazene (2020) Methoxyacetylfentanyl (2018) Cylcopropylfenatnyl (2018) Carfentanil (2017) Tetrahydrofuranylfentanyl (2017) 4-Fluoroisobutyrylfentanyl (2017) furanylfentanyl (2017) Acryloylfentanyl (2017) Acetylfentanyl (2016) MT-45 (2014) AH-7921 (2014)
Synthetic cannabinoid	4F-MDMB-BICA (2020) MDMB-4en-PINACA (2020) ADB-CHMINACA (2017) cumyl-4CN-BINACA (2017) 5F-ADB (2017) AB-CHMINACA (2017) MDMB-CHMICA (2016)
Synthetic cathinone	3-Methylmethcathinone (2021) 3-Chloromethcathinone (2021) alpha-PVP (2015) MDPV (2014) mephedrone (2010)
Amphetamine	4,4′-DMAR (2014) 5-IT (2013) 4-Methylamphetamine (2012)
Piperazine	benzylpiperazine (2007) mCPP [1-(3-chlorophenyl)piperazine] (2005)
Phenethylamine	25I–NBOMe (2014)
Arylcyclohexylamine	Methoxetamine (2014)

exponentially until it reached 100 in 2014 and 2015. Since 2017, it has been steady at about 50. The variety of classes also increased: 4 or 5 in 2004 and 2005, 6 or 7 in 2006–2009, 11 in 2010, and 8 or more since then, with a high of 13 in 2013.

Synthetic cannabinoids were the most frequently reported starting in 2009, peaking in 2012–14, declining in 2016–19, increasing again in 2020. In 2021, they remained the most frequent; 224 are currently monitored. Synthetic cathinones were the second largest group, starting in 2008, peaking in 2014 and 2015, and declining since 2016, but remaining in second place in 2021 with 162. NSO started to be consistently reported in 2012, leapt in 2016, and has remained third or second since 2017; 73 are

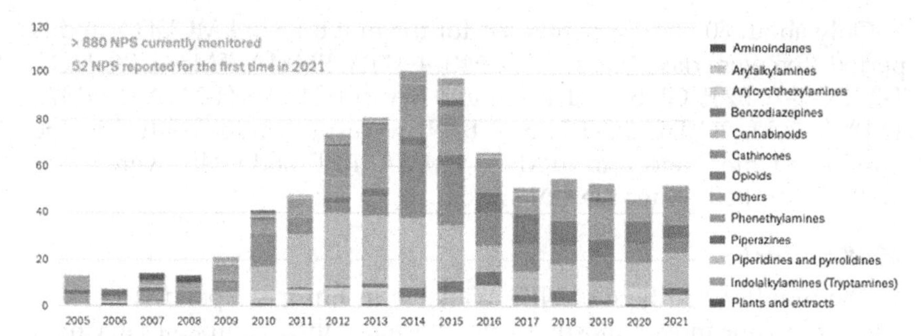

FIGURE 8.2 Annual number of NPS reported for the first time to EMCDDA. *Reproduced from NPS: 25 years of early warning and response in Europe (EMCDDA, 2022b).*

monitored currently. Designer benzodiazepines have increased steadily since 2011; 33 are currently monitored. New tryptamines and phenethylamines have been consistently notified since 2005, the former predominating from 2005 and the latter in 2012 and 2013. Arylcyclohexylamines and piperazines have been reported in multiple years, though not consistently.

Three periods of the EWS

2022 marks the 25th year of the EWS. A recent EMCDDA report divided those years into three periods based on the characteristics of the NPS market and threat and the activities conducted by the EWS (EMCDDA, 2022b).

1997 to 2007

This period marks the creation and expansion of the EWS network. It saw the development of the EWS's approach to detection, operating guidelines, and risk assessment. In 2005, the European Union passed legislation to strengthen the focus on NPS and its response, and EMCDDA formally adopted the term "new psychoactive substances" to replace "new synthetic drugs." A relatively small number of NPS were reported. Aside from *Salvia divinorum* and magic mushrooms, most were synthetic stimulants, psychedelics, and hallucinogens produced in small illicit laboratories in Europe. Amphetamines, phenethylamines, tryptamines, and piperazines were the major classes. Paramethoxymethamphetamine (PMMA) was popularly sold as "Ecstasy" to unsuspecting rave- and party-goers, causing fatalities. Benzylpiperazine "party pills" became the first NPS widely sold online (EMCDDA, 2009a). Around the middle of 2000s, companies from China began producing bulk quantities of NPS, marking a transition away from small-scale local production.

Only about 60 NPS were notified for the first time to EMCDDA in this period. Ten were risk-assessed: MDMB, 4-MTA, PMMA, TMA-2, 2C-I, 2C-T-2, 2C-T-7, BZP, GHB, and ketamine. Seven of these (4-MTA, PMMA, TMA-2, 2C-I, 2C-T-2, 2C-T-7, and BZP) were put under control in the European Union, and four (4-MTA, PMMA, BZP, and GHB) were eventually controlled under the UN system.

2008 to 2015

The number and type of NPS available in Europe expanded rapidly. The supply line moved mostly to big pharmaceutical companies in China exporting bulk amounts. The market was dominated by synthetic cannabinoids ("Spice") sold as legal replacement for cannabis and synthetic cathinones sold as legal replacement for MDMA, amphetamine, and cocaine (EMCDDA, 2009b). They were initially sold in colorful packages in smoke shops, convenience stores, and the internet in hundreds of brands marketed as "legal highs" and "research chemicals." Monitoring of the online shops tracked their increase from 170 in January 2011 to 693 in January 2012. The number of NPS first notified to EMCDDA each year grew exponentially to 100 in 2014 and 2015. NSO, designer benzodiazepines, arylcyclohexylamines, and other classes started to be reported.

In response, the EMCDDA developed and applied new methods to bolster its ability to detect, track, and understand emerging trends, such as the E-POD studies that triangulated information from a wide range of formal and informal sources, analysis of wastewater and injecting equipment to detect and track use, and monitoring of online sales. New methods were also developed to strengthen the EWS's ability to detect, assess, prioritize, and respond to threats in a timely manner: (1) a toxicovigilance system that harmonized the way information on acute poisonings and deaths was reported and analyzed; (2) a monitoring method for open source information on the Internet to detect urgent and serious threats such as outbreaks and other relevant events; (3) a signal management approach with a better method of detecting and assessing emerging threats based on their type and seriousness, and identifying response options; and (4) a stronger risk communication system to improve alert sharing in the EWS network.

The rapid increase in NPS along with the improved methods by the EWS resulted in almost 480 NPS first notified to EMCDDA in the period. Ten were risk assessed: 4-MT, 5-IT, 4,4-DMAR, mephedrone, MDPV, alpha-PVP, 25I−NBOMe, methoxetamine, MT-45, and AH-7921, all of which were put under control in the European Union.

2016 to 2022

The period is characterized by a decrease in the number of NPS notified for the first time each year (~50) and a shift away from the so-called legal highs and research chemicals. However, this has been accompanied by a growing complexity in the market characterized by more potent substances, integration of supply sources with the established illicit drug market, and targeting of the marginalized and chronic drug-using population such as the homeless and prisoners. Although synthetic cannabinoids and cathinones persisted, it was NSO and designer benzodiazepines that became increasingly important and threatening. Highly potent NSO such as carfentanil and later the nitazenes such as isotonitazene were associated with outbreaks and poisonings, only to disappear quickly and be replaced by other opioids. Designer benzodiazepines are being used to produce fake prescription Valium and Xanax in the illicit market. Fentanyl and other fentalogs are likewise fakes for Oxycontin and Norco. Control restrictions in China on the manufacture of NPS such as the fentalogs led to a greater diversification of supply chains. India became a prominent source of synthetic cathinones from around 2019, and illicit production increased in Europe.

By the end of 2018, legislation to further strengthen the EWS and allow it to respond faster had been passed by the European Union (European Parliament and Council of the European Union, 2017a,b). Revised operating guidelines for EWS and risk assessments were published along with common reporting tools (EMCDDA, 2019, 2022c). A redesigned European Database on New Drugs was launched in 2019 to provide a next-generation information system capable of electronic data management.

A little over 320 NPS were reported in this time period (until 2021), of which 17 were risk assessed. Most are synthetic cannabinoids and NSO, including MDMB-CHMICA, AB-CHMINACA, ADB-CHMINACA, 5F-MDMB-CHMINACA, 4F-MDMB-BICA, MDMB-4en-PINACA, Cumyl-4CN-BINACA, acryloylfentanyl, furanylfenatnyl, 4F-iBF, THF-F, carfentanil, cylcoprolylfentanyl, metoxyacetylfentanyl, isotonitazene, 3-methylmethcathinone, and 3-chloromethcathinone. Twelve of these were controlled in the European Union.

STRIDA

Apart from the EWS, some European countries have surveillance programs that contributed valuable information on local NPS trends and threats. One such was STRIDA in Sweden mentioned before, among the earliest to be used proactively in clinical cases. It ran nationwide from 2010 to 2016 and analyzed urine and serum or plasma samples from emergency department and intensive care unit patients referred to the program after exposure to NPS and related products, or suspected of NPS use after clinical investigation (Helander et al., 2013). All the cases were referred to the program after establishing contact with the PIC where a case code was assigned and clinical and treatment information from deidentified hospital medical records were collected and later paired with analytical results. In some cases, the substances associated with the case were also forwarded for laboratory analysis. The biological samples received by the laboratory were subjected to comprehensive toxicological investigation, which involves routine immunoassays for traditional recreational drugs and LC-MS/MS and LC-HRMS to confirm NPS, traditional recreational drugs, plant- and mushroom-derived psychoactive substances, and prescription drugs. Nuclear magnetic resonance spectroscopy also was used (Helander and Backberg, 2018).

STRIDA enrolled 6526 cases, which comprised 40% of the total NPS consultations to the PIC. Only 95 cases (18%) were referred in the first year, but the number grew annually and peaked at 756 (45%) in 2014. As STRIDA became popular, the total number of NPS consultations at the PIC grew proportionately, peaking at 1669 in 2014. When this valuable project was terminated for lack of funding in 2016, the number of NPS consultations went back to the same level it was in 2010. Most of the cases enrolled were men (74%) with a mean age of 27; about 57% were 25 or younger. The proportion of women grew over time from 20% in 2010 to 24%–29% in the following years. The mean age of women was 24.5. The age range of patients was 8–71 years.

Overall, 81% of the biological samples submitted for analysis tested positive for psychoactive drugs, often multiple ones (70%). They included NPS, traditional recreational drugs, ethanol, and prescription medications misused for their psychoactive effects. A total of 159 NPS or less common psychoactive substances (e.g., ketamine and psilocybin) were detected (Fig. 8.3); 140 of these (88%) were notified as NPS by the EWS in 2004–2016%, and 75% were not yet classified in Sweden when they were first detected. Synthetic cannabinoids and cathinones were the major NPS classes detected in STRIDA in the first 2 years. Each class accounted for 20% of NPS cases in 2010. By 2012, designer benzodiazepines (etizolam

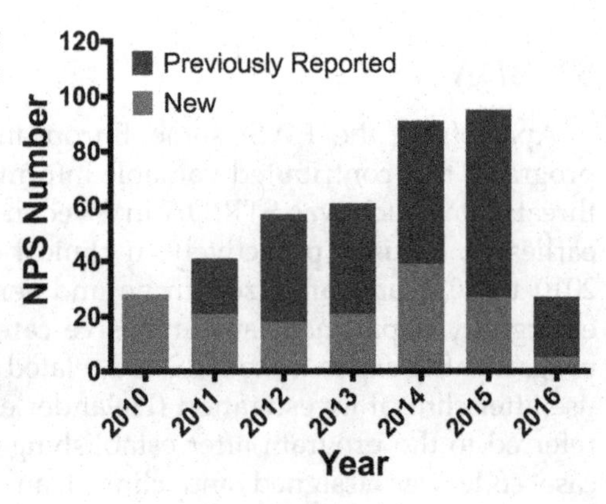

FIGURE 8.3 Annual number of NPS detected by the STRIDA project between 2010 and 2016.

and metizolam) and hallucinogens (25C- and 25I-NBOMe) started to appear, in addition to synthetic cannabinoids, synthetic cathinones, amphetamines, and phenethylamines. A powerful indole amphetamine, 5-(2-aminopropyl)indole (5-IT), was associated with 15 deaths in the first half of that year. NSO (AH-7921 and MT-45) and dissociative arylcyclohexylamines (3-MeO-PCP) started to appear in 2013 and persisted through 2016. Of note, the highly potent MDMB-CHMICA and 3-fluorophenmetrazine were linked to serious toxicity, while fentalogs (4-fluorobutyrfentanyl, 4-methoxybutyrfentanyl, acetylfentanyl, and furanylfentanyl) and escaline and its analogs appeared in 2015 (Helander et al., 2020).

Between 18 and 39 NPS (mean = 26) were detected for the first time each year from 2010 to 2015, of which 76%–96% (mean = 87%) were also reported to the EWS. The highest number was in 2014, when 39 were first reported to STRIDA, of which 35 (90%) were also reported to EWS in 2008–2014 (Helander et al., 2020). Most appeared for a relatively short time and usually disappeared after they were classified as illicit. The average delay between the first observation in STRIDA and the legal response was 1.6 years. However, a few persisted despite being classified as narcotics. Most were stimulants and synthetic cathinones, notably MDPV and alpha-PVP. Alpha-PVP was first reported to STRIDA in 2011 and scheduled after first quarter of 2013, but detections persisted until the end of STRIDA (Beck et al., 2016).

The STRIDA clinical data helped to establish the harms caused by NPS and complemented the statistics on NPS-related deaths. Its empirical monitoring and laboratory data also confirmed the previous anecdotal claims. Through notifications to EMCDDA and numerous scientific presentations and publications, it contributed to an understanding of NPS trends and harm internationally. At the time it was operating, it effectively and efficiently identified emerging NPS in Sweden, served as the primary source of data on emerging NPS that have significant potential to cause public harm in Europe, and was often first to publish numerous laboratory-confirmed intoxication cases on multiple NPS classes (Helander et al., 2014, 2017; Backberg et al., 2015a,b, 2019; Meyer et al., 2016; Beck et al., 2018). Most importantly, it demonstrated the value of collaborative work between the laboratory, medical professionals, and the PIC.

Euro-DEN

The EC set up the European Drug Emergencies Network (Euro-DEN) in 2013 to record emergency room cases from both traditional recreational drugs (TRD) and NPS. The acute toxicity data are meant to fill a gap in EMCDDA's collection (Wood et al., 2014). In the first year, it collected baseline data through an online survey of the EMCDDA's 30 Réseau

Européen d'Iformation sur les Drogues et les Toxicomanies national focal points. Only 10 countries reported systematic collection of such information on NPS at the national level, only six at the regional level. Euro-DEN set out to collect standardized data on key demographic and predefined clinical and outcome variables that had features consistent with TRD and NPS toxicity. Unlike STRIDA, however, no laboratory confirmation was required in the survey. Instead, the patient's self-report of the substances and/or the physician's interpretation of clinical features was relied upon. Data were gathered from October 2013 to September 2014 from 16 sentinel centers, which are urban centers in 10 European countries spread evenly between Western and Eastern Europe (Denmark, Estonia, France, Germany, Ireland, Norway, Poland, Spain, Switzerland, and the United Kingdom). Data were collated and processed in Euro-DEN's coordinating center in London (Dines et al., 2015).

In this baseline, there were 5529 presentations with TRD and/or NPS. About 75% were male, with average age 31; female average age was 28. Interestingly, there were variations in the number of presentations by month and day of the week: greatest in August and on Saturdays and Sundays, fewest in January and on Tuesdays. TRD comprise the highest number of reported drugs. However, without laboratory confirmation, the accuracy of this finding is very difficult to assess. Only 484 reports (5.6%) were associated with NPS, with synthetic cathinones (378, 4.3%) being most reported, especially mephedrone (245, 2.8%) and methedrone (92, 1.1%). Synthetic cannabinoids were the second most frequent NPS class. NPS presentations were concentrated in Gdansk (Poland), London and York (United Kingdom), Dublin, and Munich. There were none in Parnu and Tallin (Estonia) and Drogheda (Ireland). The most frequently reported clinical features were agitation/aggression (1467, 26.5%), anxiety (1040, 18.8%), and dyspnea (501, 9.1%). Most patients (~70%) were brought to the hospital by ambulance. A majority were either medically discharged from the emergency department (3148, 56.9%) or self-discharged; only 332 (6%) were admitted to critical care and 284 (5.1%) to psychiatry. The percentage accounted for by NPS alone for these clinical presentations and outcomes was not reported.

Euro-DEN expanded to Euro-DEN Plus after March 2015 and continued its systematic survey. The project has now 36 centers in 24 European countries. To address the concern on self-reports, Euro-DEN Plus performed a retrospective study on 10,956 cases reported to its network between October 2013 and September 2015 (Liakoni et al., 2018). Only 834 had drug confirmation. The primary reason for this low number is that toxicological screening is not routinely performed in the Euro-DEN Plus centers at the time of data collection. Data were obtained from only 8 of the 16 centers, which have limited or no capability to analyze for NPS. Only 40 cases were either self-reported or analytically confirmed for NPS.

Of these, only six self-reports were confirmed by analysis, and analytically confirmed were not self-reported. A majority (29, 72.5%) of the self-reports were not confirmed by laboratory analysis. These results further highlight the importance of the STRIDA project and the significant challenge in getting data for analytically confirmed clinical cases associated with NPS. Euro-DEN Plus has continued to collect clinical data on TRD and NPS acute toxicity, and their findings are summarized in more than a dozen publications.

The United States

Unlike in Europe where the EMCDDA was able to organize a surveillance system even before the NPS surge in 2008, no system was established in the United States right away. Instead, there was an evolution of surveillance strategies over time.

Poison control center reports

A recognition that synthetic cannabinoids and cathinones were becoming public health threats was first highlighted by the American Association of Poison Control Centers. In 2010, they reported that more than 2500 calls were associated with synthetic cannabinoids, 50 times more than in the preceding year (53) (Bronstein et al., 2011). In the next year, the number more than doubled (Fig. 8.4). The same trend occurred for synthetic cathinones starting in 2011 (Bronstein et al., 2012). The monthly statistics in their National Poison Data System and its annual reports became a source for trend data on potential harm in the early years of NPS surge in the United States. Because the information was mostly from self-reports, specific class could not be determined nor accuracy of the number of calls associated with NPS ascertained.

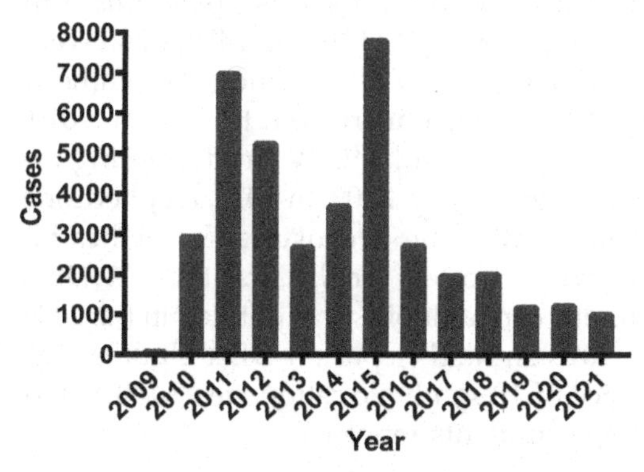

FIGURE 8.4 Annual number of poison control center calls associated with synthetic cannabinoids.

National Forensic Laboratory Information System

The other source of initial data on NPS was the NFLIS, created in 1997 by the Diversion Control Division of the DEA. It started as a data collection system for drug chemistry analysis from local, state, and federal laboratories, which submit analysis results of substances seized by law enforcement. About 283 laboratories from the 50 state systems and 108 local or municipal laboratories currently participate (US DEA, 2022). With the expansion of NFLIS in 2018, the data on drug reports became NFLIS-Drug. Two other systems were established: NFLIS-Tox collects data from public and private laboratories on toxicological findings from antemortem and postmortem drug testing, and NFLIS-MEC collects medical examiner and coroner office data on death cases in which drugs were identified.

NFLIS data have the advantage over poison control center calls of being analytically confirmed. Hence, the specific NPS and their numbers were accurately known. NFLIS issues annual and midyear reports providing collated summaries of all drugs reported by participating laboratories. It issues special reports and briefs on drug classes, including NPS, and topics relevant to drug use and harm, testing, and surveys. It also publishes Snapshots, which are 2 or 3 pages highlighting new and emerging drugs and public data tables that show drug reports by state since 2007. NFLIS-Tox and NFLIS-MEC also issue reports on lab surveys. NFLIS is not focused on NPS, as it collects data on all types of drugs. The extracted data on NPS are simply a by-product of the ongoing trend in drug use. Nevertheless, since NPS has caused mass outbreaks and thousands of individual intoxications in the past decade, NFLIS has published various types of collated data on specific NPS and NPS classes.

NFLIS-drug reports

A special report on synthetic cannabinoids and cathinones reported to NFLIS in 2009 and 2010 was published in 2011 (US DEA, 2011). It identified two synthetic cannabinoids (JWH-018 and JWH-073) in 15 instances in 2009 and 13 in 2977 instances in 2010. JWH-018 was predominant in 2010, but JWH-250 overtook JWH-073, and JWH-081 and JWH-200 were the two others with over 50 instances. Only 34 instances of synthetic cathinones were reported in 2009, which increased to 628 in 2010. Mephedrone, methcathinone, methylone, and MDPV (in decreasing order) were the synthetic cathinones reported in 2009. In 2010, mephedrone was still leading, but MDPV and methylone had overtaken methcathinone, and a new one, 4-MEC, was reported for the first time. A DEA Update on eight emerging synthetic cannabinoids was notified in the 2009 NFLIS Annual Report. In the 2010 Annual Report, a DEA Update was included notifying the emergence of 10 synthetic cathinones. MDPV and mephedrone were among the 15 stimulants reported.

An update on synthetic cannabinoids and cathinones reported from 2010 to 2013 was issued in 2014 (US DEA, 2014). This documented the rapid evolution of their molecular composition. By 2011, AM-2201 was the predominant cannabinoid, replacing JWH-018. In the second half of 2011, there was a precipitous drop in the number of JWH compounds. This was maintained until the first half of 2012, at which point AM-2201 was replaced by UR-144 and XLR-11. XLR-11 became predominant in 2013 as reports of UR-144 declined, which coincided with the time they were temporarily scheduled by the DEA. For synthetic cathinones, MDPV overtook mephedrone in 2011, with methylone second. Methylone became the most reported in 2012 and 2013. MDPV started declining in 2012; alpha-PVP had emerged in late 2011 and increased. Both are pyrovalerone derivatives. Pentedrone emerged along with alpha-PVP and persisted in 2013. The persistence of methylone until the end of 2013 despite being temporarily scheduled in 2011 is an example of a case where an NPS was controlled but continued to be sold in the recreational drug market. This is more of an exception than the rule and is true for only a handful of NPS. Both mephedrone and MDPV were temporarily scheduled with methylone in 2011, and reports on both drugs waned subsequently.

Issuance of Special Reports indicates trends in the United States. Other Special Reports published since 2011: in 2012, 2C, piperazines, and tryptamines reported in 2006–2011 (US DEA, 2012); in 2015, opiates and related drugs reported in 2009–2014, which documented the emergence of acetylfentanyl, AH-7921, and MT-45 in 2013 (US DEA, 2015); in 2016, designer benzodiazepines reported in 2009–2014, which documented the emergence of etizolam in 2014 (US DEA, 2016); in 2019, fentanyl and fentalogs reported in 2016 and 2017 (US DEA, 2019).

Survey of toxicology laboratories

NFLIS conducted its first toxicology testing laboratories survey in 2017 to collect information on caseloads, policies, and practices among public and private laboratories. Results were reported as an NFLIS-Tox publication in 2018 (US DEA, 2018a). About 231 out of 392 labs completed the full survey. Although most of the data were not specific to NPS, the survey revealed that of the 212 that responded to a query on their normal course of action for toxicology analysis requests for NPS, 46.2% (98) sent specimens to reference testing labs and 25.9% (55) conducted testing in-house. About 5.2% (11) screened in-house but sought reference labs for confirmation, while 14.2% (30) received no requests for NPS testing. Of the 218 labs that provided information on the platform used for testing, more than 90% used immunoassay for screening and 89% use mass spectrometry. Of these, 24 used LC-MS/MS and 10 used LC-(Q)TOF/MS. About 73% used GC-MS/MS and 84% used LC-MS/MS for confirmation, while 5% (10) used LC-TOF/MS. This is encouraging, as a quarter of the labs had some capability for NPS testing. At the same time though, the results highlight the scarcity of labs that can perform nontargeted data acquisition that is more compatible with NPS testing. A follow-up survey was conducted in 2021; results were expected in the summer of 2022.

In the same year as the NFLIS-Tox survey, NFLIS conducted an MEC survey that also collected baseline information on caseloads, policies, and practices (US DEA, 2018b). Of 2128 eligible MECs, 971 (45.6%) completed the full survey. About 906 provided information on their course of action when NPS testing was requested. Of these, 66% submitted the requests to a reference testing lab, 6.2% to a state lab or MEC, 8.7% had no NPS analysis requested, and only 2.1% of medical examiner and 0.6% of coroners had an in-house lab for NPS. The survey also gathered information on the testing frequency (always, sometimes, never) of specific drugs and classes and their frequency of quantitating these analytes. Of the NPS classes, less than 50% of the MECs could do toxicology testing for phenethylamines, piperazines, synthetic cannabinoids, and synthetic cathinones, while 51%–75% always did fentalog testing. Quantitative analysis was always conducted by 49.8%, 51.5%, and 57.8% of the MECs for synthetic cathinones, synthetic cannabinoids, and fentanyl/fentalogs, respectively. While it is encouraging to see that NPS testing is performed by at least half of the MECs, in-house capability is still very scarce for them in the United States. The next NFLIS-MEC survey was to be launched in 2022.

Multidisciplinary NPS testing

The scarcity of surveillance data in the early years of the NPS surge is tied to the lack of testing capability in clinical and forensic laboratories. Hence, very few specialty research laboratories with LC-HRMS capability were poised to fill the gap. One successful collaboration of experts from multiple disciplines contributed significantly to synthetic cannabinoid surveillance between 2012 and 2017. The UCSF Clinical Toxicology and Environmental Biomonitoring (CTEB) Laboratory started working on NPS-related cases with various poison control centers in California in 2010. In 2012, the laboratory was contacted by researchers from the DEA, CDC, and several poison control centers to identify the synthetic cannabinoid associated with acute kidney injury that resulted from K2 use in the Pacific Northwest (Murphy et al., 2013; Buser et al., 2014). The effort identified XLR-11, which started the informal synthetic cannabinoid surveillance collaboration among UCSF, DEA, and CDC.

The surveillance was mostly on individual and mass intoxications involving synthetic cannabinoids that were gathered and referred to the CTEB lab by a pharmacologist at the DEA, Dr. Jordan Trecki. The UCSF

CTEB lab, directed by Dr. Roy Gerona, worked on the analytical testing of biological samples by LC-HRMS, while a collaborator from the CDC, Dr. Michael Schwartz, lent his expertise in medical toxicology to find leads for nontargeted testing. These researchers often worked collaboratively with other clinical and medical toxicologists, forensic scientists, epidemiologists, public health experts, pharmacists, pharmacologists, and clinical chemists in solving the cases, especially if they were part of a mass intoxication (Trecki et al., 2015). This approach was very successful and led to providing NPS and synthetic cannabinoid data on more than 1700 samples from clusters of intoxications and outbreaks across the United States between 2012 and 2017 (Fig. 8.5).

Select data from the analysis became the basis for the DEA scheduling of at least six compounds: XLR-11, ADB-PINACA, AB-CHMINACA, AMB-CHMINACA, 5F-AMB, and AMB-FUBINACA. The 15 publications generated became the major source of surveillance data for synthetic cannabinoids in the United States and facilitated the association of a number of toxidromes to specific synthetic cannabinoids: XLR-11 to acute kidney injury, ADB-PINACA to severe delirium and seizures, and AMB-FUBINACA to strong depressant "zombie-like" behavior.

Prophetic synthetic cannabinoids library

By 2015, Dr. Samuel Banister, a synthetic organic chemist from Stanford University, joined the collaborative team and introduced a unique aspect

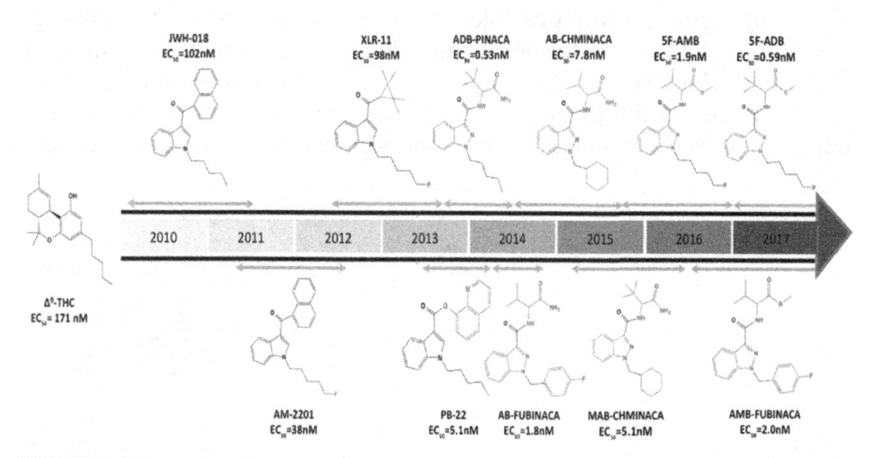

FIGURE 8.5 Molecular evolution of synthetic cannabinoids (using a specific example from each generation) and in vitro EC_{50} values. Orange arrows indicate the time periods in which our laboratory observed the cannabinoids over the course of surveillance. Red structural moieties indicate new structural motifs within the synthetic cannabinoid NPS class. Note that the EC_{50} values are from in vitro fluorometric assays of membrane potential in cells transfected with the human CB1 receptor and do not necessarily correspond to the potency of these agents in humans. For interpretation of the references to color in this figure legend, please refer online version of this title.

to the surveillance. A major challenge in NPS testing is the rapid molecular evolution of classes. With the sheer number of possible targets that can be produced by clandestine labs, testing labs often lack the reference standard to confirm a previously unreported NPS when it becomes a suspect compound in LC-HRMS analysis of NPS-related cases. The reactive nature of NPS testing labs to what comes out of the recreational drug market has put them at a disadvantage, causing long delays in confirming a newly released substance. Dr. Banister started working on a "prophetic synthetic cannabinoid" library by preemptively synthesizing compounds that are anticipated to be targeted by clandestine labs. This works well with synthetic cannabinoids, as the core structure of the class is defined by four subunits or pharmacophores (Fig. 8.6). By studying the functional groups that were used for each subunit to produce the synthetic cannabinoids that were released in the past and those that are currently trending, one can predict future functional group targets that are plausible for synthesis. By mixing and matching the functional groups possible for each subunit, one can synthesize a library of future or anticipated compounds that can be kept on standby until suspect screening by LC-HRMS identifies the compound in an NPS product or clinical sample.

This approach was successfully implemented in the well-publicized "zombie" mass outbreak in New York in the summer of 2016. In the analysis of biological samples from these cases, the CTEB lab identified a metabolite of a synthetic cannabinoid that had no commercially available reference standard. Because the parent cannot be detected in the cases, the only drug confirmation possible for the cases is that of the suspect metabolite. Custom synthesis of the reference standard by commercial labs could take weeks or months. However, because the reference standard had been in the prophetic library since February 2016, the lab was able to confirm the metabolite a few hours after it was mined as a suspect

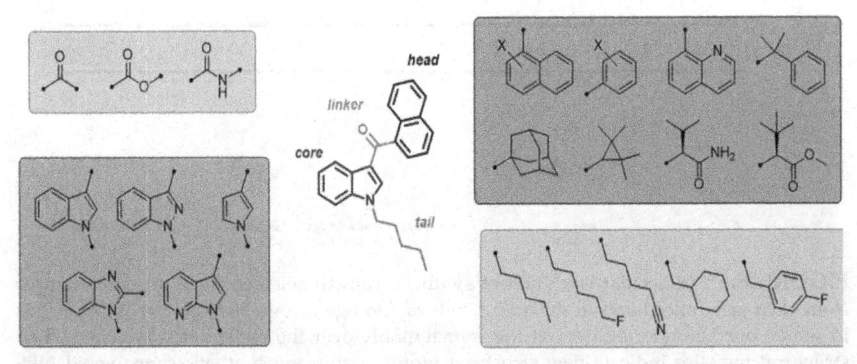

FIGURE 8.6 Generic representation of synthetic cannabinoids obtained by combination of its four subunits (pharmacophores) using JWH-018 as a template.

in the cases (Adams et al., 2017). This feature is unique to the surveillance team that set it apart from the few NPS monitoring surveillances at the time. The synthesis of prophetic libraries has now been adopted by a few surveillance groups, including the UNODC and EMCDDA.

DEA TOX

The synthetic cannabinoids surveillance project was terminated in 2017 for lack of funding. It was revived and expanded in mid-2019 through the DEA TOX project. This surveillance analyzes clinical and forensic cases that might have NPS etiology and is a formal collaborative project between the DEA and the UCSF CTEB lab funded by the US Department of Justice. Using an expanded LC-HRMS comprehensive toxicology testing panel that consists of more than 1200 drugs, including more than 950 NPS, the project screens and quantifies NPS, TRD, prescription drugs, dietary supplement stimulants, and common drug precursors, adulterants, and impurities in biological samples (serum, plasma, whole blood, urine) and in some cases, drug products and paraphernalia. The project issues quarterly reports summarizing its latest findings on top of NPS Discovery (a brief on newly discovered NPS) and Tox Alerts (a brief on currently spreading NPS threat). Aside from using prophetic NPS libraries, a unique feature of the surveillance is the publication of the ranges of quantitative levels of NPS and TRD. This is quite informative, as data on toxic levels of NPS are scarce, and therefore helpful in assessing clinical cases that may require a reference range for NPS toxicity in deciding the appropriate therapeutic regimen for an intoxicated patient. Table 8.2 gives the ranges of the top 10 identified NPS from DEA TOX in 2021.

DEA TOX collected its first full year of surveillance data only in 2021 due to delays in case collection imposed by COVID-19. The program received and analyzed 495 cases from 24 states across the United States. Of those, 90 were from fatalities, 202 from overdoses, and 203 from emergency department drug screens. Close to 70% (341) were males. NPS and/or TRD were detected in 404 of the cases (82%); in 193 of these (39%), one or more NPS were detected. Notably, NPS were detected in 71 of the 90 fatalities. Sixty-four NPS from 12 classes were confirmed and quantified; NSO and synthetic cannabinoids have the most with 15 drugs each. Synthetic cathinones, designer benzodiazepines, and amphetamines round out the top five NPS classes with the most in number. NSO were overwhelmingly the most frequently detected, followed by synthetic cannabinoids and synthetic cathinones (Fig. 8.7). Five NPS were detected in 10 or more cases: para-fluorofentanyl (NSO), mitragynine (NSO), ADB-BUTINACA (synthetic cannabinoid), metonitazene (NSO), and etizolam

TABLE 8.2 Quantitative level ranges in various biological matrices of the top 10 NPS detected in DEA TOX for 2021.

Drug	Freq	Class	Confirmed levels (ng/mL)			
			Serum	Plasma	Whole blood	Urine
para-fluorofentanyl	53	NSO	0.3–4.8	NQ	0.1–82	20.3–399
Mitragynine	22	NSO	17–185	0.4–0.8	0.8–110	2–1421
Metonitazene	16	NSO			0.7–16.5	
ADB-BUTINACA	14	Synthetic cannabinoid	1.4–4.2	4–4.5	0.4–34	
Etizolam	10	Designer benzodiazepine	0.4–55.6		1–13.4	
Tianeptine	7	NSO	56.9			14.5–18000
Alpha-methyltryptamine	7	Tryptamine				NQ
Acetyl fentanyl	6	NSO			0.9–3.4	24.7–1580
Eutylone	5	Synthetic cathinone	5.8–21		16.9–50.6	
11-Nor-9-carboxy-delta-8-THC	5	Synthetic cannabinoid	31.6	747–929		109–292

Freq, detection frequency; *NQ*, not quantified.

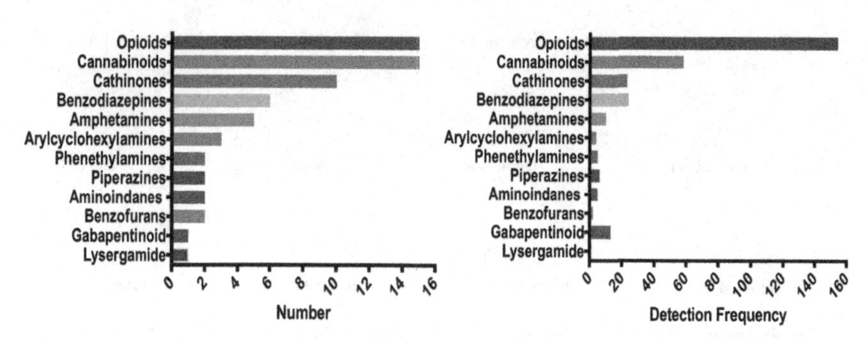

FIGURE 8.7 Number and detection frequencies of NPS classes confirmed by DEA TOX in 2021.

(designer benzodiazepine). Polydrug use is evident, especially in fatalities, with more than half of the cases having two or more NPS and/or TRD. There were 1506 NPS/TRD detections in 404 cases. The most frequently detected combinations were fentanyl and para-fluorofentanyl with or without a nitazene. Thus, the DEA TOX surveillance data in 2021 suggest that the opioid crisis is still ongoing in the United States. Whereas before, fentanyl, fentanyl-laced heroin, or fentalog-laced heroin was more common, the new feature is the deadly combination of fentanyl and a fentalog that is sometimes further mixed with a nitazene.

The trend in the domination by fentanyl and other NSO in drugs detected by the surveillance continued in 2022 (DEA, 2023). About 65% of all NPS were NSO. Designer benzodiazepines, at 20%, were the second most detected. The top five NPS were all depressants and included para-fluorofentanyl (NSO), bromazolam (designer benzodiazepine), clonazolam (designer benzodiazepine), mitragynine (NSO), and metonitazene (NSO). Fentanyl, at 35%, topped the TRD; it was confirmed in 174 of the 370 cases (47%). The majority also contained para-fluorofentanyl; the nitazenes were detected in place of para-fluorofentanyl in some cases. The mean concentration of fentanyl significantly increased in overdose cases, from 2.5 to 7.5 ng/mL, implying either increase in opioid tolerance of users or increase in the concentration of fentanyl in drug products or both. The mean concentration in death cases also increased from 14.3 to 17.6 ng/mL.

The cooccurrence of fentanyl with other drugs was demonstrated in the surveillance (Fig. 8.8) and provided additional confirmation that fentanyl is added to just about every drug product in the illicit market. The stimulants methamphetamine and cocaine, interestingly, had the highest codetection frequency with fentanyl. However, higher proportions of death cases were observed when fentanyl cooccurred with another depressant (NSO or designer benzodiazepine). The mean concentration of fentanyl in these cooccurrences in death cases ranges between 16.6 and

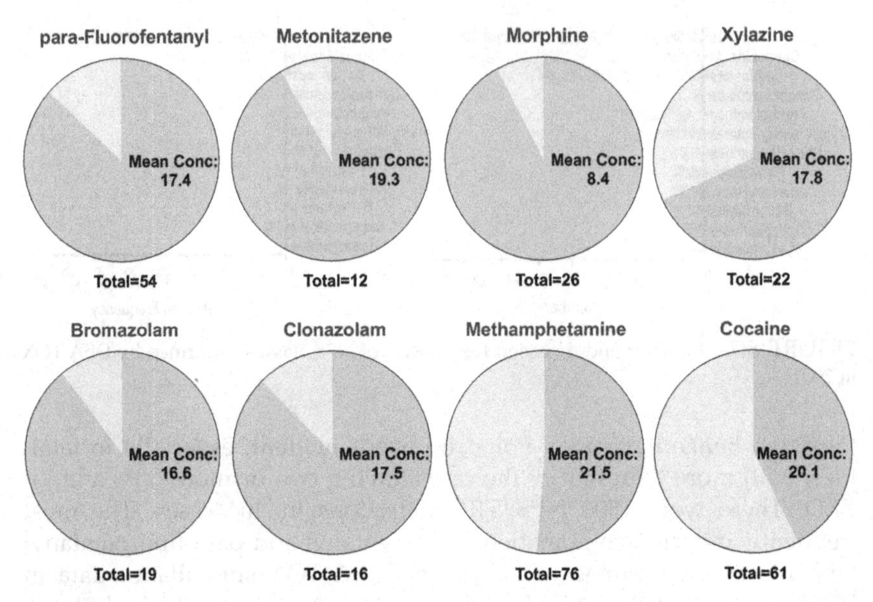

FIGURE 8.8 Distribution between deaths (*pink* [gray in print]) and overdoses (*blue* [light gray in print]) in drug cases with highest occurrence detection frequency with fentanyl among cases referred to DEA TOX in 2022. The total codetection frequency in 147 fentanyl cases (below the pie chart) and the mean concentration of fentanyl in ng/mL (in the pie chart) in death cases are shown.

21.5 ng/mL except for morphine, where the mean was 8.4 ng/mL. The use of the veterinary tranquilizer xylazine as an adulterant in fentanyl products spread. DEA TOX first detected xylazine in mass intoxication cases referred from Philadelphia in September 2021. In 2022, DEA TOX confirmed xylazine in 4 of the 24 states (Alabama, Kentucky, Pennsylvania, Tennessee) that referred cases to the program.

There are other surveillance programs in the United States that publish data on NPS. CFSRE (https://www.cfsre.org/) and NDEWS (https://ndews.org/) are two of the more popular. Limitation in space does not allow discussion of their activities; the interested reader is referred to their websites.

Other surveillance systems

Aside from surveys, poison center calls, clinical and forensic toxicology testing, and analysis of drug seizures, there are other surveillance approaches. Wastewater analysis gained traction in the mid-2000s for monitoring TRD. Wastewater analysis offers a complementary source of data for drugs that are missed by drug seizures and those that are not potent enough to cause death or emergency department visits. Thus it has the potential to provide a more comprehensive profile of drugs circulating in the community. Data on less potent NPS are also valuable as sometimes a potent isomer is a few molecular tweaks away from its less potent isomer. However, the lack of information on how some NPS are metabolized, their stability, and whether their levels in wastewater are compatible with the sensitivity of available analytical platforms presented initial challenges to this approach. Nevertheless, wastewater analysis has been applied to NPS surveillance more recently. In one study, NPS in influent wastewater from 14 sites in eight countries (Australia, China, Estonia, Italy, Netherlands, Norway, New Zealand, and United States) was monitored for 1 week at the New Year between 2019 and 2020 (Bade et al., 2021). This period was chosen because it is characterized by celebrations and parties when NPS may be more likely consumed. Results revealed geographic differences in prevalent NPS among the countries. Ten NPS, of which nine were synthetic cathinones, were semiquantified. Seven, including amphetamines, synthetic cathinones, NSO, and arylcyclohexylamine, were qualitatively detected or confirmed. Ketamine was found in all eight countries. Methcathinone was quantified in all sites except Norway. 4-Fluoroamphetamine, 4-fluoromethamphetamine, 4-methylethcathinone, and MDPV were found only in the Netherlands. Acetylfentanyl, mitragynine, and pentylone were found only in the United States. 4-Chloromethacathinone was found only in Italy. This analysis was limited by the panel used, which monitored only 200 NPS.

Similar to wastewater analysis is the use of pooled samples from portable urinals. Pooled urine sampling has the advantage of being anonymized, which frees up the requirement for individual consent. It allows collection of community-wide instead of individual patient or subject data and can capture NPS of widely varying potencies, which might not be possible for clinical cases (Archer et al., 2013a). It also makes possible the targeting of a specific geographic location or subpopulation such as rave-goers. One of the first users was Archer et al. in London, who collected samples from public stand-alone urinals in Westminster, a borough in central London with a wide variety of nighttime economy venues such as bars, late-night cafes, nightclubs, and discotheques (Archer et al., 2013b). With 12 urinals, they collected samples from 6 p.m. to 6 a.m. on a Saturday night in March 2012. LC-HRMS detected nine TRD and four NPS, including mephedrone, methiopropamine, methoxetamine, and methylhexeneamine. Extension of the study to samples collected from eight other UK cities in April 2014 detected 10 TRD, 9 NPS, and 4 anabolic steroids (Archer et al., 2015). Mephedrone was the most common NPS, detected in five cities. The others were methiopropamine (3 cities), methylhexeneamine (2), methoxetamine (2), and TFMPP, BZP, pentedrone, methylone, and glaucine (1). There were geographic differences for NPS. For example, piperazines and pentedrone were in only Birmingham, which has the highest number of NPS detected (5), methylhexanamine only in Bristol and London, and methylone only in London. NPS were detected also in all cities except Edinburgh, Leeds, and Brighton. Interestingly, methoxetamine was detected in London in 2012 but not in 2014, while methylone and glaucine were detected only in 2014.

Since benzylpiperazine became the first NPS popularly sold online about the turn of the century, the Internet has played a significant and defining role not only in marketing but equally so in the dissemination and free exchange of information on use and experiences (EMCDDA, 2009a). The increasing popularity of drug forum sites where practical information on use is openly discussed has encouraged the sale and use in the general population (Kruithof et al., 2016). Social networks and smartphone applications added fuel to the fire. And the availability of online sites on the surface and deep webs created an epidemic posing a global health threat. The dark net, where buyers and sellers can do business anonymously by paying with a virtual wallet, has created an environment in which securing highly psychoactive substances cheaply and at low legal risk is just a few clicks away (EMCDDA, 2016).

Another surveillance system that was applied for the first time to NPS before any other drug class is surveillance of Internet-based sales and websites that discuss chemistry, use, and harm. An example is the Psychonaut 2002 project funded by the European Commission (Psychonaut Web Mapping Research Group, 2005). It did qualitative and quantitative assessment of the online supply in a time-specific context or snapshot. The I-TREND (Internet Tools for Research in Europe for New Drugs) is another (Cadet-Tairou and Martinez, 2017). It monitored the evolution of online shops and user fora, conducted an online survey of users, and performed chemical analysis of products bought from online shops to produce a "top list" of NPS at the national level.

Van Hout and Hearne explored cryptomarket forum members' views and perspectives on NPS vendors and products within the context of hidden web community dynamics using Alphabay and Valhalla, two cryptomarkets popular with NPS vendors and hosting fora (Van Hout and Hearne, 2017). The study revealed that users placed high value on generation of trust, honesty, and excellence of service. They appeared well-informed, and harm reduction and vendor information exchange were top priorities. GABA-activating substances seemed to be the most popular among buyers when the study was conducted in 2016.

Changing motivations and patterns of use

Interest in NPS started almost exclusively with "psychonauts," individuals who explore altered states of consciousness, typically through hallucinatory drugs, and usually have extensive knowledge of the chemistry and pharmacology of synthetic drugs. The motivation for NPS use is to achieve altered states and access faculties of the brain that are otherwise inaccessible in a conscious state. Alexander Shulgin, who synthesized innumerable phenethylamines, amphetamines, and tryptamines and personally tested the experiences evoked, is an example. This changed with the surge of synthetic cannabinoids and cathinones in the late 2000s, especially when the products became more accessible and users discovered they were legal and usually not detected by drug testing. This brought with it a new scenario characterized by an increased number of users among young males and the consumption of drugs with unknown effects and no safety profiles.

In the first few years of the resurgence, there was a false perception of low risk because the agents were of herbal origin (specific for synthetic cannabinoids). The primary users are young and curious males such as high school and college students and young professionals who frequent the rave or party scene, as well as those who are subject to regular drug testing such as the military and other professionals but want to maintain recreational drug use. This demographic pattern is evident in early surveys in Europe and the United States. In the 2011 Eurobarometer survey, 4.8% of the 12,000 randomly selected individuals aged 15–24 (young adults, the age group with the highest drug use) responded that they had experimented with legal substances that imitate the effects of illicit drugs (EMCDDA and Europol, 2011). This trended up in succeeding years. The first formal Eurobarometer survey on NPS, in 2014, showed 8% lifetime and 3% past-year prevalence among that age group (UNODC, 2017). From Monitoring the Future, an annual nationally representative survey of US high school seniors, 10.1% reported past year use of synthetic cannabinoids from 2011 to 2013 (Palamar and Acosta, 2015). With reports on intoxications and outbreaks associated with synthetic cannabinoids in succeeding years, there was a steady decline of prevalence of past-year use among high school seniors, from 11.4% in 2011 to 3.6% in 2016 (Palamar et al., 2017). The popularity of synthetic cannabinoids and synthetic cathinones also infiltrated the military. More than 100 dischargees reported NPS use between June 2011 and January 2012 (Johnson et al., 2013), and 2.5% of 10,000 US Army drug tests that had initially been negative turned positive for synthetic cannabinoids in 2012 (Brantley, 2012).

As more online shops became available, access became easier, prices came down, and use rose. This attracted regular drug users into the mix of NPS users, including those that use drugs to enhance sexual experience, common in men having sex with men (MSM) and the transgender population. This contributed to higher morbidity and mortality with NPS use, as regular use is associated with higher risk-taking behavior. In a German study, for example, the prevalence of harmful use of recreational drugs such as cannabis and dissociative anesthetics is higher among MSM than men in the general population. Higher risk-taking behavior such as unprotected sex in MSM is significantly correlated with use of drugs (Dirks et al., 2012). In another study in Spain, MSM were found to have significantly higher lifetime and past year drug use than men in the general population. Notably, 17.4% of MSM surveyed were heavy polydrug users compared with 1.8% for men in the general population (Guerras et al., 2021).

Around the latter part of the past decade, the growing integration of the NPS supply chain with the established illicit drug market targeted more chronic and heavier drug users and the marginalized population such as the homeless and prisoners (EMCDDA, 2022b). With this phenomenon, the motivations for NPS use have become more integrated with and similar to the motivations for TRD use.

Closeup: The analytical chemistry workflow for NPS identification and surveillance

Surveillance of NPS posts numerous challenges to the analytical chemist. This was especially true during the initial proliferation of NPS in the 2010s. Analytical and reference labs all over the world were struggling to keep up with the emergence of new compounds. Numerous strategies were employed—some successful, others not. By trial and error, the Clinical Toxicology and Environmental Biomonitoring Laboratory at the University of California San Francisco developed some successful approaches that aided in the rapid resolution of NPS outbreaks.

The first of these challenges was that the federal agencies that publish reports on the prevalence of NPS, such as STRIDA, EMCDDA, and UNODC, were often describing a drug landscape that was several months outdated. This data lag became even longer in the published literature. By the time an article was reviewed for a medical or toxicologic journal (a process that might take 6 months to a year), the compound in question may have become no longer relevant! Even if you were the most up-to-date on the NPS literature, you were behind the current state of available NPS.

The second challenge to the analytical chemist is the derivative nature of NPS. Many of these compounds started as obscure structures in patent literature or research papers. Users interested in a "legal high"—intoxication that does not involve a DEA restricted substance, or a substance that might not screen positive on a drug test—would pursue these compounds. In the early days, they were available online as "research chemicals." As law enforcement caught on and began scheduling some of them, chemists in clandestine labs would derive new ones by altering the chemical structure—adding new substituent groups—as a way of

Closeup: The analytical chemistry workflow for NPS identification and surveillance (*cont'd*)

bypassing federal restriction. At the same time, chemists and users with an understanding of medicinal chemistry would also alter them to enhance their potency or activity. In this way, the number of compounds in a given class of NPS ballooned over the course of a few years.

The third challenge is that these compounds turned over rapidly within the drug supply. Some were more popular among users and would linger, but often compounds would enter the drug supply and be supplanted by the next wave of derivatives. Many users did not know what they were consuming, and even if they were purchasing "pure compound" from an online seller, it was likely to be contaminated with other compounds. Drugs purchased on the street were even more nebulous as they entered the supply under vague names such as "K2," "Spice," "Bath Salts," "Purple heroin," and "Blue pills."

Why is this a challenge to the analytical chemist? To fully comprehend, you must have a basic understanding of the workflow of an analytical lab. You receive a urine, blood, or drug sample from a hospital or agency. If it is a biologic sample, you treat it to precipitate out the proteins and other complex large molecules to minimize analytical interference. If it is a drug sample, preparation is easier, and you are often able to perform a simple extraction and dilution. You then submit the sample to your analytical method. The most common methods used for this type of analysis are gas chromatography–mass spectrometry (GC-MS) and liquid chromatography–mass spectrometry (LC-MS). LC has several benefits that are beyond the scope of this brief article and is the gold standard. The way these technologies work, in broad terms, is that they use high energy to "shatter" a molecule into multiple fragments, and these fragments are separated based upon their mass and their charge. At the end of the analysis, you receive a report—a chromatogram—that shows the abundance of the different fragments in the sample. Fortunately, a given compound fragments in a predictable way, and its chromatogram is similar to a fingerprint. Unfortunately, in a complex sample such as human blood, there can be thousands and thousands of fragments because you are shattering not just the compound of interest but also all the other small molecules present. As a result, a full chromatogram for a human sample can be a giant file consisting of multiple gigabytes of data. The way to find the signal in the noise is to compare the chromatogram to one generated by a library of reference standards. These are pure compounds that have been analyzed, and the computer software searches for matches. Of course,

continued

Closeup: The analytical chemistry workflow for NPS identification and surveillance (*cont'd*)

there is always a twist; in this case, drug metabolism. The compound you are looking for may be chemically altered in the body by common processes such as sulfonylation, glucuronidation, demethylation, and hydrolysis, and you must accommodate for this either during your treatment of the sample or by running the sample against metabolized reference standards. In many cases, you receive a sample, analyze it, and receive a confirmation based upon your reference library. You searched for an unknown in the sample and were able to confirm it with a known unknown.

This workflow poses the fourth challenge: the cost of maintaining a large reference library. Each analytically pure reference standard costs hundreds to thousands of dollars for around 1 mg. Standards degrade and must be replaced on a regular basis. And a comprehensive library may consist of 1000 reference standards. This becomes incredibly expensive to maintain, and many analytical laboratories simply cannot afford it.

The challenge for the analytical chemist is due to the unknown unknowns. The NPS from the drug supply might never have been described in an agency report or the scientific literature, or there might be no reference standard for it. In these cases, you can (and we have) spend hours to days manually combing through the chromatogram, trying to find a pattern in a sea of fragment data. And even if you do identify a possible culprit, you have to find a synthetic laboratory that will produce the reference standard, a process that is expensive and takes weeks to months, before you can definitively say that this is the unknown unknown you have been looking for. An analogy would be looking for a lost pet in a city based upon the size and weight of all the pets in the city, and you have a registry of all the pets in the city from 6 months ago, and you only have a loose idea of what type of pet it is, and when you think you have found the pet, it takes 3 months to hear back from the owner whether that was the correct pet or not. So where to begin?

To solve these wicked problems and try to resolve NPS outbreaks in real time, we took a stepwise approach.

With regard to the data lag of reporting new NPS circa 2014, we created an extensive Google Alerts search strategy based upon hundreds of compounds from agency reports, patents, and the scientific literature. This resulted in a constant inflow of media reports and articles related to NPS outbreaks. We also manually surveyed online drug forums such as BlueLight and Erowid, among others, and discussed cases with medical toxicologists and pharmacists operating the country's Poison Control call centers. This provided us with a qualitative survey of what the

Closeup: The analytical chemistry workflow for NPS identification and surveillance (*cont'd*)

psychonauts were discussing online and what people were taking when they reported to the emergency department. In turn, this helped address the issue of rapid turnover within the drug supply and of disentangling the street names for NPS products. This was how we were able to become aware of the "Zombie Outbreak" in the Bedford–Stuyvesant neighborhood of New York City, despite being in San Francisco, before the DEA reached out to us. It has also helped us to proactively reach out to healthcare facilities to obtain biologic samples before they are disposed of.

Next, we partnered with synthetic chemists such as Dr. Samuel Banister who provided us with libraries of synthetic cannabinoids—NPS that did not exist yet but might be created by clandestine labs based upon likely variations to different NPS classes—as well as likely metabolites for these compounds. This allowed us to rapidly confirm compounds before analytical standards became commercially available. And by partnering with research scientists, it also significantly reduced the cost of maintaining a comprehensive reference standard library. In this way, we were able to confirm AMB-FUBINACA as the causative agent in the Zombie Outbreak in record time (Adams et al., 2017).

Since 2015, the Clinical and Environmental Biomonitoring Laboratory at the University of California San Francisco has systematized collaborative networks with regional toxicology centers across the United States as well as the DEA in the DEA–TOX project to help resolve further outbreaks on a time scale that, while still not instantaneous, is much more rapid than the historical alternative. NPS surveillance remains complex and continues to evolve amid the addiction and mental health crises affecting much of the world. While strenuous, maintaining a wide collaborative network ranging from analytical chemistry to medicinal and synthetic chemistry to law enforcement to industry to medical toxicologists and pharmacists has been the most successful approach to date.

Axel Adams, MD, MS

Department of Emergency Medicine, University of Washington, Seattle, WA, United States

References

Adams AJ, Banister SD, Irizarry L, Trecki J, Schwartz M, Gerona R. "Zombie" outbreak caused by the synthetic cannabinoid AMB-FUBINACA in New York. N Engl J Med January 19, 2017;376(3):235–42. https://doi.org/10.1056/NEJMoa1610300. Epub December 14, 2016. PMID: 27973993.

Archer JR, Hudson S, Wood DM, Dargan PI. Analysis of urine from pooled urinals - a novel method for the detection of novel psychoactive substances. Curr Drug Abuse Rev 2013a; 6(2):86–90. https://doi.org/10.2174/1874473706666131205144014. PMID: 24308525.

Archer JR, Dargan PI, Hudson S, Wood DM. Analysis of anonymous pooled urine from portable urinals in central London confirms the significant use of novel psychoactive substances. QJM 2013b;106(2):147–52. https://doi.org/10.1093/qjmed/hcs219. Epub November 22, 2012. PMID: 23178933.

Archer JR, Hudson S, Jackson O, Yamamoto T, Lovett C, Lee HM, et al. Analysis of anonymized pooled urine in nine UK cities: variation in classical recreational drug, novel psychoactive substance and anabolic steroid use. QJM December 2015;108(12):929–33. https://doi.org/10.1093/qjmed/hcv058. Epub March 13, 2015. PMID: 25770158.

Australian Institute of Health and Welfare. National Drug Strategy Household Survey detailed report 2013. Drug statistics series no. 28. Cat. no. PHE 183. Canberra: AIHW; 2014.

Bäckberg M, Beck O, Jönsson KH, Helander A. Opioid intoxications involving butyrfentanyl, 4-fluorobutyrfentanyl, and fentanyl from the Swedish STRIDA project. Clin Toxicol 2015a;53(7):609–17. https://doi.org/10.3109/15563650.2015.1054505. Epub June 17, 2015. PMID: 26083809.

Bäckberg M, Beck O, Helander A. Phencyclidine analog use in Sweden–intoxication cases involving 3-MeO-PCP and 4-MeO-PCP from the STRIDA project. Clin Toxicol November 2015b;53(9):856–64. https://doi.org/10.3109/15563650.2015.1079325. Epub August 21, 2015. PMID: 26295489.

Bäckberg M, Pettersson Bergstrand M, Beck O, Helander A. Occurrence and time course of NPS benzodiazepines in Sweden - results from intoxication cases in the STRIDA project. Clin Toxicol March 2019;57(3):203–12. https://doi.org/10.1080/15563650.2018.1506130. Epub October 22, 2018. PMID: 30348014.

Bade R, White JM, Chen J, Baz-Lomba JA, Been F, Bijlsma L, et al. International snapshot of new psychoactive substance use: case study of eight countries over the 2019/2020 new year period. Water Res April 1, 2021;193:116891. https://doi.org/10.1016/j.watres.2021.116891. Epub February 3, 2021. PMID: 33582495.

Beck O, Franzén L, Bäckberg M, Signell P, Helander A. Toxicity evaluation of α-pyrrolidinovalerophenone (α-PVP): results from intoxication cases within the STRIDA project. Clin Toxicol August 2016;54(7):568–75. https://doi.org/10.1080/15563650.2016.1190979. PMID: 27412885.

Beck O, Bäckberg M, Signell P, Helander A. Intoxications in the STRIDA project involving a panorama of psychostimulant pyrovalerone derivatives, MDPV copycats. Clin Toxicol April 2018;56(4):256–63. https://doi.org/10.1080/15563650.2017.1370097. Epub September 12, 2017. PMID: 28895757.

Brantley CL. Spice, bath salts, salvia divinorum, and huffing: a judge advocate's guide to disposing of designer drug cases in the military. Army Lawyer; 2012. p. 16–37.

Bronstein AC, Spyker DA, Cantilena Jr LR, Green JL, Rumack BH, Dart RC. 2010 annual report of the American Association of Poison control centers' National Poison Data System (NPDS): 28th annual report. Clin Toxicol December 2011;49(10):910–41. https://doi.org/10.3109/15563650.2011.635149. Erratum in: Clin Toxicol (Phila). December 2014;52(10):1285. PMID: 22165864.

Bronstein AC, Spyker DA, Cantilena Jr LR, Rumack BH, Dart RC. 2011 annual report of the American Association of Poison control centers' National Poison Data System (NPDS): 29th annual report. Clin Toxicol December 2012;50(10):911–1164. https://doi.org/10.3109/15563650.2012.746424. Erratum in: Clin Toxicol (Phila). December 2014; 52(10): 1286–7. PMID: 23272763.

Buser GL, Gerona RR, Horowitz BZ, Vian KP, Troxell ML, Hendrickson RG, et al. Acute kidney injury associated with smoking synthetic cannabinoid. Clin Toxicol August 2014; 52(7):664–73. https://doi.org/10.3109/15563650.2014.932365. PMID: 25089722.

Cadet-Tairou A, Martinez M. I-TREND project overview. June 2017.

Commission on Narcotic Drugs, CND. Resolution 56/4. Enhancing international cooperation in the identification and reporting of new psychoactive substances. 2013. http://undocs.org/E/2013/28-E/CN.7/2013/14.

Dines AM, Wood DM, Yates C, Heyerdahl F, Hovda KE, Giraudon I, et al. Acute recreational drug and new psychoactive substance toxicity in Europe: 12 months data collection from the European Drug Emergencies Network (Euro-DEN). Clin Toxicol November 2015; 53(9):893–900. https://doi.org/10.3109/15563650.2015.1088157. Erratum in: Clin Toxicol (Phila). November 2015; 53(9):930. Liechti ME [added]; Markey, Gerard [added]; Mégarbane, Bruno [added]; Miro, Oscar [added]; Moughty, Adrian [added]. PMID: 26503789.

Dirks H, Esser S, Borgmann R, Wolter M, Fischer E, Potthoff A, et al. Substance use and sexual risk behaviour among HIV-positive men who have sex with men in specialized outpatient clinics. HIV Med October 2012;13(9):533–40. https://doi.org/10.1111/j.1468-1293.2012.01005.x. Epub March 21, 2012. PMID: 22435363.

European Monitoring Centre for Drugs and Drug Addiction. Joint action 97/396/JHA: 1997 joint action on new synthetic drugs. Luxembourg: Publications Office of the European Union; 1997.

European Monitoring Centre for Drugs and Drug Addiction. Council decision 2005/387/JHA on the information exchange, risk assessment and control of new psychoactive substances. Luxembourg: Publications Office of the European Union; 2005.

European Monitoring Centre for Drugs and Drug Addiction. Report on the risk assessment of BZP in the framework of the Council decision on new psychoactive substances. Luxembourg: Publications Office of the European Union; 2009a. www.emcdda.europa.eu/publications/risk-assessments/bzp_en.

European Monitoring Centre for Drugs and Drug Addiction. Understanding the 'spice' phenomenon. Luxembourg: Publications Office of the European Union; 2009b. https://www.emcdda.europa.eu/publications/rapidcommunications/synthetic-cannabinoids-europe-review_en.

European Monitoring Centre for Drugs and Drug Addiction. The internet and drug markets (EMCDDA Insights 21). Luxembourg: Retrieved from Publications Office of the European Union; 2016. http://www.emcdda.europa.eu/publications/insights/internet-drug-markets.

European Monitoring Centre for Drugs and Drug Addiction. EMCDDA operating guidelines for the European Union Early Warning System on new psychoactive substances. Luxembourg: Publications Office of the European Union; 2019. http://www.emcdda.europa.eu/publications/guidelines/operating-guidelines-for-the-european-union-earlywarning-system-on-new-psychoactive-substances_en.

European Monitoring Centre for Drugs and Drug Addiction. Early warning system. 2022. https://www.emcdda.europa.eu/publications/topic-overviews/eu-early-warning-system_en#section2. [Accessed 30 July 2022].

European Monitoring Centre for Drugs and Drug Addiction. New psychoactive substances: 25 years of early warning and response in Europe. An update from the EU Early Warning System. Luxembourg: Publications Office of the European Union; June 2022b.

European Monitoring Centre for Drugs and Drug Addiction. European Commission adopts measures to control two harmful new drugs amidst health concerns and surge in supply. Lisbon: EMCDDA; March 18, 2022c. https://www.emcdda.europa.eu/news/2022/3/european-commissionadopts-measures-control-two-harmful-new-drugs_en.

European Monitoring Centre for Drugs and Drug Addiction and Europol. EMCDDA–Europol 2011 annual report on the implementation of council decision 2005/387/JHA.

Lisbon: EMCDDA; 2011. https://www.emcdda.europa.eu/publications/implementation-reports/2011_en.

European Parliament and Council of the European Union. Regulation (EU) 2017/2101 of the European Parliament and of the Council of 15 November 2017 amending Regulation (EC) No 1920/2006 as regards information exchange on, and an early warning system and risk assessment procedure for, new psychoactive substances. Off J Eur Union L 2017a;305: 1–7. http://data.europa.eu/eli/reg/2017/2101/oj.

European Parliament and Council of the European Union. Directive (EU) 2017/2103 of the European Parliament and of the Council of 15 November 2017 amending Council Framework Decision 2004/757/JHA in order to include new psychoactive substances in the definition of "drug" and repealing Council Decision 2005/387/JHA. Off J Eur Union L 2017b;305:12–8. http://data.europa.eu/eli/dir/2017/2103/oj.

Evans-Brown M, Sedefov R. Responding to new psychoactive substances in the European union: early warning, risk assessment, and control measures. Handb Exp Pharmacol 2018;252:3–49. https://doi.org/10.1007/164_2018_160. PMID: 30194542.

Feng LY, Li JH. New psychoactive substances in Taiwan: challenges and strategies. Curr Opin Psychiatr July 2020;3 3(4):306–11. https://doi.org/10.1097/YCO.0000000000000604. PMID: 32167950.

Guerras JM, Hoyos J, García de Olalla P, de la Fuente L, Herrero L, Palma D, et al. Comparison of polydrug use prevalences and typologies between men who have sex with men and general population men, in Madrid and Barcelona. Int J Environ Res Publ Health 2021 Nov 4;18(21):11609. https://doi.org/10.3390/ijerph182111609. PMID: 34770122; PMCID: PMC8583212.

Health Canada. Summary of results of 2010–11 youth smoking survey. Waterloo, Canada: Health Canada; May 2012 (Controlled Substances and Tobacco Directorate). Available from:http://www.hc-sc.gc.ca/hc-ps/tobac-tabac/research-recherche/stat/_survey-sondage_2010-2011/result-eng.php.

Helander A, Bäckberg M, Hultén P, Al-Saffar Y, Beck O. Detection of new psychoactive substance use among emergency room patients: results from the Swedish STRIDA project. Forensic Sci Int October 2014;243:23–9. https://doi.org/10.1016/j.forsciint.2014.02.022. Epub March 6, 2014. PMID: 24726531.

Helander A, Bäckberg M, Signell P, Beck O. Intoxications involving acrylfentanyl and other novel designer fentanyls - results from the Swedish STRIDA project. Clin Toxicol July 2017;55(6):589–99. https://doi.org/10.1080/15563650.2017.1303141. Epub March 28, 2017. PMID: 28349714.

Helander A, Beck O, Hägerkvist R, Hultén P. Identification of novel psychoactive drug use in Sweden based on laboratory analysis–initial experiences from the STRIDA project. Scand J Clin Lab Invest August 2013;73(5):400–6. https://doi.org/10.3109/00365513.2013.793817. Epub May 22, 2013. PMID: 23692208.

Helander A, Bäckberg M. Epidemiology of NPS based confirmed overdose cases: the STRIDA project. Handb Exp Pharmacol 2018;252:461–73. https://doi.org/10.1007/164_2018_134. PMID: 30135990.

Helander A, Bäckberg M, Beck O. Drug trends and harm related to new psychoactive substances (NPS) in Sweden from 2010 to 2016: experiences from the STRIDA project. PLoS One April 23, 2020;15(4):e0232038. https://doi.org/10.1371/journal.pone.0232038. PMID: 32324788; PMCID: PMC7179898.

Institute of Environmental Science Research. Border to grave. https://www.esr.cri.nz/our-research/our-science-in-action/border-to-grave/. (Accessed 30 July 2022).

Johnson LA, Johnson RL, Portier R-B. Current "legal highs". J Emerg Med 2013;40(6): 1108–15.

Khaled SM, Hughes E, Bressington D, Zolezzi M, Radwan A, Badnapurkar A, et al. The prevalence of novel psychoactive substances (NPS) use in non-clinical populations: a systematic review protocol. Syst Rev November 21, 2016;5(1):195. https://doi.org/10.1186/s13643-016-0375-5. PMID: 27871332; PMCID: PMC5117541.

Kruithof K, Aldridge J, Décary-Hétu D, Sim M, Dujso E, Hoorens S. Internet-facilitated drugs trade – an analysis of the size, scope and the role of The Netherlands. 2016. Retrieved from Santa Monica, CA; Cambride, UK, http://www.rand.org/pubs/research_reports/RR1607.html.

Liakoni E, Yates C, Dines AM, Dargan PI, Heyerdahl F, Hovda KE, et al. Acute recreational drug toxicity: comparison of self-reports and results of immunoassay and additional analytical methods in a multicenter European case series. Medicine (Baltim) February 2018;97(5):e9784. https://doi.org/10.1097/MD.0000000000009784. PMID: 29384873; PMCID: PMC5805445.

Madras BK. The growing problem of new psychoactive substances (NPS). Curr Top Behav Neurosci 2017;32:1–18. https://doi.org/10.1007/7854_2016_34. PMID: 27571747.

Meyer MR, Bergstrand MP, Helander A, Beck O. Identification of main human urinary metabolites of the designer nitrobenzodiazepines clonazolam, meclonazepam, and nifoxipam by nano-liquid chromatography-high-resolution mass spectrometry for drug testing purposes. Anal Bioanal Chem May 2016;408(13):3571–91. https://doi.org/10.1007/s00216-016-9439-6. Epub April 12, 2016. PMID: 27071765.

Murphy T, Van Houten C, Gerona R, Moran J, Kirschner R, Marraffa J, et al. Acute kidney injury associated with synthetic cannabinoids use, multiple states, 2012. Morb Mortal Wkly Rep 2013;62(6):93–8.

Palamar JJ, Acosta P. Synthetic cannabinoid use in a nationally representative sample of US high school seniors. Drug Alcohol Depend April 1, 2015;149:194–202. https://doi.org/10.1016/j.drugalcdep.2015.01.044. Epub February 11 2015. PMID: 25736618; PMCID: PMC4361370.

Palamar JJ, Barratt MJ, Coney L, Martins SS. Synthetic cannabinoid use among high school seniors. Pediatrics October 2017;140(4):e20171330. https://doi.org/10.1542/peds.2017-1330. Epub September 11, 2017. PMID: 28893851; PMCID: PMC5613996.

Peacock A, Bruno R, Gisev N, Degenhardt L, Hall W, Sedefov R, et al. New psychoactive substances: challenges for drug surveillance, control, and public health responses. Lancet November 2, 2019;394(10209):1668–84. https://doi.org/10.1016/S0140-6736(19)32231-7. Epub October 23, 2019. PMID: 31668410.

Pirona A, Bo A, Hedrich D, Ferri M, van Gelder N, Giraudon I, et al. New psychoactive substances: current health-related practices and challenges in responding to use and harms in Europe. Int J Drug Pol February 2017;40:84–92. https://doi.org/10.1016/j.drugpo.2016.10.004. Epub December 9, 2016. PMID: 27956184.

Psychonaut Web Mapping Research Group. The psychonaut 2002 project - final reportRetrieved from Londres. 2005. http://ec.europa.eu/health/phprojects/2002/drug/drug_2002_09_en.htm.

Tanaka R, Kawamura M, Uchiyama N, Segawa K, Nakano T, Saito Y, et al. Data search system for new psychoactive substances provided by the national institute of health sciences in Japan. 2016. p. 73–8.

Tettey JNA, Crean C, Ifeagwu SC, Raithelhuber M. Emergence, diversity, and control of new psychoactive substances: a global perspective. Handb Exp Pharmacol 2018;252:51–67. https://doi.org/10.1007/164_2018_127. PMID: 29896655.

Trecki J, Gerona RR, Schwartz MD. Synthetic cannabinoid-related illnesses and deaths. N Engl J Med July 9, 2015;373(2):103–7. https://doi.org/10.1056/NEJMp1505328. PMID: 26154784.

UNODC Global SMART update Volume 6; October 2011.

UNODC Global SMART Programme. The challenge of new psychoactive substances. 2013. Austria.

United Nations Office on Drugs and Crime. Categories of NPS sold in the market. 2014. Austria.

United Nations Office on Drugs and Crime, World drug report 2017 (ISBN: 978-92-1-148291-1, eISBN: 978-92-1-060623-3, United Nations Publication, Sales No. E.17.XI.6).
United Nations Office on Drugs and Crime (UNODC). Early warning advisory toxicology highlights. Current NPS threats, Volume I. March 2019. p. 1–4.
United Nations Office on Drugs and Crime (UNODC). Early warning advisory toxicology highlights. Current NPS threats, Volume II. January 2020. p. 1–4.
UNODC Global SMART Programme. The role of drug laboratories in early warning system. 2020. Austria.
United Nations Office on Drugs and Crime (UNODC). Early warning advisory toxicology highlights. Current NPS threats, Volume III. October 2020. p. 1–6.
United Nations Office on Drugs and Crime (UNODC). Early warning advisory toxicology highlights. Current NPS threats, Volume IV. November 2021. p. 1–6.
United Nations Office on Drugs and Crime, World drug report 2021b (United Nations publication, Sales No. E.21.XI.8).
United Nations Office on Drugs and Crime. World drug report 2022. United Nations Publication; 2022.
U.S. Drug Enforcement Administration, Office of Diversion Control. National forensic laboratory information system special report: synthetic cannabinoids and cathinones reported in NFLIS, 2010–2011. Springfield, VA: U.S. Drug Enforcement Administration; 2011.
U.S. Drug Enforcement Administration, Office of Diversion Control. National forensic laboratory information system special report: emerging 2C-phenethylamines, piperazines, and tryptamines in NFLIS, 2006–2011. Springfield, VA: U.S. Drug Enforcement Administration; 2012.
U.S. Drug Enforcement Administration, Office of Diversion Control. National forensic laboratory information system special report: synthetic cannabinoids and synthetic cathinones reported in NFLIS, 2010–2013. Springfield, VA: U.S. Drug Enforcement Administration; 2014.
U.S. Drug Enforcement Administration, Office of Diversion Control. National forensic laboratory information system special report: opiates and related drugs reported in NFLIS, 2009–2014. Springfield, VA: U.S. Drug Enforcement Administration; 2015.
U.S. Drug Enforcement Administration, Office of Diversion Control. National forensic laboratory information system special report: benzodiazepines reported in NFLIS, 2009–2014. Springfield, VA: U.S. Drug Enforcement Administration; 2016.
U.S. Drug Enforcement Administration, Diversion Control Division. 2017 toxicology laboratory survey report. Springfield, VA: U.S. Drug Enforcement Administration; 2018a.
U.S. Drug Enforcement Administration, Diversion Control Division. 2017 medical examiner/coroner office survey report. Springfield, VA: U.S. Drug Enforcement Administration; 2018b.
U.S. Drug Enforcement Administration, Office of Diversion Control. National forensic laboratory information system special report: tracking fentanyl and fentanyl-related substances reported in NFLIS, 2016-2017. Springfield, VA: U.S. Drug Enforcement Administration; 2019.
U.S. Drug Enforcement Administration, Office of Diversion Control. National Forensic Library Information System. https://www.nflis.deadiversion.usdoj.gov/drug.xhtml?jfwid=jLA86tVJBh1WLCNtnz_gsKtWcJCepQQZ10oEUPSf:1. (Accessed 30 July 2022).
U.S. Drug Enforcement Administration, Diversion Control Division. DEA TOX: 2022 annual report. VA: Springfield; 2023.
Van Hout MC, Hearne E. New psychoactive substances (NPS) on cryptomarket fora: an exploratory study of characteristics of forum activity between NPS buyers and vendors. Int J Drug Pol 2017;40:102–10. https://doi.org/10.1016/j.drugpo.2016.11.007.
Wood DM, Heyerdahl F, Yates CB, Dines AM, Giraudon I, Hovda KE, et al. The European drug Emergencies network (Euro-DEN). Clin Toxicol 2014 Apr;52(4):239–41. https://doi.org/10.3109/15563650.2014.898771. Epub March 21, 2014. PMID: 24654801.

CHAPTER

9

NPS regulation and legislation

Challenges

The same challenges that new psychoactive substances (NPS) impose on public health play critically in their regulation and control. The sheer number of compounds released in the past decade and the rapid evolution of their molecular structures have relegated control measures and legislation to reactionary at best (Peacock et al., 2019; Zaami, 2019). Thus, it is often the case that a potent NPS causes mass outbreaks and/or claims the lives of users before it is scheduled and regulated. The fast pace of molecular evolution makes most NPS less relevant by the time they become targets for regulation.

Not only is the number large but the classes are quite varied. There are at least 13 major chemical and pharmacological classes, and they have widely differing ranges of potency that correlate with their ability to cause harm (Zapata et al., 2021; UNODC, 2019). Some also have promising therapeutic potential, such as synthetic cathinones, tryptamines, amphetamines, designer benzodiazepines, and new synthetic opioids. Control measures that can be appropriately implemented for each class should take a balancing approach between potential harm and therapeutic promise. A one-size-fits-all formula will not do. Closely related is the scarcity of information on pharmacology and toxicology. Unlike traditional recreational drugs, many NPS have been studied little or not at all. So an informed control measure that takes into account the potential for harm and benefit is often impossible to make.

The ability to detect and identify NPS as they are released in the recreational drug market is also necessary to implement timely, relevant, effective control measures, especially given the rapid evolution of chemical structures (Negrei et al., 2017). This necessitates monitoring methods that are capable of nontargeted analysis. There is still limited

Designer Drugs
https://doi.org/10.1016/B978-0-12-811764-4.00011-2

availability of laboratories with these analytical capabilities and the expertise to evaluate results.

NPS are trafficked on a global scale by multiple organized drug cartels linked to big commercial drug manufacturing companies. Products are marketed on both the clear web and the darknet (EMCDDA, 2022a). A crucial part of any drug control measure is the ability to regulate supply. With a globalized supply and market and a borderless Internet, the biggest challenge in mounting an effective strategy is the ability to cross state borders and jurisdictions.

An innovative approach will necessarily run contrary to existing drug control and regulatory measures for some countries. It must be evidence-based and multidisciplinary so that it can morph and evolve with scientific developments and technological progress. Real-time proactive NPS monitoring that can identify previously unreported NPS and forecast future molecular targets for drug manufacturers should be part of its core components. And it will need to be integrated with regional and international policies to respond easily to global developments (Zaami, 2019).

Control of psychoactive drugs: International drug laws

As introduced in Chapter 1, the international control of drugs is governed by three conventions:

- the 1961 Single Convention on Narcotic Drugs, focused on cannabis, cocaine, and opioids (UN, 1961);
- the 1971 Convention on Psychotropic Substances, focused on synthetic drugs such as MDMA and LSD (UN, 1971); and
- the 1988 Convention Against Illicit Traffic in Narcotic Drugs and Psychotropic Substances, covering police suppression of illicit markets (money laundering) and control of drug precursor chemicals (UN, 1988).

From these, three lists of controlled substances are defined (Table 9.1). The 1961 Convention further categorizes psychoactive substances into four schedules (I–IV) depending on their potential for abuse and addiction. The 1971 Convention has four schedules depending on their potential for abuse, addiction, and therapeutic value. The 1988 Convention includes precursors and reagents used in illicit manufacture and categorizes substances into two tables depending on whether they are precursors of psychoactive substances or reagents or solvents used in production (Zapata et al., 2021).

As described in Chapter 1, addition of NPS to the 1961 and 1971 conventions is a lengthy process, the final step of which is a vote by the member states of the UN Commission on Narcotic Drugs (CND) at an

TABLE 9.1 Controlled substance classifications under the three United Nations conventions on drugs.

Schedule of narcotic drugs (1961 Convention)			
Schedule I	**Schedule II**	**Schedule III**	**Schedule IV**
Addictive substances with serious risk of abuse Very strict degree of control Examples: cannabis and derivatives, cocaine, opium, heroin, morphine	Medically used substances with lowest risk of abuse Less strict degree of control Example: codeine	Preparation of substances listed in Schedule II as well as cocaine preparations Example: codeine preparation	Most dangerous substances in Schedule I that are harmful and with extremely limited medical or therapeutic value Example: heroin
Schedule of psychotropic substances (1971 Convention)			
Substances with high risk of abuse, serious threat to public health, and very little or no therapeutic value Very strict control except for scientific or limited medical purposes Examples: THC, LSD, MDMA, psilocybin, mescaline	Substances with risk of abuse, serious threat to public health, and low or moderate therapeutic value Less strict control Examples: methamphetamine, amphetamine, amphetamine-type stimulant	Substances with risk of abuse, serious threat to public health, and moderate or high therapeutic value Available for medical purposes Examples: buprenorphine, barbiturates	Substances with risk of abuse, minor threat to public health, and high therapeutic value Available for medical purposes Examples: benzodiazepines, analgesics, sedatives

Drug precursors, reagents, and solvents (1988 Convention)	
Table 1	Table 2
Precursors of psychotropic substances and reagents for converting and extracting narcotics Examples: ephedrine, safrole, lysergic acid, acetic anhydride (converting agent), potassium permanganate (extracting agent)	Industrial reagents and solvents that can be used in the production of narcotics and psychotropic substances Examples: acetone, toluene, ethyl ether, sulfuric acid

annual meeting. Since the World Health Organization's Expert Committee on Drug Dependence first examined the potential harm and abuse of individual NPS in 2014, only 71 have been scheduled under the international conventions (UNODC, 2023). This is only 6% of the 1182 NPS that have been reported by the United Nations Office on Drugs and Crime as of 2022.

Investigation by the Expert Committee takes time, as the process of recommendation needs to be evidence-based. This is quite difficult when there is a scarcity of data and evidence on the propensity of most NPS to cause adverse effects. Most of these data come from case reports or series on single or mass intoxications (CND, 2015). Only a few NPS also have pharmacological or toxicological data from animal studies readily available when they are first identified. The process is also slowed because CND decisions are made annually. Hence, a year or more typically passes after the NPS is initially released before it can be added to the conventions. This is too late for most countries, especially if the substance is potent and causing fatal intoxication.

Only individual NPS, not classes, can be added to the conventions because of the concern that adding entire classes will have a negative impact on research and discovery of therapeutic agents. Moreover, not all member states have the capability to monitor a large number of novel compounds, a requirement imposed on each member state once a drug is scheduled under the international conventions. In a practical sense, evaluation of a large number of NPS for addition to the conventions will also take away too much time from the CND to attend to other drug issues. So the CND prioritizes a smaller number of NPS for scheduling based on their propensity to cause harm and the prevalence of their use.

Control of psychoactive drugs: Scheduling in the United States

The Comprehensive Drug Abuse Prevention and Control Act of 1970, commonly known as the Controlled Substance Act (CSA), created the US Drug Enforcement Administration, an arm of the Department of Justice charged with enforcement and prevention of abuse. The CSA established the scheduling and regulatory framework for existing controlled substances and the process by which new drugs are evaluated and scheduled. The schedules (I–V) were based on potential for abuse, accepted medical use, safety, and potential for addiction (CSA, 1970).

Permanent scheduling is conducted under the authority of the Department of Human Health Services based on data provided by the DEA, the National Institute on Drug Abuse, the Food and Drug Administration, and the scientific and medical community who have relevant information on any of the eight criteria used in scheduling. The criteria are

- the substance's actual and relative potential for abuse;
- scientific evidence on its pharmacological effect if known;
- the state of current scientific knowledge of the substance;
- the scope, duration, and significance of abuse;

- its history and current pattern of abuse;
- its psychological or physiological dependence liability;
- its risk to public health if any; and
- whether it is an immediate precursor of a controlled substance.

The rise in the number of designer drugs in the late 1970s and early 80s, especially the more potent opioids, called for a faster scheduling response. The process of securing a permanent schedule is long and slow. The Comprehensive Control Act of 1984 granted the US attorney general, through the DEA, the emergency authority to place a substance temporarily into Schedule I for a year with an option to extend for 6 months if the following criteria are met: action is necessary to avoid imminent hazard to public safety, the substance is not listed in any other schedule under Section 202 of the CSA, no exemption or approval is in effect under 21 U.S.C 355, and the DEA administrator must consider factors 4, 5, and 6 in Section 201(c) of the CSA.

With the continued evolution of structures, albeit at a slower rate than what was observed for NPS in the past decade, the Federal Analogue Act of 1986 addressed their increasing manufacture, distribution, and use on the basis of chemical similarity to known controlled substances of abuse (Federal Analogue Act, 1986). This act defined a controlled substance analog under Schedule I or II if one of the following criteria is met:

- a chemical structure that is substantially similar to the chemical structure of an existing controlled substance in Schedule I or II;
- a stimulant, depressant, or hallucinogenic effect on the central nervous system that is substantially similar to that of a controlled substance in Schedule I or II;
- with respect to a particular person, which such person represents or intends to have a stimulant, depressant, or hallucinogenic effect on the central nervous system that is substantially similar to the stimulant, depressant, or hallucinogenic effect on the central nervous system of a controlled substance in Schedule I or II.

Although the definitions of "chemical and functional similarity" have always been ambiguous, the Federal Analogue Act was not really challenged until the NPS surge started to gain more grounds in 2009. Synthetic cannabinoid is a good example. Although they bind the same CB1 and CB2 receptors that THC binds, most of their structures are significantly different from THC. Furthermore, they have effects that are not observed in THC. Although there was no formal claim done, it was generally recognized that the Federal Analogue Act did not apply to most NPS in the past decade. Instead, the DEA used temporary scheduling as a means of emergency control over potent synthetic cannabinoids and synthetic cathinones that had caused substantial public harm. The first temporary placement of five synthetic cannabinoids into Schedule I occurred in March 2011. Temporary placement of the first three synthetic cathinones followed in October 2011. These temporary schedulings become permanent in 12 or 18 months.

A reboot of the Federal Analogue Act was passed in 2012 through the Synthetic Drug Abuse Prevention Act. It placed 15 synthetic cannabinoids and five classes of synthetic cannabinoids defined as cannabimimetic agents into Schedule I. The Act went further to specify the structural substitutions to the pharmacophores of these classes that would generate substances covered by the temporary schedule as cannabimimetic agents (Synthetic Drug Abuse Prevention Act, 2012). Specifying the exact structural requirement for what qualifies as an analog was unprecedented and initially seemed a clever move. However, NPS manufacturers reacted by simply changing the entire functional group in one of the pharmacophores that resulted in a different class that does not fall under the five classes, rendering the scheduling inapplicable. Changing the entire functional group in a pharmacophore then became a trend in generating later generations of synthetic cannabinoid classes. Some of the resulting new classes proved to be much more potent than the previous classes, an unfortunate unintended consequence of the SDAPA.

Another unprecedented step by the DEA was the class-wide scheduling of fentanyl analogs. Temporary and permanent scheduling are generally done on an individual drug basis. However, the proliferation of potent fentanyl analogs that started in 2016 worsened the opioid crisis and claimed even more fatalities from overdose. The reactionary temporary scheduling of individual fentanyl analogs did not help in alleviating the crisis, as there are thousands of potential fentanyl analogs that manufacturers can synthesize to replace those that are already scheduled, and some of them have higher potency than previous fentalogs. This finally prompted the temporary class-wide scheduling of fentanyl analogs in February 2018 (Lewis, 2018). It spurred a similar action in China, the predominant source of fentalogs. As a result, fentalogs fell dramatically in the marketplace, only to be replaced by other classes of new synthetic opioids such as the benizimidazoles (nitazenes). Class-wide scheduling has never been applied to any other NPS class in the United States.

The European framework for NPS control

The EMCDDA has led the response to NPS in the European Union. In 1997, legislation known as the Joint Action on New Synthetic Drugs defined a three-step framework: information exchange, risk assessment, and control measures (EMCDDA, 1997). Information exchange is facilitated through the network created by the early warning system (EWS) that paved the way for an effective surveillance system in Europe, especially after its focus on NPS was strengthened in 2005 (EMCDDA, 2005). Details on the operation of the EWS were described in Chapter 8.

The legal framework starts with the EWS, through which an alert to all member states is issued by a formal notification after the EMCDDA has reviewed data on newly discovered NPS reported by a member state's national EWS. Then the EMCDDA begins monitoring the substance for reports of harm. Based on the level of harm observed, the EWS can mount a response ranging from placing the substance under intensive monitoring to issuing risk communications, including public health alerts, to preparation of an initial report that may lead to a risk assessment. Since 2005, the EWS has issued initial reports on 30 NPS (Table 9.2) (EMCDDA, 2022b). Issuance of the initial report is the final stage of early warning.

The initial report is used by the European Commission and Council to determine whether a formal risk assessment is required. If so, the scientific committee of the EMCDDA conducts it. In general, the process reviews the possible health and social risks and the implications of placing it under control. Both the probability that some harm may occur from using the substance (risk) and the degree of seriousness of such harm (hazard) based on its origins and type are assessed.

Because scientific data on the pharmacology and toxicology of a newly reported NPS are usually scarce, the committee uses a wide range of available evidence from case reports gathered by the EMCDDA network and its other monitoring systems. This usually includes reports from forensic and toxicology laboratories and law enforcement agencies; sometimes, anecdotal reports are also considered. These reports are weighted based on their reliability and relevance. Hence, a peer-reviewed publication reporting intoxication from an analytically confirmed NPS will be given more weight than an unpublished laboratory report. The latter may have lower reliability, but its relevance is still significant to risk assessment. To get around the lack of empirical data, an analysis of the possible nature and risks of the substance is extrapolated from established data on known substances such as similar controlled drugs or substances that have similar chemical characteristics, pharmacology, and psychological and behavioral effects. A risk–benefit ratio is also assessed. Potential therapeutic benefits, industrial use, or other uses of significant economic value are considered. Substances with established therapeutic value or those that can be used in the preparation of medicinal products may be exempted from risk assessment.

A completed risk assessment is documented on a report written by the scientific committee that contains an analysis of the available scientific

TABLE 9.2 NPS that were issued initial report by the EMCDDA.

Year	Drug	Class
2005	meta-Chlorophenyl piperazine (mCPP)	Piperazine
2007	Benzylpiperazine (BZP)	Piperazine
2010	Mephedrone	Synthetic cathinone
2012	4-methylamphetamine (4-MA)	Amphetamine
2013	5-(2-aminopropyl)indole (5-IT)	Amphetamine
2014	AH-7921	New synthetic opioid
2014	3,4-methylenedioxy pyrovalerone (MDPV)	Synthetic cathinone
2014	Methoxetamine (MXE)	Arylcyclohexylamine
2014	25I-NBOMe	Phenethylamine
2014	4,4′-DMAR	Amphetamine
2014	MT-45	New synthetic opioid
2015	α-Pyrrolidinovaler ophenone (α-PVP)	Synthetic cathinone
2016	Acetylfentanyl	New synthetic opioid
2016	MDMB-CHMICA	Synthetic cannabinoid
2017	Acryloylfentanyl	New synthetic opioid
2017	Furanylfentanyl	New synthetic opioid
2017	AB-CHMINACA	Synthetic cannabinoid
2017	5F-MDMB-PINACA	Synthetic cannabinoid
2017	4-fluoroisobutyryl fentanyl (4F-iBF)	New synthetic opioid
2017	Tetrahydrofu ranylfentanyl (THF)	New synthetic opioid
2017	Cumyl-4-cyano-BINACA	Synthetic cannabinoid
2017	ADB-CHMINACA	Synthetic cannabinoid
2017	Carfentanil	New synthetic opioid
2018	Methoxyacetylfentanyl	New synthetic opioid
2018	Cyclopropylfentanyl	New synthetic opioid

TABLE 9.2 NPS that were issued initial report by the EMCDDA.—cont'd

Year	Drug	Class
2020	Isotonitazene	New synthetic opioid
2020	MDMB-4en-PINACA	Synthetic cannabinoid
2020	4F-MDMB-BICA	Synthetic cannabinoid
2021	3-chloromethcathinone (3-CMC)	Synthetic cathinone
2021	3-methylmethcathinone (3-MMC)	Synthetic cathinone

and law enforcement information. The report includes information on the following:

- physical and chemical properties;
- mechanisms of action and medical value;
- health risks associated with the NPS;
- social risks associated with the NPS;
- information on manufacture and seizures or detections by authorities;
- information on the level of involvement of organized crime in manufacture and distribution;
- assessment in the UN systems;
- description of control measures in member states if applicable;
- options for control and possible consequences of control measures;
- chemical precursors used in the NPS manufacture.

Since 1997, the EMCDDA has issued risk assessments on 35 NPS, of which 26 were done between 2009 and 2022 (EMCDDA, 2022c): 10 new synthetic opioids, 7 amphetamines, 6 synthetic cannabinoids, 4 synthetic cathinones, 4 phenethylamines, 2 arylcyclohexylamines, 1 piperazine, and 1 gabapentinoid (Table 9.3).

Once a risk assessment report is issued, the council decides if control measures will be submitted. If submitted, the EMCDDA member states will have one year to implement the recommended control measures and criminal penalties in accordance with their national law.

Scheduling strategies in Asia and Oceania

Legal attitudes toward illegal drug use are similar in most countries in Asia. They are generally more punitive than in the United States and Europe, where illegal use of Schedule I or II substances can be punished with imprisonment of 3–10 years. Despite the similarity, the legislative

TABLE 9.3 NPS subjected to formal risk assessment by the EMCDDA.

Class	Drug
New synthetic opioid	Isotonitazene (2020) Methoxyacetylfentanyl (2018) Cylcopropylfenatnyl (2018) Carfentanil (2018) Tetrahydrofuranylfentanyl (2018) 4-fluoroisobutyrylfentanyl (2018) furanylfentanyl (2017) Acryloylfentanyl (2017) MT-45 (2014) AH-7921 (2014)
Synthetic cannabinoid	4F-MDMB-BICA (2021) MDMB-4en-PINACA (2021) 5F-MDMB-PINACA (2018) ADB-CHMINACA (2018) Cumyl-4CN-BINACA (2018) MDMB-CHMICA (2017)
Synthetic cathinone	3-methylmethcathinone (2022) 3-chloromethcathinone (2022) alpha-PVP (2016) Mephedrone (2011)
Amphetamine	4,4′-DMAR (2015) 5-IT (2014) 4-methylamphetamine (2014) TMA-2 (2004) PMMA (2003) 4-MTA (1999) MBDB (1998)
Phenethylamine	25I-NBOMe (2014) 2C-I (2004) 2C-T-2 (2004) 2C-T-7 (2004)
Arylcyclohexylamine	Methoxetamine (2014) Ketamine (2002)
Piperazine	Benzylpiperazine (2009)
Gabapentinoid	GHB (2002)

criteria and response for NPS control vary widely (Feng et al., 2020). Although newly released NPS appear later in Asian countries than in Europe and North America, a broad selection also reaches the recreational drug market in Asia. Between 2007 and 2015, for example, 940 NPS were reported in Northeast Asian countries (Japan, South Korea, Taiwan, and China). Of these, 882 were reported in at least one country and 96 were not under control (6%). Japan regulates 41% of the controlled NPS, while

China, South Korea, and Taiwan regulate 28%, 21%, and 10%, respectively (Lee et al., 2017).

Because there is hardly any governing body like the EMCDDA in Asia, it is impossible to cover all the different regulation and control measures there. We take three as examples: Taiwan, South Korea, and Japan. These were chosen because the status of NPS distribution, spread, and use are best known. All three have complied with the UN drug conventions by enacting corresponding domestic laws. In Taiwan, there are the Statute of Controlled Drugs for legal purposes and the Statute for the Prevention and Control of Illicit Drugs for illegal purposes (Taiwan FDA, 2018). In South Korea, the Narcotic Control Act regulates narcotics, psychotropic substances, temporary drugs, and precursors (Korean Supreme Prosecutors' Office, 2017). In Japan, there is a control law for each of the three pharmacological classes—opium, cannabis, and amphetamines (stimulants)—in addition to the Narcotic and Psychotropic Substances Control Law (Tanaka et al., 2016).

The scheduling criteria and legislative progress on control vary widely in these three countries. As of 2019, Taiwan had not stipulated a new law or provisions in response to NPS emergence and surge. In South Korea, a temporary narcotic designation for NPS was added to the Narcotic Control Act in 2011. In Japan, a new system of dangerous drugs was implemented to cope with the psychotoxicity of new substances. This includes the Temporary Designation System and Analog Control System implemented in 2011, the JHW-018 Analogue Control enforced from 2013, the Dangerous Drug Regulation promulgated in 2014, and the Cathinone Analog Control enforced in 2014 and 2015. The promulgations of the legislative controls in both South Korea and Japan resulted in higher numbers of NPS scheduled as controlled substances. As of 2015, there were 358 in Japan and 245 in South Korea. In contrast, the lack of legislation in Taiwan resulted in the lowest number of NPS scheduled among the three countries, 91 as of 2015 (Li et al., 2020).

Among the three countries, the effect of legislation on the NPS situation is best exemplified by Japan. It led to the rapid decline of head shops, clearly affecting supply. After the tightening of regulations, difficulty in procurement caused a majority of patients with NPS-related psychiatric disorders to change their drug of choice.

NPS control and legislation vary by jurisdiction in Australia. Since 2013, Queensland, New South Wales, South Australia, and Western Australia have introduced blanket bans on possessing or selling any nonprescription substance that has a psychoactive effect except alcohol, tobacco, and food. In other jurisdictions, specific NPS are scheduled, and others are regularly added.

New Zealand has a unique legislation on NPS control. In July 2013, the Psychoactive Substances Act was passed. Manufacturers and conveyers of any psychoactive substance, including energy pills, party pills, and herbal products, are legally bound to provide to the consumer scientific evidence that their product is "low risk" (Psychoactive Substances Act, 2013). This regulatory framework is similar to the approval of medications by the US Food and Drug Administration. Manufacturers, sellers, and importers of NPS in New Zealand are required to have their product approved or rejected by the government based on the company's preclinical and clinical data. This is quite an innovative approach that constitutes an evidence-based policy on NPS safety for distribution and use.

Generic legislation and blanket ban

As is evident in various requirements by international, regional, and national governing bodies, scheduling NPS on an individual basis is lengthy and reactionary. Often a potent NPS is allowed to claim dozens if not hundreds or thousands of lives before control is implemented. Tension exists on both legislative timing and level of control. Time is needed to build evidence, and this provides opportunity for expansion of availability, problematic use, and harm. On the other hand, excessive restriction hinders exploration of therapeutic benefit.

Temporary scheduling helps with the legislative timing, but the level of regulatory restriction has remained a strongly debated subject. Countries more concerned about harm have adopted policies that control broad classes of substances through generic (simultaneous control of clusters), analog, and neurochemical legislation. In generic legislation, specific variations of a core molecular structure are defined, by which substances (including those that might be targeted in the future) can be controlled without being specifically referred to in the legislation. Analog legislation is much broader; it encompasses more general similarity to a parent drug structure. Neurochemical legislation invokes similarity of effects on the brain, on top of chemical similarity (Tettey et al., 2018). Although these types of legislation can proactively ban all possible members of an NPS class or chemical family that are deemed to be too potent to risk public safety, they can also impede research on therapeutic benefits. In practical terms, this creates confusion about which specific NPS is covered. For example, what is the extent covered by chemical similarity that defines an analog? This is evident in the US Federal Analog Act of 1986, where, depending on which lawyer you consult, a given substance can be either an analog of a controlled substance or not. The issue here of course is that the last word on what comprises an analog should be the chemist's even in a court of law. Often the legislation itself is vague on the extent of chemical similarity that is within the definition of an analog.

A far broader approach is to impose a blanket ban on all psychoactive substances, as in the United Kingdom (Reuter and Pardo, 2017). Although this can dramatically limit the presence of NPS in the recreational market, it is also the most stifling to new drug development. Moreover, it is the most difficult to enforce, and it bans drugs that have very low risk of harm.

It remains to be seen which of these control measures can effectively respond to the NPS challenge. Perhaps in another decade or so when we have collected more data on use and harm following more years after implementation, a clear picture will emerge.

Closeup: Fire and ice, the two faces of blanket bans on new psychoactive substances

In one of his poems, Robert Frost used fire and ice as metaphor for perceptions of desire and hatred that are bound to doom the human race to self-destruction. In a way, banning new psychoactive substances (NPS) shares both elements. Fire is the desire to explore the therapeutic potential of various NPS. Ice is the hatred that NPS engenders for every human life it claims; policy-wise, it can also represent the ice-cold hatred for psychoactive substances of the prohibition movement. These elements have fueled the long-running debate over banning NPS and other psychoactive substances. But rather than causing the end of NPS, retrocausality seems to apply more. Will banning NPS put an end to either or both? Let us look at some enacted regulatory policies to catch a glimpse of the emerging answer.

Ireland is the first country to issue a blanket ban on NPS, through the Criminal Justice (Psychoactive Substances) Act of 2010. It targeted primarily NPS vendors, particularly head shops where the majority of mephedrone and synthetic cannabinoids were being sold early in the NPS resurgence. Following this legislation, the vast majority of head shops in Ireland selling NPS closed, and the remaining ones stopped NPS sales. NPS use and harms declined in parallel. The National Drug Prevalence Survey in 2010/11 reported past year use among people 15–24 and 25–34 years old to be 9.7% and 4.6%, respectively. The next time the survey was conducted, in 2014/15, a dramatic decline to 1.9% and 1.3% was reported. In a small study of adolescents presenting to addiction treatment services in Dublin, one in three involved problematic use of NPS in early 2010. By early 2011, the number was zero (Smyth, 2023).

continued

Closeup: Fire and ice, the two faces of blanket bans on new psychoactive substances (*cont'd*)

At first glance, these data yield a rosy picture of success for the legislation. A decline in NPS use, however, was also observed in the United States, where no blanket ban was imposed. From Monitoring of the Future surveys, for example, the past year use prevalence of synthetic cannabinoids among high school seniors was 11.4% in 2011. With reports on intoxications and outbreaks in the succeeding years, this rate was down to 3.6% in 2016 (Palamar et al., 2017). The United States issued neither a blanket ban on NPS nor a generic ban on synthetic cannabinoids between 2011 and 2016. Self-reports of problematic use of a drug also become unreliable once individuals are threatened directly or indirectly by punitive action that could be imposed by a legislative act. Unless a self-report of not taking a drug after legislation was enacted can be backed up by an objective pharmacological measure such as segmental hair analysis before and after the legislative ban, one must be careful in taking self-reports at face value. So the rosy picture is suspect.

Poland and the United Kingdom followed Ireland's lead on blanket ban. The UK passed its Psychoactive Substances Act (PSA) in 2016 that criminalized the production, supply, or possession with intent to supply of any substance with psychoactive effects. Included in the criminalization is the import and export of a psychoactive substance via a foreign website even for personal use. Critics of the bill before it passed pointed out concern over the exclusion of harm or dependence as criteria by which a psychoactive substance is either included or excluded from its scope. The all-inclusive ban for psychoactive substances without regard for whether it may cause harm or not is feared to impede the discovery and fulfillment of potential benefits. It can lead to people facing disproportionate sanctions for offenses associated with substances that pose very little harm. The inability to predict whether a new compound is psychoactive or not based solely on its molecular structure makes it very difficult to prove that the manufacturer, producer, or importer was aware of its psychoactive properties. This will be problematic for prosecution (Stevens et al., 2015).

About 5 years after the UK bill passed, a group assessed the effectiveness of the ban using measurable health outcomes that interventional public health laws like the PSA aim to address—NPS use, acute intoxications, and mortality associated with NPS (Neicun et al., 2022). Immediately after the ban, NPS use in England and Wales among those aged 16–24 years declined (past year and lifetime use prevalence rates declined from 2.8% to 1.2% and 6.1% to 4.2% between 2014/15 and

Closeup: Fire and ice, the two faces of blanket bans on new psychoactive substances (*cont'd*)

2016/17). These levels were maintained up to 2018/19, the last period presented in the study. In Scotland, no change in lifetime use prevalence among adults aged 16–59 years (only data available) between 2014/15 and 2017/18 (1.6% vs. 1.8%) was observed. No data on hospital admissions caused by NPS were available for England and Wales, so admissions from other and unspecified narcotics were used as a surrogate. These admissions increased by 27% between 2014/15 and 2019/20. The number of deaths from NPS remained essentially the same between 2015 and 2019 in England and Wales (114 vs. 125) while it increased by eight times in Scotland (112 vs. 802) in the same period. These numbers clearly suggest that, at best, the blanket ban marginally lowered NPS use. However, NPS intoxications and deaths paradoxically increased despite the blanket ban. This observation is not unique to blanket ban, however. Data on two other regulatory strategies, individual listing and generic control, showed similar results in health outcomes from other countries in Europe that adopted these strategies.

It is hard to quantify the effect of blanket bans on scientific exploration of potential therapeutic benefits from some NPS or other drug targets related to NPS. However, we know from history that LSD and MDMA started as therapeutic agents before they were banned. Upon their prohibition, research stopped for 40 years, only to resume recently (Negrei et al., 2017). The same fate is bound to happen to some contemporary NPS. As pointed out in the beginning of this book, the FDA-approved antidepressant bupropion (Wellbutrin) is a synthetic cathinone. Other members of this NPS class may have antidepressant use, but a blanket ban will make it impossible or very difficult to explore that potential. Various compounds from cannabis have demonstrable analgesic properties by targeting the CB2 receptor. The same receptor is targeted by innumerable synthetic cannabinoids. A few of these compounds might produce potent analgesia without psychoactive effects. The race is already on to find therapeutic applications for tryptamines in alleviating symptoms of various mental health conditions and chronic pain, while we may yet find therapeutic benefits from other amphetamines and phenethylamines. Other NPS can be added to this list. The list may not be as long as those that have been exploited in recreational drug markets, but a significant number may have legitimate health benefits. All this can be lost to blanket bans.

continued

Closeup: Fire and ice, the two faces of blanket bans on new psychoactive substances (*cont'd*)

The data so far come from only a couple of countries, so it may be premature to judge the effectiveness of the blanket ban. But the emerging picture does suggest that banning may not so much melt the ice as extinguish the fire that fuels the search for therapeutic benefit. We can fairly say that after more than a decade since the NPS resurgence, the right regulatory strategy has not been found.

Roy Gerona, PhD

Associate Professor, Department of Obstetrics, Gynecology and Reproductive Sciences, University of California San Francisco, San Francisco, CA, United States

References

Commission on Narcotic Drugs, CND. UNODC-WHO expert consultation on new psychoactive substances, Vienna, 9–11 December 2014. Report by LSS/RAB/DPA/UN. 2015. http://undocs.org/E/CN.7/2015/CRP.2.

Controlled Substance Act, 21 U.S.C. § 812; 1970.

European Monitoring Centre for Drugs and Drug Addiction. Joint action 97/396/JHA: 1997 Joint action on new Synthetic drugs. Luxembourg: Publications Office of the European Union; 1997.

European Monitoring Centre for Drugs and Drug Addiction. Council decision 2005/387/JHA on the information exchange, risk assessment and control of new psychoactive substances. Luxembourg: Publications Office of the European Union; 2005.

European Monitoring Centre for Drugs and Drug Addiction. New psychoactive substances: 25 years of early warning and response in Europe. An update from the EU Early Warning System. Luxembourg: Publications Office of the European Union; June 2022a.

European Monitoring Centre for Drugs and Drug Addiction. Early warning system. 2022. https://www.emcdda.europa.eu/publications/topic-overviews/eu-early-warning-system_en#section2. [Accessed 30 July 2022].

European Monitoring Centre for Drugs and Drug Addiction. Risk assessment of new psychoactive substances. 2022. https://www.emcdda.europa.eu/publications/topic-overviews/risk-assessment-new-psychoactive-substances-nps_en. [Accessed 2 August 2022].

Federal Analogue Act, 21 U.S.C. § 813. 1986.

Feng LY, Wada K, Chung H, Han E, Li JH. Comparison of legislative management for new psychoactive substances control among Taiwan, South Korea, and Japan. Kaohsiung J Med Sci February 2020;36(2):135–42. https://doi.org/10.1002/kjm2.12140. Epub October 23, 2019. PMID: 31643137.

Lee J, Yang S, Kang Y, Han E, Feng LY, Li JH, et al. Prevalence of new psychoactive substances in Northeast Asia from 2007 to 2015. Forensic Sci Int 2017;272:1–9.

Lewis MJ, DEA. Schedules of controlled substances: temporary placement of fentanyl-related substances in schedule I. Doc. No. 2018-02319. Fed Regist 2018;83(25):5188–92.

Neicun J, Roman-Urrestarazu A, Czabanowska K. Legal responses to novel psychoactive substances implemented by ten European countries: an analysis from legal spidemiology. Emerg Trends Drugs Addiction Health 2022;2:10044.

Negrei C, Galateanu B, Stan M, Balalau C, Dumitru MLB, Ozcagli E, et al. Worldwide legislative challenges related to psychoactive drugs. DARU June 2, 2017;25(1):14. https://doi.org/10.1186/s40199-017-0180-2. PMID: 28578694; PMCID: PMC5455135.

Palamar JJ, Barratt MJ, Coney L, Martins SS. Synthetic cannabinoid use among high school seniors. Pediatrics October 2017;140(4):e20171330. https://doi.org/10.1542/peds.2017-1330. Epub September 11, 2017. PMID: 28893851; PMCID: PMC5613996.

Peacock A, Bruno R, Gisev N, Degenhardt L, Hall W, Sedefov R, et al. New psychoactive substances: challenges for drug surveillance, control, and public health responses. Lancet November 2, 2019;394(10209):1668–84. https://doi.org/10.1016/S0140-6736(19)32231-7. Epub October 23, 2019. PMID: 31668410.

Psychoactive Substances Act 2013 (NZ) 2013 No 53.

Reuter P, Pardo B. New psychoactive substances: are there any good options for regulating new psychoactive substances? Int J Drug Pol February 2017;40:117–22. https://doi.org/10.1016/j.drugpo.2016.10.020. Epub November 23, 2016. PMID: 27889115.

Smyth BP. Head shops and new psychoactive substances: a public health perspective. Ir J Psychol Med 2023 Mar;40(1):89–96. https://doi.org/10.1017/ipm.2020.131. Epub January 22, 2021. PMID: 33478611.

Stevens A, Fortson R, Measham F, Sumnall H. Legally flawed, scientifically problematic, potentially harmful: the UK Psychoactive Substance Bill. Int J Drug Pol December 2015;26(12):1167–70. https://doi.org/10.1016/j.drugpo.2015.10.005. Epub October 27, 2015. PMID: 26525856.

Supreme Prosecutors' Office, Republic of Korea. Drug white paper 2017. Seoul, Republic of Korea: Supreme Prosecutor's Office, Republic of Korea. 2017. Available from: http://antidrug.drugfree.or.kr/page/?mIdx=190&mode=view&idx=12433&retUrl=mIdx%3D190.

Synthetic Drug Abuse Prevention Act, 21 U.S.C. § 812; 2012.

Tanaka R, Maiko Kawamura N, Uchiyama K, Segawa T, Nakano YS, et al. Data search system for new psychoactive substances provided by the national institute of health sciences in Japan. Bull Natl Inst Health Sci 2016;134:73–8.

Taiwan Food and Drug Administration. Annual report of drug abuse statistics in Taiwan 2018. Taipei, Taiwan: Taiwan Food and Drug Administration. Available from: http://www.fda.gov.tw/TC/site.aspx?sid=1578.

Tettey JNA, Crean C, Ifeagwu SC, Raithelhuber M. Emergence, diversity, and control of new psychoactive substances: a global perspective. Handb Exp Pharmacol 2018;252:51–67. https://doi.org/10.1007/164_2018_127. PMID: 29896655.

UN. Single convention on narcotic drugs. 1961. https://www.unodc.org/pdf/convention_1961_en.pdf. [Accessed 1 August 2022].

UN. Convention on psychotropic substances. 1971. https://www.unodc.org/pdf/convention_1971_en.pdf. [Accessed 1 August 2022].

UN. Convention against the illicit traffic in narcotic drugs and psychotropic substances. 1988. https://www.unodc.org/pdf/convention_1988_en.pdf. [Accessed 1 August 2022].

United Nations Office on Drugs and Crime (UNDOC). Early warning advisory on new psychoactive substances. Categories of NPS sold in the market. UNODC ITS; 2019.

UNODC. Early warning advisory on new psychoactive substances. 2023. https://www.unodc.org/LSS/Page/NPS. [Accessed 28 April 2023].

Zaami S. New psychoactive substances: concerted efforts and common legislative answers for stemming a growing health hazard. Eur Rev Med Pharmacol Sci November 2019;23(22):9681–90. https://doi.org/10.26355/eurrev_201911_19529. PMID: 31799633.

Zapata F, Matey JM, Montalvo G, Garcia-Ruiz C. Chemical classification of new psychoactive substances. Microchem J April 2021;163:1–13. https://doi.org/10.1016/j.microc.2020.105877.

Index

'*Note:* Page numbers followed by "f" indicate figures, "t" indicate tables, and "b" indicate boxes'.

O

P